ACS SYMPOSIUM SERIES 486

Phosphorus Chemistry

Developments in American Science

Edward N. Walsh, EDITOR
Consultant

Edward J. Griffith, EDITOR
Monsanto Company

Robert W. Parry, EDITOR
University of Utah

Louis D. Quin, EDITOR
University of Massachusetts

Developed from a symposium sponsored
by the Divisions of Inorganic Chemistry
and Industrial and Engineering Chemistry, Inc.,
at the Fourth Chemical Congress of North America
(202nd National Meeting of the American Chemical Society),
New York, New York
August 25–30, 1991

American Chemical Society, Washington, DC 1992

Library of Congress Cataloging-in-Publication Data

American Chemical Society. Meeting (202nd: 1991: New York, N.Y.)
  Phosphorus chemistry: developments in American science   Edward
N. Walsh, editor . . .[et al.].
    p.   cm.—(ACS Symposium Series, 0097–6156; 486).

"Developed from a symposium sponsored by the Divisions of
Inorganic Chemistry and Industrial and Engineering Chemistry at the
202nd National Meeting of the American Chemical Society, New York,
New York, August 25–30, 1991."
  Includes bibliographical references and index.

  ISBN 0–8412–2213–4

  1. Phosphorus—Congresses. 2. Phosphorus compounds—Congresses.
I. Walsh, Edward N., 1925–   . II. American Chemical Society.
Division of Inorganic Chemistry. III. American Chemical Society.
Division of Industrial and Engineering Chemistry. IV. Title. V. Series.

QD181.P1A54   1991
546'.712—dc20   92–5966
                                                   CIP

The paper used in this publication meets the minimum requirements of American National
Standard for Information Sciences—Permanence of Paper for Printed Library Materials, ANSI
Z39.48–1984. ∞

PRINTED IN THE UNITED STATES OF AMERICA

Foreword

THE ACS SYMPOSIUM SERIES was founded in 1974 to provide a medium for publishing symposia quickly in book form. The format of the Series parallels that of the continuing ADVANCES IN CHEMISTRY SERIES except that, in order to save time, the papers are not typeset, but are reproduced as they are submitted by the authors in camera-ready form. Papers are reviewed under the supervision of the editors with the assistance of the Advisory Board and are selected to maintain the integrity of the symposia. Both reviews and reports of research are acceptable, because symposia may embrace both types of presentation. However, verbatim reproductions of previously published papers are not accepted.

Arthur Dock Fon Toy—Scientist, Colleague, Friend

Arthur Toy is a man of real accomplishment, a forthright and strictly honest individual, an excellent scientist, and a wonderful human being. It is our great pleasure to dedicate Phosphorus Chemistry: Developments in American Science *to him.*

Contents

viii

INDEXES

ix

Preface

PHOSPHORUS IS UBIQUITOUS IN OUR WORLD. It is the key element in the genetic tape that guides the reproduction of all species; it is essential to agriculture and to all forms of life; it is one of the major elements found in many solid-state metal phosphide materials such as GaP and InP. Its salts leaven our food and clean our teeth; its organic derivatives reduce the flammability of materials around us; and it plays an important role in keeping us clean. On a more basic level, its chemistry, which is so important in life, provides many conceptual challenges in topics ranging from unusual bonding patterns and stereochemistry to reaction kinetics that control important chemical processes. Many important catalysts owe their effectiveness to phosphorus ligands.

In planning a book based on a two-day symposium on phosphorus and its compounds, one is faced with an embarrassment of riches. It is important to select and focus on a limited number of important topics, but this is done at the expense of other important topics. Of necessity, we have made those choices for this volume.

Frank Westheimer presents a thought-provoking overview of why nature chose phosphates to make the genetic tape. Even the youngest students have heard of DNA, and most have seen models of the famed double helix wherein hereditary information is encoded, but the current question, *Why are phosphates in that helix?*, is usually passed over. The answer is in this volume. Other significant biochemical concerns, such as hydrolysis mechanisms for phosphate compounds and the ^{31}P NMR spectroscopy of duplex oligonucleotides and DNA complexes, are also addressed.

Chapters by Heinz Harnisch and Arthur Toy provide rare insight into the field—from describing the circuitous paths scientists travel in developing industrial products to how basic phosphorus chemistry can be harnessed for the good of humankind. These uncommon reports were an interesting feature of the symposium and provide an introduction to the book. More detailed information on applied phosphorus chemistry is found in chapters 7, 17, 19, and 20.

Fundamental chemical questions have not been neglected. The chapters on phosphorus atoms in unusual stereochemical patterns and phosphorus atoms with high or low coordination numbers expand our views of phosphorus bonding and theory. Important questions of mechanism and kinetics for phosphorus compounds are considered by experts in the field. Basic phosphorus chemistry provides more than its share of challenges!

What had to be omitted from this volume? The large and important area of phosphorus solid-state chemistry has been omitted, not because it is not important, but because it could constitute a volume in itself. For that same reason, the related area of volatile phosphorus precursors for solid-state synthesis has received only passing attention (*see* Chapter 4). Military applications of phosphorus were deliberately passed up for more positively focused topics.

We recognize that while all topics have not been addressed here, we are confident that the work presented is an appropriate cross section of the bulk of American phosphorus chemistry.

EDWARD N. WALSH
33 Concord Drive
New York, NY 10956

EDWARD J. GRIFFITH
Monsanto Company
St. Louis, MO 63167

ROBERT W. PARRY
University of Utah
Salt Lake City, UT 84112

LOUIS D. QUIN
University of Massachusetts
Amherst, MA 01003

January 22, 1992

A Tribute to Arthur D. F. Toy

A Life of Dedication to Research in Industrial Chemistry

H. Harnisch

Hoechst Aktiengesellschaft, D−6230 Frankfurt
am Main 80, Germany

The paper describes the outstanding contribution of
Dr. A.D.F. Toy to phosphorus chemistry to which he
devoted much of his professional life. Dr. Toy's
worldwide recognized research work includes basic
investigations such as the development of new
methods to form phosphorus-carbon bonds as well as
the development and commercialization of many new
production processes and phosphorus-based industrial products. Besides an overview of his contributions to phosphorus chemistry, the professional
career of Dr. Toy is appreciated, which shows that
an extraordinary scientifc qualification in
connection with the ability to create a stimulating
and challenging atmosphere for an effective
scientific cooperation within a research community
is an essential prerequisite for a successful
industrial research leader.

Education

Dr. Toy was born in Canton, China, in 1915. At the age of 12 he
emigrated to the United States, where he first went to the
elementary school in Chicago. He then went to Joliet, south-west
of Chicago and enrolled in Joliet Township High School where he
was made cadet commandant of the ROTC in his senior year.

Arthur Toy studied chemistry at the University of Illinois.
In 1939 he gained his bachelor's degree and in 1940 his master's
degree. In 1942 he acquired his PhD. His thesis supervisor was
Professor Ludwig Frederick Audrieth, who, until his death,
remained a close friend of Dr. Toy and with whom Arthur had many
fruitful discussions on scientific matters.

Professional Career

After completing his studies, Arthur Toy started his professional
career in the laboratory of Victor Chemical Works in Chicago

0097−6156/92/0486−xiii$06.00/0

Heights. Dr. Toy soon became a group leader, was then promoted to director of organic research, associate director of research, and in 1963 became director of research at Victor Chemical Works.

Some years after the acquisition of Victor Chemical Works, Stauffer combined the Victor research with a section of its own research in the Eastern Research Center in the New York area. As Stauffer wanted to retain Arthur's scientific expertise and experience for the company, he was offered a sabbatical year as a visiting scientist under Professor Harry Julius Emeléus in Cambridge, and could then rejoin the Stauffer team as a Senior Scientist.

An influential personality like Arthur Toy, with his scientific background, human integrity, critical intelligence and powers of judgement is almost predestined, even when not possessing a formal title, to take on a management role of some kind within a research community. So, after returning from Britain to the Eastern Research Center, he was soon entrusted with management responsibilities in addition to his duties as Senior Scientist. He became the head of various research departments in Dobbs Ferry, then Director of the Eastern Research Center and finally in 1979 became Director of Research of the Stauffer Chemical Company.

Scientific Accomplishments

Dr. Toy is certainly one of the most prominent pioneers of industrial phosphororganic chemistry as recognized by his numerous publications and more than 80 U.S. patents. His publications and patents are inspired to a great extent by his steady efforts to arrange even complicated chemical processes as simply as possible in order to enable their usage and commercialization on a technical scale.

Furthermore, as a crucial condition for any kind of chemical innovation, his scientific work has never lacked an imaginative, creative and unconventional element.

He has applied his knowledge to the research programmes of the two companies to which he belonged during his working life. These programmes led to a broad variety of important new products and processes; under his leadership new, commercially important areas have been opened up for phosphorus chemistry.

His extensive career in phosphorus chemistry has emphasized topics like flame retardants for textiles and plastics, metal extractants, chelating agents, surfactants, lubricants, stabilizers for plastics, food additives, economical processes for the synthesis of phosphorus insecticides and a considerable number of useful intermediates such as phenylphosphonous dichloride and derivatives as well as reactions leading to the formation of phosphorus-carbon bonds.

In the field of industrial chemicals, among others, Dr. Toy and his associates pioneered Stauffer Chemical Company's Fyrol flame retardants and were instrumental in the generation of new flame retardant plasticizers like the Phosflex types and new industrial hydraulic fluids such as the Fyrquel and Fyrgard series. Most of these products have attained annual production volumes in excess of one million pounds.

Since the applications of his outstanding scientific work impact virtually every area of "everyday living", it is almost impossible to single out the most important specific contribution. Accordingly, for today I would like to highlight three areas of his creative research which, in my opinion, combine both interesting reactions from a scientific point of view as well as the implementation of new and valuable technical innovations in a unique way.

Selected Scientific Contributions

Derivatives of Phosphoric Acid. Because of their application as agrochemicals and intermediates for organophosphorus chemistry, derivatives of pyrophosphoric acid have attracted considerable attention. In this area, Dr. Toy substantially developed the synthesis of precursors and improved processes for the commercial preparation of pyrophosphate insecticides.

As demonstrated in Figure 1, the reaction of dimethyl chlorophosphate, preferably with excess trimethyl phosphate at 105-125°C, leads after the elimination of chloromethane to tetramethyl pyrophosphate in 85 % yield (1).

The direct synthesis of tetraalkyl pyrophosphates can be achieved via hydrolysis of dialkyl chlorophosphates (2) which, if necessary, is conducted in the presence of a base like pyridine (Figure 2).

Thus, tetraethyl pyrophosphate (TEPP), which has gained substantial interest as a contact insecticide can be obtained almost quantitatively. Likewise, this pathway also makes accessible tetraamino pyrophosphates, such as the insecticide OMPA (3, 4).

As a further method, which is also suitable for the synthesis of mixed tetraalkyl pyrophosphates (5), the reaction of dialkyl chlorophosphates with dialkyl hydrogen phosphates in the presence of molar amounts of a tertiary amine has been established (Figure 3).

In the context of agricultural application, tetraalkyl dithiodiphosphates are especially interesting. According to the above described procedure, the base-facilitated hydrolysis of dialkyl chlorothiophosphates leads to the corresponding tetraalkyl dithiodiphosphates (6).

From this class of compounds, products like Sulfotep or ASPON, as shown in Figure 4, emerged. ASPON, which exhibits a very low mammalian toxicity, has been applied as a selective insecticide for the chinch bug until recently.

Dr. Toy has also made remarkable contributions to the development of the insecticides Parathion and Methylparathion (Figure 5).

The corresponding dialkyl chlorothiophosphates, as crucial precursors, are obtained in good yields and purity by the reaction of the easily accessible dialkyl dithiophosphates with phosphorus pentachloride (7).

As a further improvement, he discovered that the formation of Parathion, starting from diethyl chlorothiophosphate, is favored by the addition of catalysts such as tertiary amines and

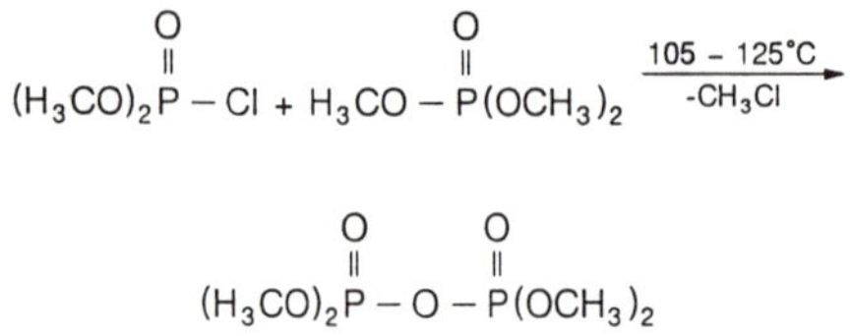

$$(H_3CO)_2 \overset{\overset{O}{\|}}{P} - Cl + H_3CO - \overset{\overset{O}{\|}}{P}(OCH_3)_2 \xrightarrow[-CH_3Cl]{105 - 125^\circ C}$$

$$(H_3CO)_2 \overset{\overset{O}{\|}}{P} - O - \overset{\overset{O}{\|}}{P}(OCH_3)_2$$

Figure 1. Synthesis of Tetramethyl pyrophosphate

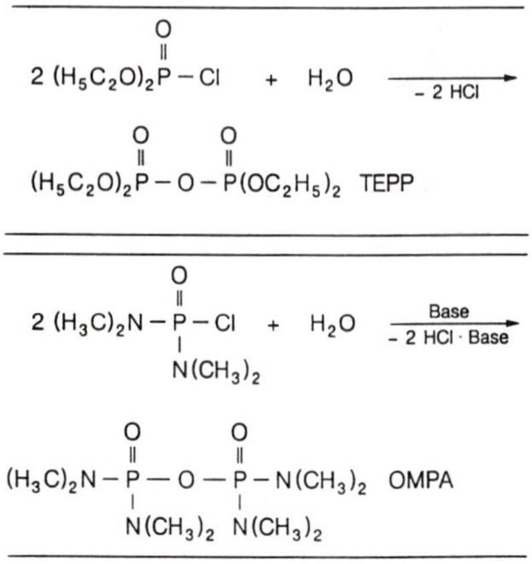

$$2 (H_5C_2O)_2 \overset{\overset{O}{\|}}{P} - Cl + H_2O \xrightarrow[-2 HCl]{}$$

$$(H_5C_2O)_2 \overset{\overset{O}{\|}}{P} - O - \overset{\overset{O}{\|}}{P}(OC_2H_5)_2 \quad TEPP$$

$$2 (H_3C)_2 N - \underset{\underset{N(CH_3)_2}{|}}{\overset{\overset{O}{\|}}{P}} - Cl + H_2O \xrightarrow[-2 HCl \cdot Base]{Base}$$

$$(H_3C)_2 N - \underset{\underset{N(CH_3)_2}{|}}{\overset{\overset{O}{\|}}{P}} - O - \underset{\underset{N(CH_3)_2}{|}}{\overset{\overset{O}{\|}}{P}} - N(CH_3)_2 \quad OMPA$$

Figure 2. Pyrophosphate insecticides

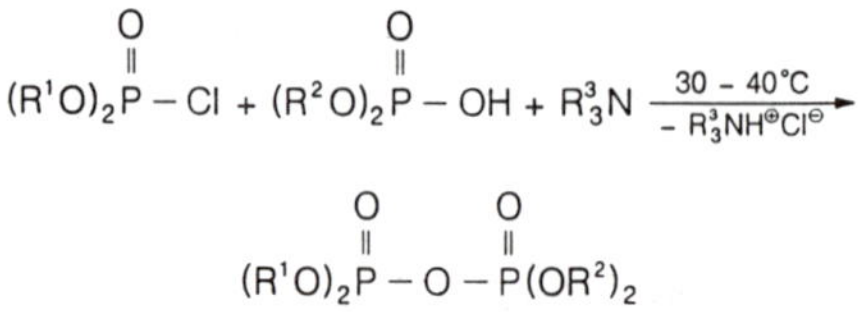

$$(R^1O)_2 \overset{\overset{O}{\|}}{P} - Cl + (R^2O)_2 \overset{\overset{O}{\|}}{P} - OH + R_3^3 N \xrightarrow[-R_3^3 NH^\oplus Cl^\ominus]{30 - 40^\circ C}$$

$$(R^1O)_2 \overset{\overset{O}{\|}}{P} - O - \overset{\overset{O}{\|}}{P}(OR^2)_2$$

Figure 3. Tetraalkyl pyrophosphates

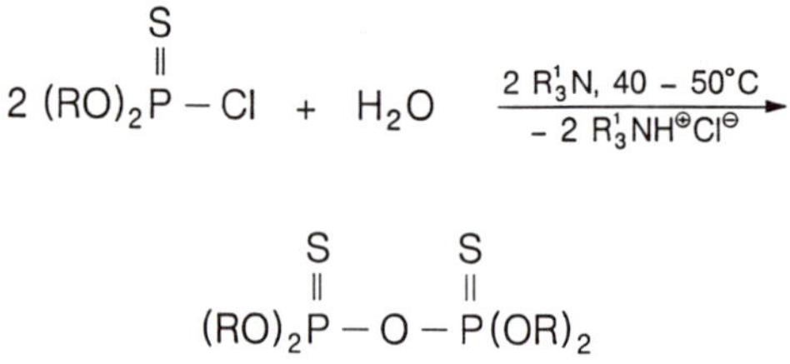

$$2 (RO)_2 \overset{\overset{S}{\|}}{P} - Cl + H_2O \xrightarrow[-2 R_3^1 NH^\oplus Cl^\ominus]{2 R_3^1 N, 40 - 50^\circ C}$$

$$(RO)_2 \overset{\overset{S}{\|}}{P} - O - \overset{\overset{S}{\|}}{P}(OR)_2$$

Sulfotep (R = C_2H_5)

ASPON (R = nC_3H_7)

Figure 4. Tetraalkyl dithiodiphosphates

xvi

phosphines. Using the sodium salt of p-nitro-phenol, the synthesis of Parathion furthermore can be so accomplished at moderate temperatures with high yields (8).

Because of an equivalent effectiveness but a significantly decreased mammalian toxicity, nowadays the sales volume of Methylparathion (about $ 220 million annually) well exceeds that of Parathion.

As Dr. Toy found, not only dialkyl dithiophosphates but also tetraphosphorus decasulfide is a valuable starting material for numerous syntheses leading to versatile and useful intermediates for organophosphorus syntheses (Figure 6).

For example, the reaction of alkylchlorides with tetraphosphorus decasulfide under pressure at elevated temperatures (200-270°C) leads to the formation of alkyl dichlorodithiophosphates and dialkyl chlorotrithiophosphates. The transformation of the latter into alkyl dichlorodithiophosphates by reaction with thiophosphoryl trichloride can easily be achieved. This two-step process can also be carried out as a "one-pot reaction" with satisfactory yields (9).

Starting from dialkyl sulfides, tetraphosphorus decasulfide and thiophosphoryl trichloride (Figure 7), a similar reaction, depending on the applied ratio of the starting materials, furnishes alkyl dichlorodithiophosphates and dialkyl chlorotrithiophosphates. Omitting thiophosphoryl trichloride in this type of reaction leads to trialkyl tetrathiophosphates (10). For example, triethyl tetrathiophosphate can be obtained in 90 % yield according to this method (11).

Dichloro Organo Phosphines. In 1873 A. Michaelis discovered the reaction of benzene with phosphorus trichloride leading to dichloro phenylphosphine. But the technical application could not be achieved until much later when the said process was successfully influenced by applying suitable catalysts (Figure 8).

Dr. Toy for example demonstrated that at 550°C chlorobenzene exhibits a considerable catalytic activity in this reaction (12).

Another reaction, leading to dichloro phenylphosphine starts from chlorobenzene, phosphorus and phosphorus trichloride. This process is advantageously carried out at 300-360°C under pressure. Dr. Toy's endeavor in optimizing the reaction finally led to yields of 80 % (13).

Furthermore, as outlined in Figure 9, Dr. Toy developed a process for the synthesis of dichloro chloromethyl phosphine, starting from the readily accessible chloromethyl phosphonic dichloride. This involves the treatment with thiophosphoryl trichloride yielding chloromethyl thiophosphonic dichloride (14). Subsequent desulfuration with dichloro phenylphosphine smoothly leads to dichloro chloromethyl phosphine in 85 % yield (14, 15). In this context, the chemical properties of this compound also were thoroughly investigated. Most recently this compound was found to be a very useful starting material for the synthesis of heterophospholes by Schmidtpeter et. al.

An interesting and remarkable process was created for the generation of alkane bis-dichlorophosphines which may serve as starting materials for chelating agents with catalytic properties

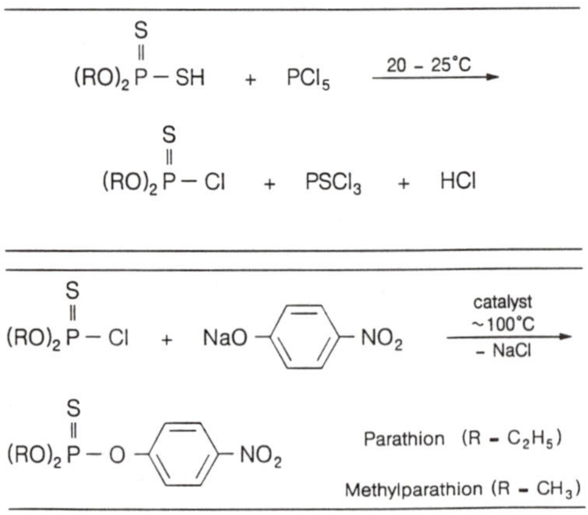

$$(RO)_2 \overset{\overset{\text{S}}{\|}}{P} - SH \quad + \quad PCl_5 \quad \xrightarrow{\;20 - 25°C\;}$$

$$(RO)_2 \overset{\overset{\text{S}}{\|}}{P} - Cl \quad + \quad PSCl_3 \quad + \quad HCl$$

$$(RO)_2 \overset{\overset{\text{S}}{\|}}{P} - Cl \quad + \quad NaO -\!\!\!\left\langle\!\!\!\bigcirc\!\!\!\right\rangle\!\!\!- NO_2 \quad \xrightarrow[- NaCl]{\overset{\text{catalyst}}{\sim 100°C}}$$

$$(RO)_2 \overset{\overset{\text{S}}{\|}}{P} - O -\!\!\!\left\langle\!\!\!\bigcirc\!\!\!\right\rangle\!\!\!- NO_2 \qquad \text{Parathion } (R = C_2H_5)$$

Methylparathion $(R = CH_3)$

Figure 5. Syntheses of Parathion type insecticides

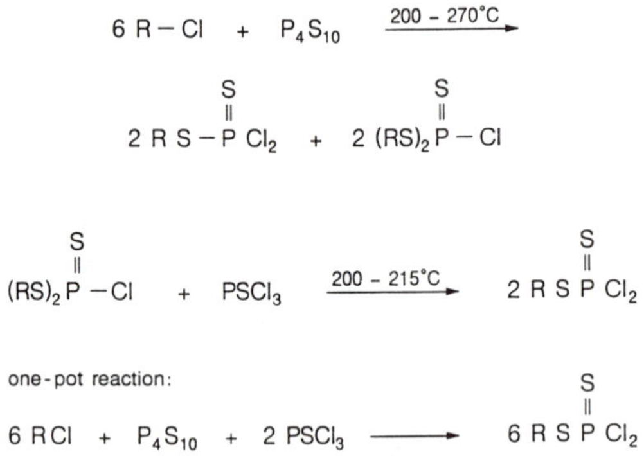

$$6 \, R - Cl \quad + \quad P_4S_{10} \quad \xrightarrow{\;200 - 270°C\;}$$

$$2 \, R \, S - \overset{\overset{\text{S}}{\|}}{P} \, Cl_2 \quad + \quad 2 \, (RS)_2 \overset{\overset{\text{S}}{\|}}{P} - Cl$$

$$(RS)_2 \overset{\overset{\text{S}}{\|}}{P} - Cl \quad + \quad PSCl_3 \quad \xrightarrow{\;200 - 215°C\;} \quad 2 \, R \, S \, \overset{\overset{\text{S}}{\|}}{P} \, Cl_2$$

one-pot reaction:

$$6 \, RCl \quad + \quad P_4S_{10} \quad + \quad 2 \, PSCl_3 \quad \longrightarrow \quad 6 \, R \, S \, \overset{\overset{\text{S}}{\|}}{P} \, Cl_2$$

Figure 6. Alkyl chlorothiophosphates from Tetraphosphorus decasulfide

$$6 \, R_2S \quad + \quad P_4S_{10} \quad + \quad 8 \, PSCl_3 \quad \xrightarrow{\;200°C\;} \quad 12 \, RS - \overset{\overset{\text{S}}{\|}}{P} \, Cl_2$$

$$6 \, R_2S \quad + \quad P_4S_{10} \quad + \quad 2 \, PSCl_3 \quad \xrightarrow{\;200°C\;} \quad 6 \, (RS)_2 \overset{\overset{\text{S}}{\|}}{P} \, Cl$$

$$6 \, (H_5C_2)_2S \quad + \quad P_4S_{10} \quad \xrightarrow{\;210 - 220°C\;} \quad 4 \, (H_5C_2S)_3 \overset{\overset{\text{S}}{\|}}{P} \; 90\%$$

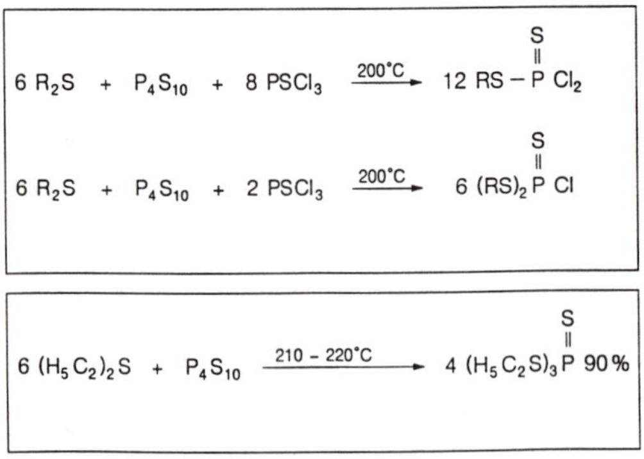

Figure 7. Syntheses with Tetraphosphorus decasulfide

xviii

(16) (Figure 10). In the simplest case, ethylene is reacted with phosphorus trichloride and elemental phosphorus under pressure at 160-200°C. Various other olefins can readily be used in this reaction (16).

Derivatives of Phosphonic Acids. A very noteworthy method for the production of arylphosphonic dichlorides starting from arene dichlorophosphines and phosphorus pentoxide leads, in the most prominent case, to phenylphosphonic dichloride. Hereby, according to Figure 11, the mixture of dichloro phenylphosphine and phosphorus pentoxide is chlorinated at 130-150°C (17).

Phosphonic acid derivatives have found widespread applications as flame retardants for plastics and textiles.

For example, the reaction of phenylphosphonic dichloride with aromatic dihydroxy compounds opens a route to polyphosphonates (18). This pioneer patent was followed by a considerable number of further applications in this area (e.g. 19). These polymers are especially useful as flame retardants for poly-(ethyleneterephthalate) molding formulations.

Starting in the 40's, Dr. Toy and his co-workers at Victor Chemical Works began to develop flame-proof polymers and later flame retardants for fibers like cotton and cellulose, based on optionally partly brominated diallyl phosphonates. (Figure 12) (20). A well known compound from this series, introduced as "Phoresins", is diallyl cyanoethylphosphonate (21). Other flame retardants for rayon have been similarly established.

A comparable, low-priced phosphorus-based monomer, suitable for the flame protection of various materials, is Bis-(β-chloroethyl) vinylphosphonate, developed commercially by Stauffer under the trade name Fyrol Bis-Beta (22).

The three-step synthesis (Figure 13) comprises the addition of ethylene oxide to phosphorus trichloride, subsequent thermal Arbusow rearrangement of tris-2-chloroethyl phosphite, and, finally, the elimination of hydrogen chloride (23). The final dehydrochlorination is advantageously carried out as a continous process using basic aluminum oxide (24).

This important product was successfully employed as a flame retardant cross-linker for curable unsaturated polyesters as well as for reinforced polyesters, epoxy resins, rigid polyurethanes and synthetic latices and binders in textile and paper manufacturing (22, 25, 26).

Furthermore Fyrol Bis-Beta serves as a precursor for the synthesis of a water-soluble vinylphosphonate oligomer, which was introduced in 1974 by Stauffer under the trade name Fyrol 76.

Fyrol 76 is a reactive, curable flame retardant for textiles, in particular for cotton fabric (27, 28), where it co-reacts with N-methylol-acrylamide in the presence of a free radical catalyst such as a persulfate. In the 70's the application of Fyrol 76 was one of the most common methods to obtain flame-proof cotton, which was utilized, for example, by Cone Mills Corporation for the manufacture of flame-proof pajamas for children.

In 1976, the US market for flame retardants applied to cotton was estimated at 2 million lb./year. Presumably, about half of this amount was attributed to Fyrol 76.

$$C_6H_6 \;+\; PCl_3 \quad \xrightarrow[-HCl]{\substack{\text{catalyst} \\ 550°C}} \quad H_5C_6 - PCl_2$$

catalyst: monochlorobenzene

$$3\,H_5C_6Cl \;+\; 2\,P \;+\; PCl_3 \quad \xrightarrow{300-360°C}$$

$$3\,H_5C_6 - PCl_2$$

Figure 8. Dichloro phenylphosphine

$$ClCH_2 - \overset{\overset{\displaystyle O}{\|}}{P}Cl_2 \;+\; PSCl_3 \quad \xrightarrow{200°C}$$

$$ClCH_2 - \overset{\overset{\displaystyle S}{\|}}{P}Cl_2 \;+\; POCl_3$$

$$ClCH_2 - \overset{\overset{\displaystyle S}{\|}}{P}Cl_2 \;+\; C_6H_5 - PCl_2 \quad \xrightarrow{165-175°C}$$

$$ClCH_2 - PCl_2 \;+\; H_5C_6 - \overset{\overset{\displaystyle S}{\|}}{P}Cl_2$$

Figure 9. Dichloro chloromethyl phosphine

$$3\,H_2C = CH_2 \;+\; 2\,P \;+\; 4\,PCl_3 \quad \xrightarrow{160-200°C}$$

$$3\,Cl_2P - CH_2CH_2 - PCl_2 \quad \sim 60\%$$

Figure 10. Ethylene bis–dichlorophosphine

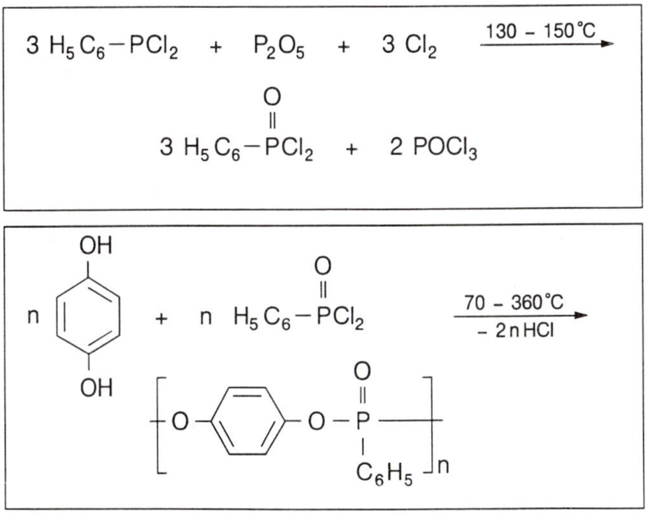

$$3\ H_5C_6-PCl_2\ +\ P_2O_5\ +\ 3\ Cl_2 \xrightarrow{130\ -\ 150\,°C}$$

$$3\ H_5C_6-\overset{\overset{\displaystyle O}{\|}}{P}Cl_2\ +\ 2\ POCl_3$$

$$n\ \underset{OH}{\overset{OH}{\bigcirc}}\ +\ n\ H_5C_6-\overset{\overset{\displaystyle O}{\|}}{P}Cl_2 \xrightarrow[-\,2n\,HCl]{70\ -\ 360\,°C}$$

$$\left[\!-O-\bigcirc-O-\overset{\overset{\displaystyle O}{\|}}{\underset{\underset{\displaystyle C_6H_5}{|}}{P}}\!-\!\right]_n$$

Figure 11. Phenyl phosphonic dichloride and Polyphosphonates

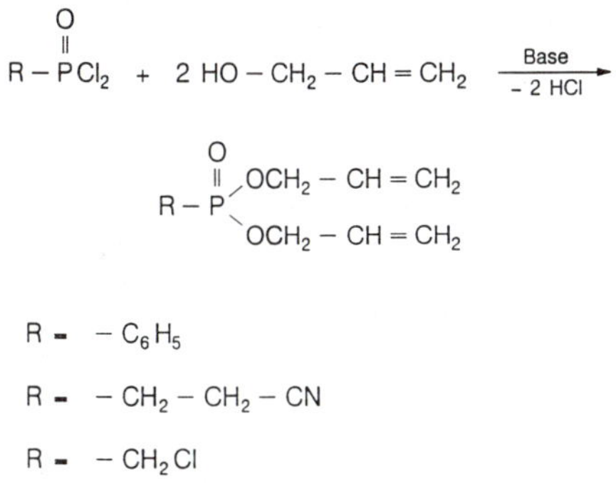

$$R-\overset{\overset{\displaystyle O}{\|}}{P}Cl_2\ +\ 2\ HO-CH_2-CH=CH_2 \xrightarrow[-\,2\,HCl]{Base}$$

$$R-\overset{\overset{\displaystyle O}{\|}}{P}\overset{\displaystyle OCH_2-CH=CH_2}{\underset{\displaystyle OCH_2-CH=CH_2}{\big\langle}}$$

$$R\ -\ \ -C_6H_5$$

$$R\ -\ \ -CH_2-CH_2-CN$$

$$R\ -\ \ -CH_2Cl$$

Figure 12. Diallyl phosphonates

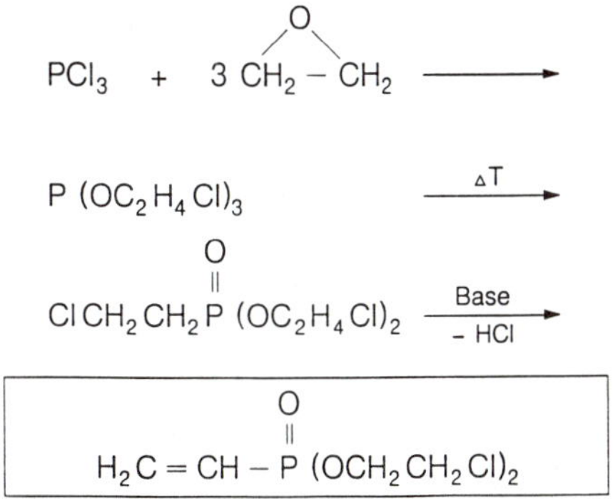

$$PCl_3\ +\ 3\ \overset{\displaystyle O}{\overset{\diagup\ \diagdown}{CH_2-CH_2}} \longrightarrow$$

$$P\,(OC_2H_4Cl)_3 \xrightarrow{\Delta T}$$

$$ClCH_2CH_2\overset{\overset{\displaystyle O}{\|}}{P}\,(OC_2H_4Cl)_2 \xrightarrow[-\,HCl]{Base}$$

$$H_2C=CH-\overset{\overset{\displaystyle O}{\|}}{P}\,(OCH_2CH_2Cl)_2$$

Figure 13. Fyrol Bis-Beta

xxi

In general, the flame protection of both rigid and flexible
polyurethane foams was an important area in the development of
phosphorus-containing flame retardants. Some of these compounds
are of the additive type, while others, containing functional
groups, as already shown above, are of the reactive type. One of
the most important flame retardants, especially for rigid urethane foam was Fyrol 6 (Figure 14). It is a reactive type and incorporates permanently in the rigid foam structure.

Further important commercial flame retardants, developed
by associates of Dr. Toy are Fyrol 99 (non-reactive for soft
polyurethane foams) as well as halogen-free compounds like Fyrol
51 e.g. for polyurethanes or Fyrol HMP for the fire-protection of
automotive air filters (29). In the application for polymer
films, in addition to flame protection, Fyrol HMP was (now
Victastab HMP) found to enhance the tensile strength of Mylar
polyester films considerably.

Finally I would like to mention Dr. Toy's significant contributions to the synthesis of various chloro-thiophosphonates
and phosphinates which have been widely used for the manufacture
of related agrochemicals such as insecticides and fungicides.

Alkanes, as well as several arenes like benzene, are reacted
with thiophosphoryl trichloride under pressure at temperatures of
280-340°C to yield thiophosphonic dichlorides and thiophosphinic
chlorides (30). Starting from ethane, depending on the reaction
conditions, ethyl thiophosphonic dichloride and diethyl thiophosphinic chloride are obtained as outlined in Figure 15.

A similar reaction can be utilized by reacting or-
ganochlorides with thiophosphoryl trichloride and elemental
phosphorus (31).

Dr. Toy further demonstrated that a combination of the
systems chloroalkane, phosphorus trichloride, tetraphosphorus
decasulfide and elemental phosphorus can be utilized for the
synthesis of thiophosphonic dichlorides (Figure 16).

According to this pathway, for example, methyl thiophosphonic dichloride is obtained in 80 % yield (32).

Optionally, this product can be generated by the reaction of
dimethyl disulfide, phosphorus trichloride and elemental phosphorus. This reaction proceeds at temperatures around 300°C, and,
depending on conditions, may produce the corresponding thiophosphinic chloride as a by-product (33).

As a last example, which completes the series of imaginative
syntheses for thiophosphonic dichlorides, in Figure 17 I would
like to highlight the cleavage of alkyl thiophosphoryl chlorides
and trialkyl tetrathiophosphates, which where discussed earlier,
with phosphorus trichloride at 300°C, which, for example, leads
to methylphosponothioic dichloride as well as thiophosphoryl
trichloride (34).

In summary, Dr. Toy in this field devised numerous new
syntheses for several key intermediates, including phenyl thiophosphonic dichloride and ethyl thiophosphonic dichloride. The
former enabled the breakthrough to be made with insecticides like
Du Pont's EPN. Ethyl thiophosphonic dichloride is an intermediate
for Stauffer's Dyfonate (Figure 18). The latter was introduced in
1967 for soil application with a high stability against

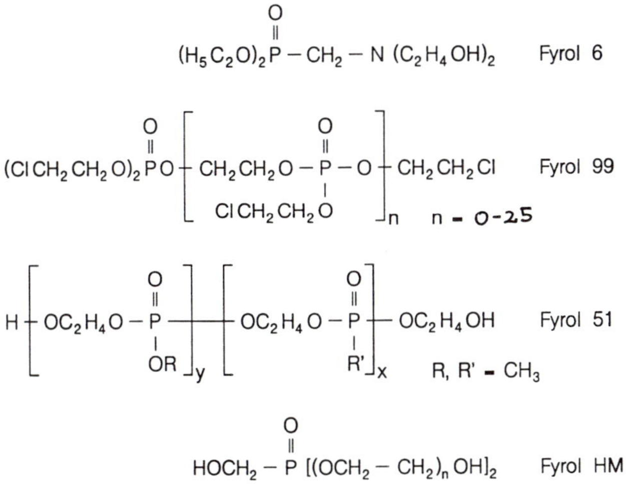

$$
(H_5C_2O)_2\overset{\overset{\displaystyle O}{\|}}{P} - CH_2 - N\,(C_2H_4OH)_2 \qquad \text{Fyrol 6}
$$

$$
(ClCH_2CH_2O)_2\overset{\overset{\displaystyle O}{\|}}{P}O \left[CH_2CH_2O - \overset{\overset{\displaystyle O}{\|}}{\underset{\underset{\displaystyle ClCH_2CH_2O}{|}}{P}} - O \right]_n CH_2CH_2Cl \qquad \text{Fyrol 99} \qquad n = 0\text{-}25
$$

$$
H \left[OC_2H_4O - \overset{\overset{\displaystyle O}{\|}}{\underset{\underset{\displaystyle OR}{|}}{P}} \right]_y \left[OC_2H_4O - \overset{\overset{\displaystyle O}{\|}}{\underset{\underset{\displaystyle R'}{|}}{P}} \right]_x OC_2H_4OH \qquad \text{Fyrol 51} \qquad R,\,R' = CH_3
$$

$$
HOCH_2 - \overset{\overset{\displaystyle O}{\|}}{P} \,[(OCH_2 - CH_2)_n\,OH]_2 \qquad \text{Fyrol HM}
$$

Figure 14. Fyrol flame retardants

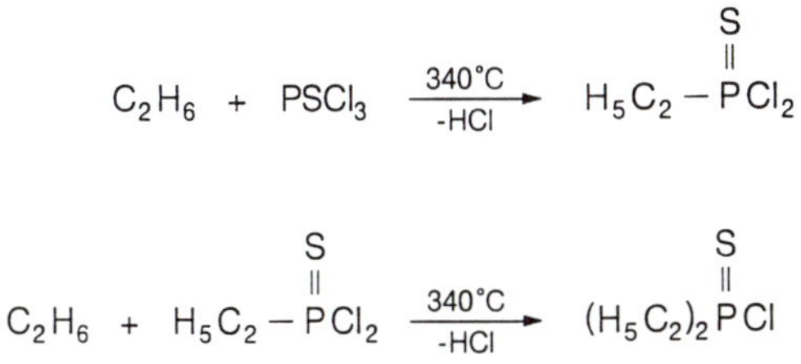

$$
C_2H_6 \;+\; PSCl_3 \quad \xrightarrow[\text{-HCl}]{340^\circ C} \quad H_5C_2 - \overset{\overset{\displaystyle S}{\|}}{P}Cl_2
$$

$$
C_2H_6 \;+\; H_5C_2 - \overset{\overset{\displaystyle S}{\|}}{P}Cl_2 \quad \xrightarrow[\text{-HCl}]{340^\circ C} \quad (H_5C_2)_2 \overset{\overset{\displaystyle S}{\|}}{P}Cl
$$

Figure 15. Ethyl thiophosphonic dichloride

$$
10\;CH_3Cl \;+\; P_4S_{10} \;+\; 4\;PCl_3 \;+\; 2.7\;P \quad \xrightarrow{300^\circ C}
$$
$$
10\;H_3C\overset{\overset{\displaystyle S}{\|}}{P}Cl_2 \;+\; 0.7\;PCl_3
$$

$$
3\;H_3CSSCH_3 \;+\; 4\;PCl_3 \;+\; 2\;P \quad \xrightarrow{300^\circ C}
$$
$$
6\;H_3C\overset{\overset{\displaystyle S}{\|}}{P}Cl_2
$$

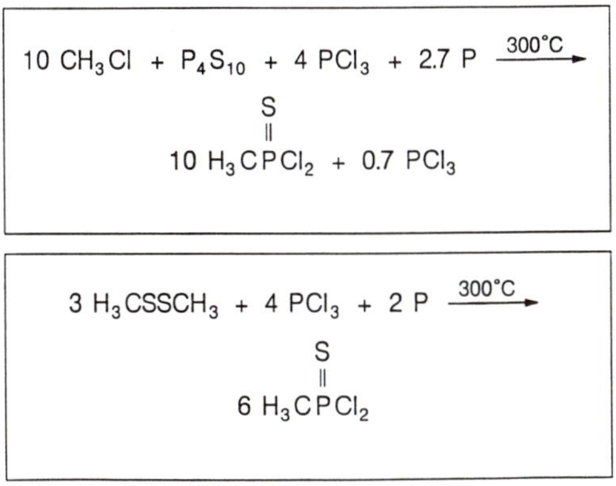

Figure 16. Syntheses of Methyl thiophosphonic dichloride

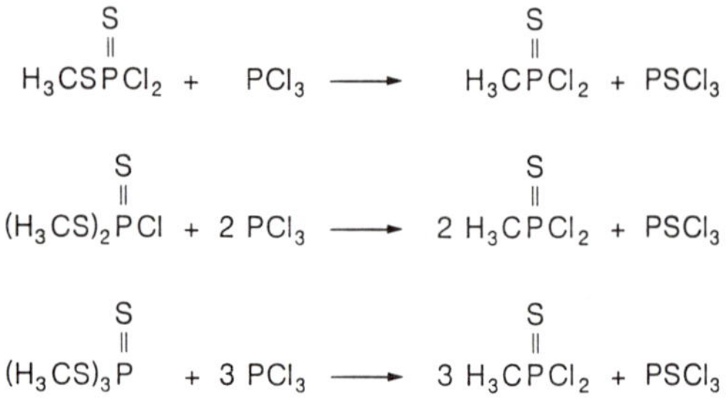

$$\underset{H_3CSPCl_2}{\overset{\overset{\text{S}}{\|}}{}} + PCl_3 \longrightarrow \underset{H_3CPCl_2}{\overset{\overset{\text{S}}{\|}}{}} + PSCl_3$$

$$\underset{(H_3CS)_2PCl}{\overset{\overset{\text{S}}{\|}}{}} + 2\,PCl_3 \longrightarrow 2\,\underset{H_3CPCl_2}{\overset{\overset{\text{S}}{\|}}{}} + PSCl_3$$

$$\underset{(H_3CS)_3P}{\overset{\overset{\text{S}}{\|}}{}} + 3\,PCl_3 \longrightarrow 3\,\underset{H_3CPCl_2}{\overset{\overset{\text{S}}{\|}}{}} + PSCl_3$$

Figure 17. Methyl thiophosphonic dichloride by cleavage reactions with PCl_3

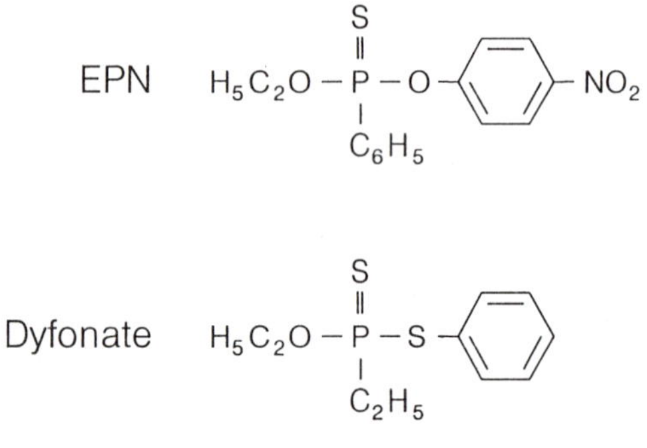

EPN $\quad H_5C_2O - \overset{\overset{\text{S}}{\|}}{\underset{C_6H_5}{P}} - O - \langle\!\!\!\!\bigcirc\!\!\!\!\rangle - NO_2$

Dyfonate $\quad H_5C_2O - \overset{\overset{\text{S}}{\|}}{\underset{C_2H_5}{P}} - S - \langle\!\!\!\!\bigcirc\!\!\!\!\rangle$

Figure 18. Insecticides EPN and Dyfonate

hydrolysis and showed a tremendous increase in sales and market penetration (35, 36).

After this survey of some main areas of Dr. Toy's scientific work, let me finally show that he has not only confined his energies to work in company research departments. His professional activities and commitment to the chemical sciences extend far beyond company requirements.

Other activities

His special affinity for, and links to, phosphorus chemistry occasioned him to write two books on the subject. His first work, "The Chemistry of Phosphorus", appeared in 1975 and was aimed at specialists working in the field of industrial phosphorus chemistry. In contrast, his second book, entitled "Phosphorus Chemistry in Everyday Living", was written for a broader range of readers with a general interest in science. The first edition was published in 1976, and the second, with Edward Walsh as co-author, in 1987.

Dr. Toy's major contribution to the activities of the American Chemical Society deserves a special mention. He has played an active role in a tremendous number of local and national ACS committees. To include them all in my address today and to duly acknowledge Arthur's involvement in them is simply not possible, so permit me to mention just a few. From 1972 to 1974 he was the Chairman-Elect and from 1974 to 1976 Chairman of the New York section of the ACS. He also served as Chairman of the Program Committee, and Juror and Chairman of the Nominating Committee for the Nichols Medal. As for the American Chemical Society nationwide, he was a councilor, a member of the Committee on International Activities, member of the Budget and Finance Committee and numerous sub-committees, Chairman of the Arrangement Committee for the National ACS Centennial Meeting in 1976, and a member of the Committee on Program Review. He also served a term in the Council Policy Committee.

Dr. Toy has also been active internationally. Apart from belonging to the Royal Society of London, the Chinese-American Chemical Society, and being a former member of the Association of German Chemists, he was for years a member of the Scientific Board of the International Conference on Phosphorus Chemistry and served as Associate Technical Director in the organisation of the NRC/ACS workshop to Egypt. After his retirement he presented a series of lectures at the Research Institute of Elemental Organic Chemistry at Nankai University in Tientsin. He also lectured at various other research institutes in China at the same time as his mother country, China, was opening up to the outside world.

For his services to the New York Section of the American Chemical Society, Dr. Toy received the Outstanding Service Award in 1978. He was the recipient of the Distinguished Alumnus Award from his school, the Joliet Junior College, in 1985, and the Eli Whitney Award of the Connecticut Patent Law Association in 1988. Dr. Toy has been elected to Phi Beta Kappa, Sigma Xi and Phi Lambda Upsilon.

Conclusion

Dr. Toy is an outstanding example of a personality who has dedicated his life to research in industrial chemistry. With his scientific talents and his ability to create a stimulating and challenging atmosphere of scientific cooperation, he has brilliantly fulfilled all the requirements expected from an industrial research leader. He has made remarkable contributions to our knowledge in the field of Phosphorus Chemistry, which have been recognized worldwide. He has served the chemical community with his untiring engagement for the American Chemical Society and other scientific institutions within and outside the United States.

Literature Cited

1) Toy, A. D. F. J. Am. Chem. Soc. 1949, **71**, 2268
2) Toy, A. D. F. USP. 2 504 165 (1950)
3) Toy, A. D. F.; Costello Jr., J. R. USP. 2 706 738 **(1955)**
4) Toy, A. D. F.; Costello Jr., J. R. USP. 2 717 249 **(1955)**
5) Toy, A. D. F. USP. 2 683 733 **(1954)**
6) Toy, A. D. F. USP. 2 663 722 **(1953)**
7) Toy, A. D. F.; McDonald, G. A. USP. 2 715 136 **(1955)**
8) Toy, A. D. F.; Beck, T. M. USP. 2 471 464 **(1949)**
9) Uhing, E. H.; Toy, A. D. F. USP. 3 879 500 **(1975)**
10) Uhing, E. H.; Toy, A. D. F. USP. 4 100 230 **(1978)**
11) Toy, A. D. F.; Uhing, E. H. USP. 4 034 024 **(1977)**
12) Toy, A. D. F.; Cooper, R. S. USP. 3 029 282 **(1962)**
13) Via, F. A.; Uhing, E. H.; Toy, A. D. F. USP. 3 864 394 **(1975)**
14) Uhing, E.; Rattenbury, K.; Toy, A. D. F. J. Am. Chem. Soc. 1961, **83**, 2299
15) Toy, A. D. F.; Rattenbury, K. H. USP. 3 244 745 **(1966)**
16) Toy, A. D. F.; Uhing, E. H. USP. 3 976 690 **(1976)**
17) Toy, A. D. F. USP. 2 482 810 **(1949)**
18) Toy, A. D. F. USP. 2 435 252 **(1948)**
19) Toy, A. D. F. USP. 2 572 076 **(1951)**
20) Toy, A. D. F. J. Am. Chem. Soc. 1948, **70**, 186, see also Toy, A. D. F.; Brown, L. V. Ind. Eng. Chem. 1948, **40**, 2276
21) Toy, A. D. F.; Rattenbury, K. H. USP. 2 735 789 **(1956)**
22) Weil, E. D. J. Fire and Flammability/Flame Retardant Chem. 1974, Vol. 1, 125
23) Leupold, E. USP. 2 959 609 **(1960)**
24) Stamm, W. USP. 3 576 924 **(1971)**, 3 725 300 **(1973)**
25) Adler, A.; Brenner, W. Nature 1970, **225** 60
26) Kraft, P.; Brunner, R. USP. 3 691 127 **(1972)** Kraft, P.; Yuen, S. DE-PS. 2 452 369 **(1975)**
27) Eisenberg, B. J.; Weil, E. D. Textile Chemist and Colorist, 1974, **6**, 180
28) Eisenberg, B. J.; Weil, E. D. Pap. Math. Techn. Conf. Am. Assoc. Text. Chem. Color. 1975, 17-26
29) Weil, E. D.; Phosphorus Based Flame Retardants, in Flame Retard. Polym. Mater. 1978, **2**, 103-133; Levin, M.; Atlas, S. M.; Pearce E. M., Ed.; Plenum Press, New York and London 1978
30) Uhing, E. H.; Toy, A. D. F. USP. 3 790 629 **(1974)**

31) Toy, A. D. F.; Uhing, E. H. USP. 3 726 918 (1973)
32) Uhing, E. H.; Toy, A. D. F. USP. 3 803 226 (1974)
33) Uhing, E. H.; Toy, A. D. F. USP. 4 000 190 (1976)
34) Toy, A. D. F.; Uhing, E. H. USP. 3 962 323 (1976)
35) Büchel, K. H. Pflanzenschutz und Schädlingsbekämpfung; Georg Thieme Verlag: Stuttgart 1977
36) Fest, C.; Schmidt, K.-J. The Chemistry of Organophosphorus Pesticides; Springer Verlag 1973

RECEIVED January 17, 1992

Research in Industrial Phosphorus Chemistry

A Reminiscence

Arthur D. F. Toy

14 Katydid Lane, Stamford, CT 06903

The research effort on industrial phosphorus chemistry by my colleagues and I have led to some technical successes. However in industrial research, the bottom line is commercial success. Fortunately, some of our technical successes have resulted in phosphorus products which were commercialized and which made a profit. For research success, much has been written on the need for good inspiration, careful planning based on theoretical considerations and hard work. Our experiences have shown that also needed is alertness to the unexpected or serendipity and a willingness to change research direction when more promising leads are uncovered. Also important is the courage to go ahead and do the work instead of being inhibited by prior knowledge which may not be true. Finally there is no substitute for being lucky. Examples will be given on the path ways some of our research efforts on phosphorus chemistry followed which led to successful commercial products.

I had thought that old chemists are like old soldiers, they just fade away. In fact, I learned first hand that chemists who got into research management faded away much faster then those who remained as active scientists. About five years before my retirement, my wife and I were attending a national ACS meeting. We stopped at an information booth manned by one of the local chemists to ask for directions. The chemist looked at my name tag and asked "Are you the Dr. Toy who used to publish on phosphorus chemistry?" I told him yes. Then he blurted out, "I thought you were dead."

NOTE: This is the text of the address given by Dr. Toy at the symposium on which this book is based.

0097–6156/92/0486–xxix$06.00/0

I haven't seen any of your publications for such a long time." My wife assured him that I was not dead, that I just got into research management.

I thought it may be of interest, especially to academic chemists, for me to discuss some of my experiences in research on phosphorus chemistry in an industrial laboratory. In thinking back on my experiences as an industrial chemist, I found that there really isn't very much difference between industrial and academic research. I know that some of my academic friends think otherwise. The other day a professor friend called to tell me about two of his former students, who are husband and wife. The professor said that the wife had found an industrial position, but the husband was having difficulty finding a tenure-track teaching position. I asked him why didn't the husband also try for an industrial position. The professor's quick reply was "But he is much too good a scientist to go into industry!"

I have found that there are very good scientists in academia and also some very good scientists working in industry. I feel that the major difference between industrial research and academic research is that in industrial research, the final goal is to make money for the organization. In other words, it is necessary to have technical success followed by commercial success. In the course of such industrial research, new knowledge may be discovered. Some of the new knowledge occasionally is even published in the scientific literature, but usually it ends up in patent literature which is not so widely read. In academic research, as you all know, the main goal is for the discovery of new knowledge and the dissemination of that knowledge through publications. The goal is technical success only. However, in many cases such new knowledge is of sufficient industrial importance that the professor involved becomes very wealthy.

My experiences have also taught me that whether in academic or industrial research, success depends a great deal on good inspiration, careful thinking, meticulous planning, and hard work, but more important is an alertness for the unexpected or serendipitous event. Finally, there is no substitute for being lucky.

I found out about the importance of good luck and serendipity in research quite early in my research career. In fact, it was serendipity that indirectly got me into research in phosphorus chemistry   When I joined Professor L. F. Audrieth's research group at the University of Illinois, he was changing direction on one of his research projects. In the late 1930's an important antibacterial was sulfanilamide.

$$H_2NSO_2\langle O \rangle NH_2$$
Sulfanilamide

Prof. Audrieth thought, based on Prof. Reynold Fuson's Principle of Vinylogy (1) which was in vogue in those days, that it may not be necessary to have the benzene ring, (which represents three vinyl groups) between the amine and the sulfonic acid group, in order for the compound to have physiological properties. To check on that hypothesis the research project was assigned to the graduate student Mike Sveda. Sveda made many sulfonic acid derivatives of organic amines. As some of you may have heard the story, Sveda accidentally dropped his cigarette on one of the many crystalline compounds he had drying on watch glasses on his desk. When he dusted off the cigarette and smoked it, he discovered that it had a sweet taste. So he tasted all the crystals he had on his desk and found that the compound responsible for the sweet taste was sodium cyclohexylsulfonamide (2):

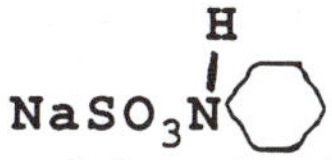

$$NaSO_3\overset{\overset{\displaystyle H}{|}}{N}\hspace{-0.3em}\bigcirc$$

Sodium cyclohexylsulfonamide,
main ingredient of Cyclamate, a sweetener

That compound eventually become the main ingredient for Cyclamate, the artificial sweetener which enjoyed commercial success in the U. S. for many years and is still used in many foreign countries. Incidentally, that is the only case I know in which the of smoking a cigarette had a beneficial effect.

Prof. Audrieth, with this new lead, decided to expand his program on synthetic sweeteners and to study the relationship between chemical structure and sweetness. Since phosphorus is the next door neighbor to sulfur in the periodic table, I was assigned the project of looking into the synthesis and properties of the organic amides of phosphoric acid. I prepared many such amides of phosphoric acid, including cyclohexylphosphoramide:

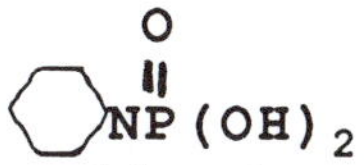

$$\bigcirc\hspace{-0.3em}N\overset{\overset{\displaystyle O}{||}}{P}(OH)_2$$

Cyclohexylphosphoramide

I tasted every one of them and none was sweet. Based on what I know today about the toxicity of some of the organic phosphorus compounds, I would hesitate to taste again any new phosphorus compound.

With this background, when I joined Victor Chemical Works in 1942, the first thing I did was to talk to some of the older chemists there to find out how they made some of their discoveries. They were very helpful in getting me introduced into industrial chemistry. One of the very important new compounds Victor was introducing into the market at that time was the new baking acid, the coated anhydrous monocalcium acid phosphate. It was given the trade name of V-90. Mr. Julian Schlaeger, the inventor of the compound, told me this story: He said that the leavening action in the baking of biscuit with monocalcium acid phosphate as the baking acid is due to the carbon dioxide gas liberated by the action of the acidic monocalcium phosphate on the sodium bicarbonate (equation 1).

$$3CaH_4(PO_4)_2 \cdot H_2O + 8NaHCO_3 \longrightarrow$$

$$8CO_2 + Ca_3(PO_4)_2 + 4Na_2HPO_4 + 11H_2O \quad (1)$$

Monocalcium acid phosphate has many desirable properties as the baking acid, but it has one undesirable property. It is too reactive. In biscuit making, its fast reaction with sodium bicarbonate caused the release of two thirds of the available carbon dioxide gas during the dough mixing stage. Some of the released gas stays in the dough, but a large part is lost to the atmosphere and is unavailable for leavening during baking.

Schlaeger thought that it might be possible to slow down the reactivity of the monocalcium acid phosphate by coating it with a thin coating of calcium dihydrogen pyrophosphate or glassy calcium polyphosphate. He thought also that such a thin coating may be formed by heating the monocalcium acid phosphate for a short period of time at the proper temperature shown in equations 2 and 3.

$$Ca(H_2PO_4)_2 \cdot H_2O \xrightarrow{\Delta} Ca^{+2} \left[\begin{matrix} O & O \\ \| & \| \\ HOPOPOH \\ | & | \\ O & O \end{matrix} \right]^{-2} + 2H_2O \quad (2)$$

Monocalcium acid phosphate   Calcium dihydrogen pyrophosphate

$$(n+2)Ca(H_2PO_4)_2 \cdot H_2O \xrightarrow{\Delta}$$

$$2Ca_{\frac{n+2}{2}}^{+2} \left[\begin{matrix} O & O & O \\ \| & \| & \| \\ HOPO(PO)_nPOH \\ | & | & | \\ O & O & O \end{matrix} \right]^{-(n+2)} + (3n+4)H_2O \quad (3)$$

Calcium polyphosphate

This inspiration came after many attempts to physically coat the fine monocalcium acid phosphate crystals failed. As shown in the above equations, calcium dihydrogen pyrophosphate has two less acidic hydrogens and is less water soluble than monocalcium acid phosphate. The calcium polyphosphate contains no acidic hydrogen except at the end of a long chain. It should also be insoluble in water. Formation of either of these compounds on the surface of the monocalcium acid phosphate crystals should make them less reactive with sodium bicarbonate during the dough mixing stage. Also since both coatings have less acidic hydrogen, they should have a less acidic or tart taste. In fact, Schlaeger used this simple but effective taste test as a guide to follow the progress of his chemical coating experiments. In those days he didn't have available to him differential thermal analysis or thermogravimetric equipment. The temperature he chose for the heating experiments was the arbitrary range of 210 - 220°. He knew that monocalcium acid phosphate loses its water of hydration at above 140°. At 210°, he was exploring new chemistry.

According to plan, he carried out the heat treatment experiments. He tasted the material periodically and found that after about 30 minutes of heating, there was a reduction in the tart taste. This discovery was made at 4:30 p.m., time to go home. He and his assistant formulated some self-rising flour with this heat treated material. As a control, he also formulated some self-rising flour containing the regular monocalcium acid phosphate. Then he took those two formulations home, made them into biscuit dough and baked them in his oven. He found that he had discovered a new baking acid.

In the manufacturing process developed subsequently, the new baking acid was made by the reaction of lime with concentrated phosphoric acid. The fine dry crystals obtained were then heated treated at approximately 200 - 220°. The baking tests with this new material gave excellent results.

One day without warning, batch after batch of the newly produced material failed. The new material was no longer superior to the ordinary monocalcium acid phosphate. This unexpected turn of events caused a big commotion in the laboratory. Subsequent microscopic studies showed that regular monocalcium phosphate dissolved almost completely with water, leaving hardly any residue. The same was observed for the unsatisfactory material.

The good material, however, dissolved much more slowly. Even more important, there were some very fine transparent empty glassy shells having the shape of the original crystals remaining on the microscope slide. These shells were then collected and analyzed. It was found that they contained in addition to phosphorus and calcium, also large amounts of potassium, aluminum, sodium and magnesium along with traces of other minor elements.

At that particular period, Victor Chemical Works was making a transition from manufacturing phosphoric acid from phosphorus produced by the blast furnace to phosphorus produced by the electric furnace. The blast furnace phosphoric acid was less pure. It contained the above elements as impurities. The unsatisfactory material was made from the more pure electric furnace acid. So they solved the problem by adding the impurities to the "pure" electric furnace acid. After that the addition of some of the impurities became part of the process.

It was fortuitous therefore that Schlaeger used the monocalcium acid phosphate prepared from blast furnace acid in his original experiments. If he had used material made from the more pure electric furnace acid, the new baking acid might not have been discovered. Later research showed that even though there was indeed a protective coating formed on the fine anhydrous monocalcium acid phosphate crystals, these coatings were not calcium acid pyrophosphate nor were they calcium polyphosphate as originally planned by Schlaeger. The coating was a glass with some crystallinity. The major component seemed to be a mixed potassium, aluminum, calcium and magnesium polyphosphate. In the manufacturing process, the monocalcium acid phosphate precipitated out first. The acid phosphate of the impurities stayed in the mother liquor. Upon drying, they coated the surface of the monocalcium acid phosphate crystals. On heating, they then formed the glassy polyphosphate coating. I thought that Schlaeger's story is very significant. On hindsight, we know that his original hypothesis on the chemistry of the coating on monocalcium acid phosphate was all wrong. However, if he hadn't had his original idea and worked hard at it, he would not have discovered the unexpected result. He really deserved the good luck he had in his successful discovery (*3*, *4*).

When I started working at Victor Chemical Works in 1942, my boss, Dr. Howard Adler, wanted me to look into the possibility of making some phosphorus-containing polymers. It is known that organic phosphorus compounds have flame retarding properties. To put things in the proper historical perspective, at that time vinyl polymers and nylon were relatively new commercial products. (Polymers such as the ethylene terephthalate polyesters and polycarbonate were not introduced to the market place until many years later.)

The largest commercial organic phosphate esters then on the market were tricresyl and triphenyl phosphate. Both of these esters were produced by the reaction of phosphorus oxychloride with the phenol or the cresol, usually as a mixture of para- and metacresol (equations 4, 5).

$$
\underset{\overset{\displaystyle \text{Cl}}{|}}{\overset{\overset{\displaystyle \text{O}}{\|}}{\text{Cl-P-Cl}}} \; + \; 3\;\bigcirc\text{-OH} \longrightarrow \bigcirc\text{O-}\underset{\overset{\displaystyle |}{\text{O}}}{\overset{\overset{\displaystyle \text{O}}{\|}}{\text{P}}}\text{-O}\bigcirc \; + \; 3\text{HCl} \quad (4)
$$

Triphenyl phosphate

and

$$
\underset{\overset{\displaystyle \text{Cl}}{|}}{\overset{\overset{\displaystyle \text{O}}{\|}}{\text{Cl-P-Cl}}} \; + \; 3\;\text{CH}_3\bigcirc\text{OH} \longrightarrow
$$

$$
\text{CH}_3\bigcirc\text{O-}\underset{\overset{\displaystyle |}{\text{O}}}{\overset{\overset{\displaystyle \text{O}}{\|}}{\text{P}}}\text{-O}\bigcirc\text{-CH}_3 \; + \; 3\text{HCl} \quad (5)
$$

Tri-p-cresyl phosphate

I thought that an analogous reaction between the difunctional phenyl phosphorodichlorodate with the difunctional hydroquinone should yield a polyphosphate ester (equation 6).

Even though I used quite pure phenyl phosphorodichlorodate, when the reactants were heated together for a long time under conditions suitable for producing a very high molecular weight polymer, the mixture become an infusible and insoluble gel after only about 70 - 80% of the theoretical amount of HCl had evolved.

Subsequently, I found that the gelling was due to a cross-linking reaction caused by a small quantity of trifunctional phosphorus oxychloride generated from the disproportionation of phenyl phosphorodichlorodate.

$$
(n+1) \; Cl\overset{\overset{\displaystyle O}{\|}}{\underset{\underset{\displaystyle \hexagon}{|}}{P}}Cl \;\; + \;\; (n+1) \; HO\!-\!\hexagon\!-\!OH \;\; \longrightarrow
$$

$$
-O\!-\!\hexagon\!-\!O\overset{\overset{\displaystyle O}{\|}}{\underset{\underset{\displaystyle \hexagon}{|}}{P}} - (O\!-\!\hexagon\!-\!O\overset{\overset{\displaystyle O}{\|}}{\underset{\underset{\displaystyle \hexagon}{|}}{P}})_n - \;\; + \;\; 2(n+1) \; HCl \quad (6)
$$

This disproportionation occurred under the heating conditions used in the polymerization reaction (equation 7).

$$
4 \; Cl\!-\!\overset{\overset{\displaystyle O}{\|}}{\underset{\underset{\underset{\hexagon}{|}}{O}}{P}}\!-\!Cl \;\; \rightleftharpoons \;\; Cl\!-\!\overset{\overset{\displaystyle O}{\|}}{\underset{\underset{\displaystyle Cl}{|}}{P}}\!-\!Cl \;\; + \;\; 2 \; Cl\!-\!\overset{\overset{\displaystyle O}{\|}}{\underset{\underset{\underset{\hexagon}{|}}{O}}{P}}\!-\!O\!-\!\hexagon
$$

$$
+ \;\; \hexagon\!-\!O\!-\!\overset{\overset{\displaystyle O}{\|}}{\underset{\underset{\underset{\hexagon}{|}}{O}}{P}}\!-\!O\!-\!\hexagon \quad (7)
$$

We all know that it takes very little trifunctional reactants to cause cross-linking. Incidentally, other combinations of products could be equally well written to balance the equation, but I did not carry out a quantitative analysis of the products. I want to point out that some of the equations still to come are also not balanced for the same reason.

I decided next to use a thermally more stable organophosphorus dichloride for my phosphorus polymer research. The compound I picked was phenylphosphonic dichloride, which contains the more stable carbon-phosphorus bond. This is a well known compound first reported in the literature in the 1870's by Michaelis. It could be made by the oxidation of phenylphosphonous dichloride (equation 9). The latter compound is made by the reaction of phosphorus trichloride with benzene in the presence of aluminum chloride catalyst or by the pyrolysis of phosphorus trichloride with benzene at a high temperature range of around 600° - 700° (equation 8). The polymer I obtained using this organophosphorus dichloride and hydroquinone (equation 10) is a true thermoplastic (5). It is a very tough and horn-like material. Upon melting, it can be pulled out into very long silky fibers. The tensile strength of these fibers may be increased by the process of cold drawing. We made many polymers of this class by varying

$$PCl_3 + C_6H_6 \xrightarrow[\text{@ }600°]{AlCl_3 \text{ or}} C_6H_5PCl_2 + HCl \quad (8)$$

$$C_6H_5PCl_2 \xrightarrow{(O)} C_6H_5\overset{O}{\overset{\|}{P}}Cl_2 \quad (9)$$

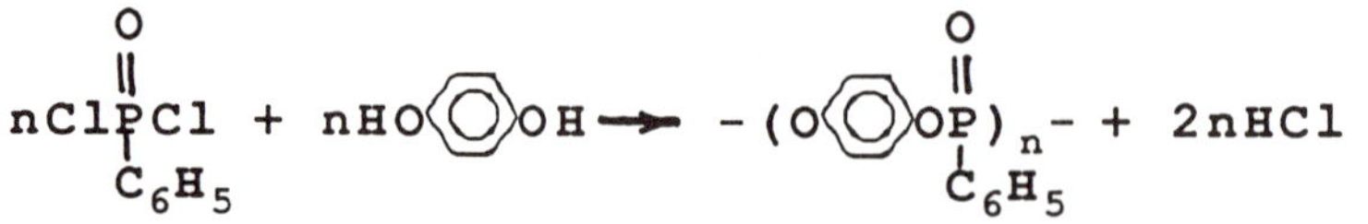

$$nCl\overset{O}{\underset{C_6H_5}{\overset{\|}{P}}}Cl + nHO\langle\bigcirc\rangle OH \longrightarrow -(O\langle\bigcirc\rangle O\underset{C_6H_5}{\overset{O}{\overset{\|}{P}}})_n- + 2nHCl \quad (10)$$

the nature of the bifunctional phosphorus reactants or by replacing the hydroquinone with other dihydroxy aromatic compounds. For example, the glass transition temperature of the polymer is greatly increased when hydroquinone is replaced with tetrachloro-bisphenol. Since these are phosphorus containing polymers, they do not burn easily.

Our original goal of making thermoplastic phosphorus containing polymers was considered a technical success. However, we weren't able to commercialize any of them. So this technical success was a commercial failure for us. About twenty some years later, the Japanese chemists from Toyobo Ltd. (6) prepared a polymer using phenylphosphonic dichloride and sulfonyl bisphenol as the dihydroxy reactant. That polymer called Heim Additive (equation 11) was considered seriously for use as an additive type of flame retardant for polyester fibers.

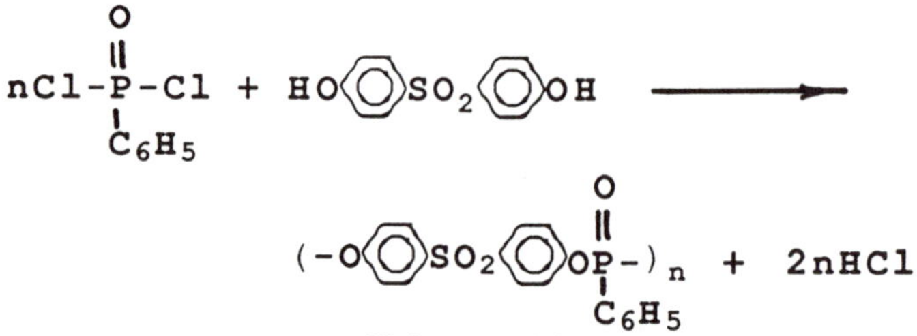

$$nCl-\overset{O}{\underset{C_6H_5}{\overset{\|}{P}}}-Cl + HO\langle\bigcirc\rangle SO_2\langle\bigcirc\rangle OH \longrightarrow$$

$$(-O\langle\bigcirc\rangle SO_2\langle\bigcirc\rangle O\underset{C_6H_5}{\overset{O}{\overset{\|}{P}}}-)_n + 2nHCl \quad (11)$$

Heim Additive

We then decided to look into another class of phosphorus containing polymers formed by the addition polymerization of diallyl organophosphonates. An example of such monomers is diallyl phenylphosphonate (7) shown in (equation 12).

$$C_6H_5\overset{O}{\overset{\|}{P}}Cl_2 + 2CH_2=CHCH_2OH + 2C_5H_5N \longrightarrow$$

$$C_6H_5\overset{O}{\overset{\|}{P}}(OCH_2CH=CH_2)_2 + 2C_5H_5N\cdot HCl \quad (12)$$

Diallyl phenylphosphonate

This monomer polymerizes when heated in an oxygen-free atmosphere in the presence of benzoyl peroxide catalyst. The polymer obtained is a hard, glassy, transparent, flame-resistant solid. The monomer is also capable of forming solid copolymers with such monomers as methyl methacrylate, vinyl acetate, diallyl phthalate and unsaturated polyesters. By using the proper ratio of methyl methacrylate and diallyl phenylphosphonate, we obtained a copolymer having the same index of refraction as that of the fiber glass fabric. A fiber glass fabric laminate made with this copolymer is almost transparent (8).

We worked with many potential customers trying to commercialize this monomer. The most promising application was in the copolymerization with unsaturated polyesters. The copolymer containing about 25% of diallyl phenylphosphonate is flame resistant. Eventually it was determined that the monomer was too expensive for the intended use and the project was shelved. This is another example of a technical success but a commercial failure.

However, in the course of working with phenylphosphonous dichloride as the intermediate for our phosphorus polymer research, we learned a great deal about the compound and its preparation. The method of preparation we decided on was by the pyrolysis of benzene with phosphorus trichloride, (equation 13). Both reactants are relatively inexpensive.

$$PCl_3 + C_6H_6 \xrightarrow{@600°} C_6H_5PCl_2 + HCl \qquad (13)$$

In our study of phenylphosphonous dichloride and its various derivatives we found that they have very interesting properties. We thought that other chemists might find uses for them. By working in close cooperation with our commercial development people, we announced the availability of the compound and some of its derivatives in pilot-plant quantities to the chemical community.

Eventually the phenylphosphonothionic dichloride derivative was found to have a large use as an intermediate for the new insecticide, EPN (A), by its inventor (9) and phenylphosphinic acid (B) was shown to be a good stabilizer against yellowing on heating for Nylon 6-6 by Nylon manufacturers (chart I) (10, 11).

There were also lesser commercial uses found for other derivatives of phenylphosphonous dichloride. The demand for phenylphosphonous dichloride became so large that it was necessary for us to go from the pilot plant to a commercial plant. The original commercial plant then had to be be expanded several times. So here is a case of a commercial success that came out unexpectedly as a by-product of our research on organic phosphorus polymers.

The pyrolysis procedure we used for phenylphosphonous dichloride was first reported by Michaelis in the 1870's. In our first small pilot plant using hot tubes, we got only 2 - 3% conversion per pass of the reactants. Even though the unconverted reactants were recycled, the low conversion rate per pass was definitely a short-coming.

Chart I. Phenylphosphonous dichloride and some derivatives

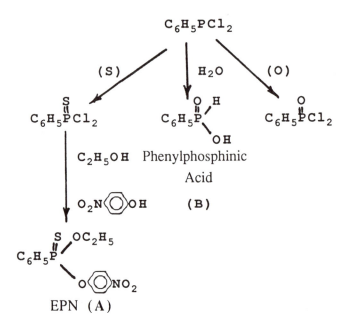

$C_6H_5PCl_2$

(S) H_2O (O)

$C_6H_5\overset{\overset{S}{\parallel}}{P}Cl_2$ $C_6H_5\overset{\overset{O}{\parallel}}{P}\overset{\displaystyle H}{\underset{\textstyle OH}{\diagup}}$ $C_6H_5\overset{\overset{O}{\parallel}}{P}Cl_2$

C_2H_5OH Phenylphosphinic
Acid

$O_2N\langle O\rangle OH$ (B)

$C_6H_5\overset{\overset{S}{\parallel}}{P}\overset{\displaystyle OC_2H_5}{\underset{\textstyle O\langle O\rangle NO_2}{\diagup}}$

EPN (A)

We therefore tried to find a catalyst which could increase the percent of conversion per pass. In thinking about what might be a good catalyst, I recalled a casual conversation with the chemical engineer, Harold Sorestokke, who worked closely with me in the original small pilot plant. In that pilot plant, we collected the unconverted reactants along with the product. After fractionation to get the product, we then recycled the unconverted reactants. Sorestokke had mentioned that when he used recycled materials, the conversion rate seemed to increase somewhat. At that time, I had analyzed the recycled reactants and found that they invariably contained 1 - 2% of by-product chlorobenzene. None of us paid much attention to that. When I thought about a possible catalyst, it then occurred to me that chlorobenzene may be a candidate. The question was how might the chlorobenzene act as a catalyst.

Finally I worked out the following hypothesis on paper. We know from phosphorus chemistry that the formation of phosphorus pentachloride from phosphorus trichloride and chlorine is a reversible reaction (equation 14).

$$PCl_3 + Cl_2 \underset{\Delta}{\overset{}{\rightleftharpoons}} PCl_5 \qquad (14)$$

Phosphorus
pentachloride

I speculated that under the conditions of our reaction, it may be possible for the phosphorus trichloride to form a transitory quaternary compound, phenyltetrachlorophosphorane, with chlorobenzene which then immediately thermally decomposes into phenylphosphonous dichloride and chlorine in a reaction analogous to the reverse reaction of PCl_5 (equation 15).

$$C_6H_5Cl + PCl_3 \rightleftharpoons [C_6H_5PCl_4] \rightleftharpoons$$
$$C_6H_5PCl_2 + Cl_2 \quad (15)$$

Tetrachloro-
phenylphosphorane

The chlorine of course would immediately undergo reaction with the benzene to form more chlorobenzene to push the reaction to the right and to start the reaction cycle again (equation 16).

$$Cl_2 + C_6H_6 \longrightarrow C_6H_5Cl + HCl \quad (16)$$

I discussed my speculation with one of my assistants, Bob Cooper. He thought I was crazy. When I suggested that he should test out my speculation in the equipment he had set up for related pyrolysis reactions, he told me that I was out of my mind. I told him that so what if it was a crazy idea. If it didn't work, all he had to lose was two or three days of work. In order to stop me from pestering him, he did carry out the reaction. However, he added sufficient chlorobenzene to the system, in an order of 20 - 30%, so that there was a meaningful amount of it in the vapor. After a few days, he came to me smiling to report that my crazy idea worked. The added chlorobenzene significantly increased the percent of conversion per pass. After optimization studies on the amount of chlorobenzene to use, he was able to increase the percent conversion per pass of phenylphosphonous dichloride by more than three hundred percent (12). He also found that if he used too much chlorobenzene, there formed as a by-product chlorophenylphosphonous dichloride. Chlorobenzene was quickly introduced as a catalyst in our commercial production plant.

In later years we worked out on paper a free radical mechanism to explain the effectiveness of chlorobenzene as a catalyst. The free radical mechanism seems to be more reasonable on theoretical grounds. The important thing to us however is that chlorobenzene does work as a catalyst. I remember when I was a graduate student, I heard a lecture by professor Homer Adkins of the University of Wisconsin. He said that in a chemical reaction, what is important is what goes in and what comes out. What goes on in between belongs to the realm of speculation and the processes visualized change from time to time.

I have no pretension of being a theoretical chemist. In fact when it comes to theoretical chemistry I always feel the same as one of my colleagues, Dr. Thomas Beck. He once told me after returning from a national ACS meeting, that he had to sit through a theoretical paper to get to the paper he was interested in. He said facetiously that the only thing he understood from the first lecture was "next slide please." Nevertheless, I always loved to speculate about possible reaction mechanisms for new routes to synthesis. I found it very satisfying when the results came out as predicted but I was even more satisfied when the results came out better than predicted. I found that sometimes it is an advantage not to be hampered by too many preconceptions or misconceptions. I was fortunate to have as a co-worker, Mr. Gene Uhing, who was of the same mind. (Unfortunately, Gene died in April, 1991.) In many cases after we discussed a possible new reaction, if it were not too big a project, he would go ahead and try it out first in the laboratory before he checked the literature. Both he and I had complete faith in his knowledge of phosphorus chemistry and especially his laboratory technique.

I would like to tell you another case in which we proposed the reaction mechanism before we carried out the reaction in the laboratory. Fortunately the result came out as predicted. We were interested in making some organophosphorus compounds containing two phosphonous acid chloride groups as shown in the following formula:

$$\overset{\overset{\displaystyle R}{|}}{Cl_2PCHCH_2PCl_2}$$

The obvious standard procedure is to use Grignard reagents. In general we don't like to use the expensive Grignard reagents unless we have no other choice.

The addition of the PCl_3 across the double bond using uv or other free radical initiators (equation 17) is well known (*13*). The thermal intiation of this reaction had not been reported. When we tried out this reaction thermally by heating an excess of PCl_3 with ethylene under pressure at 200 - 250°, only a trace amount of $ClCH_2CH_2PCl_2$ was formed.

$$PCl_3 + CH_2=CH_2 \xrightarrow{\text{uv}} ClCH_2CH_2PCl_2 \quad (17)$$

We thought that the low yield of $ClCH_2CH_2PCl_2$ under thermal conditions was due to unfavorable equilibrium or due to decomposition. We knew from the literature that the reaction of an alkyl chloride with PCl_3 and elemental phosphorus at 200 - 300° shown in equation 18 produced alkylphosphonous dichloride (*14*):

$$C_2H_5Cl + PCl_3 \rightleftharpoons [C_2H_5PCl_4] + [P] \longrightarrow$$

$$C_2H_5PCl_2 + PCl_3 \quad (18)$$

It was assumed that the reaction went through a quaternary intermediate. We thought that analogously if we could trap the $ClCH_2CH_2PCl_2$ as soon as it was formed by treating it as though it were a -PCl_2 substituted alkyl chloride as shown in the following sequence of equations (19, 20, 21), we might prevent its decomposition.

$$CH_2=CH_2 \ + \ PCl_3 \ \xrightarrow{\Delta} \ [ClCH_2CH_2PCl_2] \qquad (19)$$

$$[ClCH_2CH_2PCl_2] \ + \ PCl_3 \longrightarrow [Cl_4PCH_2CH_2PCl_2] \quad (20)$$

$$[Cl_4PCH_2CH_2PCl_2] \ + \ (P) \longrightarrow Cl_2PCH_2CH_2PCl_2$$

$$+ \ PCl_3 \qquad (21)$$

Accordingly, we heated a mixture of ethylene, excess PCl_3 and elemental phosphorus together in an autoclave at 200 - 250° for 4 - 6 hours. The yield of the desired 1,2-bis(dichlorophosphino)ethane, $Cl_2PCH_2CH_2PCl_2$, was 70%. When we applied this reaction to substituted ethylenes, such as propylene, or 1-butene, we obtained the corresponding 1,2-bis(dichlorophosphino)alkanes, but in lower yields (*15,16*) (chart II).

Chart II. Yields of $Cl_2PCH_2CH(R)PCl_2$

1-alkene	R	% yield
ethylene	H	70
propylene	CH_3	66
1-butene	CH_2CH_3	47
1-pentene	$(CH_2)_2CH_3$	41
1-octene	$(CH_2)_5CH_3$	20

Many of my academic friends told me that they have used our procedure for their synthesis of 1,2-bis(dichlorophosphino)ethane. We have even given permission to a specialty manufacturer of rare phosphorus compounds to use this process. Unfortunately for us, the products made by this simple process never took off commercially. So this is another example of a technical success but a commercial failure.

In another case the results came out better than what we had planned. In fact, the final result was entirely different from our original goal.

My original goal in this case was to make some pure diethyl phosphoric acid. I thought that it would be easy to do by using the simple procedure of cooking the diethyl phosphorochlorodate with glacial acetic acid and distill off the lower boiling by-product, acetyl chloride (equation 22).

$$(C_2H_5O)_2\overset{O}{\overset{\|}{P}}Cl \ + \ CH_3\overset{O}{\overset{\|}{C}}OH \ \xrightarrow{\qquad\diagup\qquad}$$

$$(C_2H_5O)_2\overset{O}{\overset{\|}{P}}OH \ + \ CH_3\overset{O}{\overset{\|}{C}}Cl \qquad (22)$$

I did that by cooking together one mol each of the reactants. When I distilled the reaction mixture I got almost all of the glacial acetic acid and diethyl phosphorochlorodate back. Apparently no reaction took place. There was, however, at the end of the distillation about a gram of liquid residue left in the pot. My first inclination was to clean out the apparatus and try some other reactions. However, since the distillation was carried out in a fractionation apparatus, and was connected to a vacuum pump, I decided to see whether the little residue was distillable under reduced pressure. By flaming with a Bunsen burner, I was able to collect about 1/2 g of a high-boiling colorless distillate. I thought that it might be the diethyl phosphoric acid I was looking for. To confirm my suspicion, I took the small amount of distillate in the original receiver to my good friend, Rodger Wreath, the analytical chemist. I asked him to make a simple titration for me, nothing fancy and nothing quantitative. I also told him that I expected a simple titration curve of a monobasic acid. About 30 minutes later, he called to report that in his titration, he couldn't get the monobasic acid titration curve I was looking for. He kept getting a fading end point. I realized then that I didn't get the expected diethyl phosphoric acid. I speculated that I must have gotten a little tetraethyl pyrophosphate through the controlled hydrolysis of the diethyl phosphorochlorodate. This could happen from a little moisture inadvertently getting into the reaction. The pyrophosphate bond must have been hydrolyzing while he was titrating it.

I decided to follow up on this speculation. Eventually I worked out a process by the controlled hydrolysis of diethyl phosphorochlorodate to produce a very pure tetraethyl pyrophosphate in excellent yield (equation 23).

$$2(C_2H_5O)_2\overset{O}{\underset{}{P}}Cl + H_2O + 2C_5H_5N \longrightarrow$$

$$(C_2H_5O)_2\overset{O}{\underset{}{P}}O\overset{O}{\underset{}{P}}(OC_2H_5)_2 + 2C_5H_5N \cdot HCl \qquad (23)$$

It was at this time, which was very soon after the end of World War II, when we learned that Gerhardt Schraeder of Germany had found that tetraethyl pyrophosphate was a good insecticide. Schraeder had prepared the compound in a crude form by heating one mol of phosphorus oxychloride with three mols of triethyl phosphate (equation 24).

$$OPCl_3 + 3(C_2H_5O)_3PO \xrightarrow{\Delta} (C_2H_5O)_2\overset{O}{\underset{}{P}}O\overset{O}{\underset{}{P}}(OC_2H_5)_2$$

$$(20 - 25\%)$$

$$[\text{Products Complex}]$$

$$+ nC_2H_5Cl \qquad (24)$$

Even though Schraeder reported that he obtained hexaethyl tetraphosphate, subsequent studies showed that actually various reorganization reactions occurred. The reaction mixture contained about 20 - 25% tetraethyl pyrophosphate. Tetraethyl pyrophosphate was the active insecticide ingredient. One of my laboratory colleagues discovered another procedure which also produced tetraethyl pyrophosphate in a crude form but in a 35 - 40% yield. This was accomplished by heating together 4 mols of triethyl phosphate with 0.5 mol of P_4O_{10} (equation 25). For agricultural insecticide uses, such a crude mixture prepared by such a simple method was good enough. Our elegant process of making the tetraethyl pyrophosphate in a very pure form and with a high yield was redundant and economically not competitive.

$$4(C_2H_5O)_3PO + 1/2P_4O_{10} \longrightarrow$$

$$3(C_2H_5O)_2\overset{O}{\underset{}{P}}O\overset{O}{\underset{}{P}}(OCH_2CH_5)_2 \quad (25)$$

$$(35 - 40\%)$$

However, our procedure is more versatile in making the homologs of tetraethyl pyrophosphate and we made a few of them and studied their properties (18). We were very much interested in the toxicity of these homologs to warm-blooded animals. The standard quick test in those days was by intraperitoneal injections into white mice to determine the MLD_{50}, the minimum lethal doze to kill 50% of the mice. As shown in the data in chart III, the homologs are less toxic, but for practical purposes, they are still highly toxic. We felt that they would not have any advantage over the crude tetraethyl pyrophosphate in the market at that time.

Chart III. Toxicity of tetralkyl pyrophosphates

Pyrophosphates	MLD_{50} mg/kg
Tetramethyl	1.9
Tetraethyl	0.8
Tetra-n-propyl	9.5
Tetra-i-propyl	13.3
Tetra-n-butyl	14.2

We didn't stop there. We then decided to apply our general synthetic procedure to the preparation of the tetraethyl dithionopyrophosphate and its homologs and study their properties (19). The low toxicity of the higher homologs of tetraethyl dithionopyrophosphate shown in chart IV certainly came as a surprise to us.

Chart IV. Toxicity of tetraalkyl dithionopyrophosphate

$(RO)_2\overset{S}{\underset{}{P}}O\overset{S}{\underset{}{P}}(OR)_2$	MLD_{50} mg/kg
R = C_2H_5	8
n-C_3H_7	>3000
i-C_3H_7	>3000
n-C_4H_9	>3000

Based on our knowledge of the toxicity of the homologs of tetraethyl pyrophosphate, we would have predicted that the homologs of the tetraethyl dithionopyrophosphate would be less toxic but they would still be highly toxic. We were glad that we were not hampered by prior knowledge and had not relied on our predictions but went ahead and made the homologs anyway and tested their toxicity. The low toxicity to warm-blooded animals of the higher homologs gave us the incentive to carry out extensive insecticidal evaluations. The compound we selected for such an evaluation was the tetra n-propyl dithionopyrophosphate. We found that it was a good insecticide for its days. It was especially effective for the control of Chinch bugs, a problem on golf courses and lawns along the U. S. Eastern seaboard. The compound was commercialized for many years under the trade name of Aspon.

As you will recall, the initial goal of this research was to prepare a pure sample of diethyl phosphoric acid and we ended up with the discovery of a commercial insecticide. Such is the nature of industrial research. Incidentally, we eventually made the pure diethyl phosphoric acid. We did it by the controlled hydrolysis of tetraethyl pyrophosphate.

The last example I am going to tell you about involved my good friend, Dr. Edward Walsh. We learned from our commercial development people that there was a need for a flame retardant for urethane foam.

Dr. Walsh's group was given the assignment of coming up with a good flame retardant. Since urethane foams are made by the reaction of a polyol with a polyisocyanate (equation 26), they concentrated their effort in making polyhydroxy organic phosphorus compounds. The rationale was that such organic phosphorus-containing polyols would chemically incorporate into the structure of the polyurethane foam and impart flame resistance to the foam. Several dozens of such compounds were prepared and evaluated in urethane foams in our laboratory.

$$R(NCO)_x + R'(OH)_y \rightleftharpoons -(\overset{H}{\underset{R}{N}}-\overset{O}{\overset{\|}{C}}-O-R'O-\overset{O}{\overset{\|}{C}})_n - \qquad (26)$$

For one reason or other, most of them did not have all the desired properties. However, the compound which we called Fyrol 6 (20) prepared as shown in equation 27 became a very successful commercial flame

$$(C_2H_5O)_2P(O)H + H_2CO + HN(CH_2CH_2OH)_2 \longrightarrow$$

$$(C_2H_5O)_2P(O)(NCH_2CH_2OH)_2 + H_2O \qquad (27)$$
$$\text{Fyrol 6}$$

retardant for rigid polyurethane foam. Rigid polyurethane foam flame retarded with Fyrol 6 is used extensively for packaging and insulation, particularly in household refrigerators, refrigerated cars and trucks. The roof of the huge Superdome in New Orleans is insulated with rigid polyurethane foam flame retarded with Fyrol 6. The compound which we called Fyrol 2 or Fyrol HMP (equation 28) passed our tests as a flame retardant but we didn't get much business for it as a flame retardant.

$$HP(O)(OH)_2 \; + \; H_2CO \; + \; n(\overset{O}{\overbrace{CH_2CH_2}}) \; \longrightarrow$$

$$HOCH_2P(O)[(OCH_2CH_2)_{2n}OH]_2 \quad (28)$$

Fyrol 2 or Fyrol HMP

However, one of our customers (21) found it to be an effective additive to polyethylene terephthalate in order to increase the strength and reduce the striation when the polymer was extruded into polyester film. This certainly was an unexpected commercial success. I would regard this commercial success as pure luck on our part. After all, Dr. Walsh's group designed the compound to be a flame retardant.

At my retirement roasting party, the general manager of Stauffer's Specialty Chemical Division presented me with a very large hand painted scroll entitled "Technical Success, Commercial Graveyard." On the scroll are many graves with tomb stones each having the name of a "technically success product" which failed commercially. He wanted to remind me that not all of our technical successes were also commercial successes. In fact, two of the products, Aerosafe ER and Fyrol 76 even won the I.R-100 Awards. The I.R-100 Awards are for the 100 outstanding commercial products introduced in a particular year in this country. Aerosafe ER was a flame resistant fluid for the hydraulic system of airplanes. However, our management decided to get out of that business because it was not all that profitable and the risk was quite high.

Fyrol 76 is a vinylphosphonate/methylphosphonate oligomer and is water soluble. It was developed by Dr. Edward Weil and his group (22,23). It is prepared by the thermal condensation of bis(2-chloroethyl) vinylphosphonate with dimethyl methyl-phosphonate as shown in the idealized equation 29.

Fyrol 76 is generally applied to cellulosic materials such as cotton flannel in admixture with N-methylol acrylamide along with a free radical catalyst such as potassium persulfate. The mixture in aqueous solution is padded on the cotton flannel and then dried and cured at 145 - 180°. The reactions that occur during the curing process chemically bond the Fyrol 76 to the cotton flannel. The chemical bonding is the result of the joining of the double bonds of the vinyl groups in Fyrol 76 to the double bonds of the methylol acrylamides. The methylol acrylamides are then chemically bound to the hydroxyl groups of the cellulose molecules in the cotton flannel via the methylol groups.

$$\text{(n+1) CH}_2\text{=CH-}\overset{\overset{\text{O}}{\|}}{\text{P}}\text{(OCH}_2\text{CH}_2\text{Cl)}_2 \quad + \quad \text{(n+1) CH}_3\overset{\overset{\text{O}}{\|}}{\text{P}}\text{(OCH}_3\text{)}_2$$

bis(2-chloroethyl)   methyl
vinylphosphonate   methylphosphonate

$$\xrightarrow{\text{K}_2\text{S}_2\text{O}_8}$$

$$\text{CH}_3\text{O}\overset{\overset{\text{O}}{\|}}{\underset{\underset{\text{CH}_3}{|}}{\text{P}}}\text{(OCH}_2\text{CH}_2\text{O}\overset{\overset{\text{O}}{\|}}{\underset{\underset{\text{CH=CH}_2}{|}}{\text{P}}}\text{OCH}_2\text{CH}_2\text{O}\overset{\overset{\text{O}}{\|}}{\underset{\underset{\text{CH}_3}{|}}{\text{P}}}\text{)}_n\text{OCH}_2\text{CH}_2\text{O}\overset{\overset{\text{O}}{\|}}{\underset{\underset{\text{CH=CH}_2}{|}}{\text{P}}}\text{OCH}_2\text{CH}_2\text{Cl}$$

Fyrol 76

$$+ \quad 2\text{(n+1)}\,\text{CH}_3\text{Cl} \quad (29)$$

Cotton flannels treated with the Fyrol 76 system retain acceptable softness and strength. The flame-retardant property is not adversely affected after 50 wash and dry cycles using perborate bleach. We had expected Fyrol 76 to be an assured commercial success. Unfortunately, everybody likes to have flame retarded children's pajamas to protect their children from accidental burning, but they are unwilling to pay a little extra for such protection. However, Dr. Weil informed me that recently there is revived commercial interest in Fyrol 76.

In conclusion, I hope my reminiscences give you an idea on how we carried out our industrial research in phosphorus chemistry. As I said at the beginning, in research, it is very important to have good inspiration, careful planning, hard work and an awareness of the unexpected. However, there is no real substitute for good luck and a lot of it. Not all of our efforts led to technical successes and not all of our technical successes led to commercial successes. Fortunately we did have sufficient commercial successes and profits to have our company's continued support of our research program.

I hope this discussion presents a somewhat different perspective on industrial research which is not normally found in the literature.

Literature Cited

1. Fuson, R. *Chem. Review.* **1935,** *16*, p 1.
2. Audrieth, L. F.; Sveda, M. *J. Org. Chem.* 9, **1944,** *9*, pp 89-101.
3. Schlaeger, J. R. U. S. Patent 2,160,232; May 30, 1939 (to Victor Chemical Works).
4. Toy, Arthur D. F.; Walsh, Edward N. *Phosphorus Chemistry in Everyday Living; 2nd Edition*; American Chemical Society: Washington, DC, 1987; pp 34-39.
5. Toy, A. D. F. U.S.Patent 2,435,252; Feb. 3, 1948 (to Victor Chemical Works).
6. Masai, Y.; Kato, Y; Fukui, N. U.S.Patent 3,719,727; March 6, 1973 (to Toyo Spinning).
7. Toy, A. D. F. *J. Am.Chem. Soc.* **1948,** *70*, pp 186-188.

8. Toy, A. D. F.; Brown, L. V. *Ind. Eng. Chem.*. **1948**, *40*, pp 2276-2279.
9. Jelinek, A. G. U.S. Patent 2,503,390 ; Apr. 1950 (to E.I. du Pont de Nemours and Co.).
10. Anton, Anthony U.S. Patent 3,377,314; Apr. 9, 1968 (to E.I du Pont de Nemours and Co.).
11. Ben, Victor R. U.S Patent 2,981,715; Apr. 25. 1961 (to E.I.du Pont de Nemours and Co.).
12. Toy, A. D. F.; Cooper, R. S. U.S. Patent 3,029, 282; Apr. 10, 1962 (to Stauffer Chemical Co.).
13. Fild, M.; Schmutzler, R. In *Organic Phosphorus Compounds* ; Kosolapoff, G. M., and Maier, L., Eds.; Wiley Interscience, New York, NY, 1972, Vol.4; Chapter 8.
14. Bliznyuk, N. K.; Kvasha, Z. N.; Kolomiets, A. F. *Zh. Obshch. Khim.* **1967**, *37*, p 890.
15. Toy, A. D. F.; Uhing, E. H. U.S. Patent 3,976,690; Aug. 24, 1976 (to Stauffer Chemical Co.).
16. Uhing, E. H.; Toy, A. D. F. In *Phosphorus Chemistry, Proceeding of the 1981 International Conference;* Quin, L. D.and Verkade, J. G., Eds.; ACS Symposium Series 171; Am. Chem. Soc.: Washington, DC, 1980, pp 333-336.
17. Toy, A. D. F. U.S. Patent 2,504,165; Apr. 18, 1950 (to Victor Chemical Works).
18. Toy, A. D. F.; *J. Am. Chem. Soc.* **1948**, *70*, pp 3882-6.
19. Toy, A. D. F.; *J Am. Chem. Soc.* **1951**, *73*, pp 4670-4.
20. Beck, T. M.; Walsh, E. N. U.S.Patent 3,235,517; Feb, 16, 1966.(to Stauffer Chemical Co.).
21. Eaton, E. E.; Small, J. R. U.S. Patent 3,406,153; Oct. 15, 1968 (to E.I. du Pont de Nemours and Co.).
22. Weil, Edward D. U.S. Patent 3,855,359; Dec. 17, 1974 (to Stauffer Chemical Co.).
23. Weil, Edward D. U.S. Patent 4,017,257; Apr. 12, 1977 (to Stauffer Chemical Co.).

RECEIVED December 10, 1991

Chapter 1

The Role of Phosphorus in Chemistry and Biochemistry

An Overview

F. H. Westheimer

Department of Chemistry, Harvard University, Cambridge, MA 02138

Ionized phosphate esters are ubiquitous in biochemistry principally for two reasons. First, metabolites generally must be charged, so that they will not pass through a lipid membrane and so be lost to the cell. Second, the charge must be negative, so as to repel nucleophiles and thus resist destruction by hydrolysis. Phosphates uniquely allow for these requirements.

We cannot overestimate the importance of phosphorus to chemistry and biochemistry; its central place has been quietly acknowledged, especially by the work that molecular biologists carry out daily. Figure 1, courtesy of Professor George Kenyon, illustrates the position of phosphorus in the Periodic Table.

INDUSTRIAL CHEMISTRY Phosphorus chemistry has proved a minor branch of traditional organic chemistry and of modern inorganic chemistry, but it is of great importance in industry, and especially in biochemistry and molecular biology. The greatest tonnage of industrial phosphorus chemicals is undoubtedly superphosphate, used as fertilizer; the phosphate is incorporated into growing plants. The most ubiquitous phosphorus product of industry is certainly matches, where phosphorus sesquisulfide (P_4S_3) as fuel, along with potassium chlorate as oxidizer, provides the incendiary mixture; this mixture loses none of its importance because the combination is a very old discovery. Phosphorus compounds are also important in foodstuffs — for example, phosphoric acid acidifies soft drinks — and in plasticizers, and in fire retardants; phosphorus compounds are used as war gases, a fact that many of us feared would become all too evident in the Gulf War. Arthur Toy's books, entitled Inorganic Phosphorus Chemistry and Phosphorus Compounds in Everyday Living, have been a principal source of information about industrial phosphorus chemicals for countless students. The

0097–6156/92/0486–0001$06.00/0
© 1992 American Chemical Society

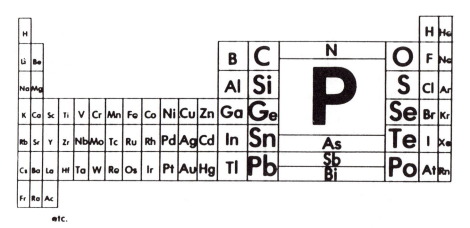

Fig. 1. The position of phosphorus in the Periodic Table.

symposium in his honor allows us to acknowledge the role he has played over the years, especially in popularizing the importance of phosphorus chemistry.

BIOLOGICAL CHEMISTRY My personal interest in phosphorus chemistry is largely directed toward its role in biochemistry. I first became acquainted with the importance of the field of nucleotides because of a series of lectures that Lord Todd gave at the University of Chicago in 1948, and I am forever grateful to him for the exciting and inciteful introduction to the field that he provided.

The role of phosphates in biochemistry was foreshadowed by the use of phosphates as fertilizer; the occurrance of phosphates in plants implied an important role of phosphates — or anyway of phosphorus — in the processes of life.*(1)*

DNA. The most spectacular role of phosphates in life processes is as the central building block in nucleic acids. The non-scientific public — that is to say, almost everybody — is well acquainted with the concept that the genetic material is DNA, or deoxyribonucleic acid, but relatively few non-scientists know that DNA is a chain of diesters of phosphoric acid. Philip Handler told the story of a young man whom he met on an airplane, who told him that he knew all about nucleic acids, and really had only one question: What is an acid?

But phosphates are also involved in ATP, that is in adenosine triphosphate, and in other compounds for the storage of chemical energy. Further, phosphate residues are attached to several coenzymes, and phosphate residues, combined with the hydroxyl groups of the serine and threonine and tyrosine residues of enzymes, control the action of these catalysts.*(2)* An understanding of the fundamental chemistry of phosphorus allows us to understand the chemistry of these biologically important materials. In 1987, I offered a view of why Nature chose phosphates for the genetic tape, and for the storage of chemical energy, and for control processes, and so on.

THE IMPORTANCE OF BEING IONIZED The most important aspect of my analysis was derived from a paper that Bernard David *(3)* published in 1958, entitled "The Importance of Being Ionized." Cells — presumably including the earliest cells on earth — are distinguished by lipid membranes. Davis pointed out that ionized compounds generally cannot pass through lipid membranes, whereas most electrically neutral compounds do so; the highly polar ions do not dissolve in the nonpolar fatty acid residues of the membrane. Therefore, most metabolites, if they are to be retained within cells, must be ionized. Of course, there are some compounds — a minority of them — that can get along without being ionized. Steroids and other compounds that are almost totally insoluble in water may dissolve in the membrane but will not pass through to a surrounding watery environment. (Figure 2)

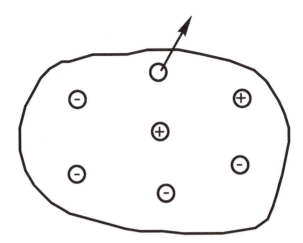

Fig. 2. The escape of unionized molecules from a cell membrane

Some sugars and other compounds with many hydroxyl groups may be so insoluble in the membranes that, once they are generated within a cell, they will remain inside, and will not escape to the outside, watery environment. But many, and perhaps most unionized compounds will dissolve to a sufficient extent in a membrane to pass through it, and be lost to the cell. Primitive organisms depended for survival on keeping important metabolites within the cell membrane, and an important way to do this, perhaps the only practical way to do this for many compounds, was to manage somehow to convert the metabolite into an ionized compound.

Phosphoric acid is a moderately strong acid; it and its mono- and diesters show an ionizable proton with a pK of about 2. It follows that mono- and diesters of phosphoric acid will be almost completely ionized in aqueous biological media where the pH is usually somewhere near neutrality. They will, therefore, be retained within a cell membrane. An effective way to convert a metabolite to an ionized compound, and thus retain it within a cell membrane, is to attach a phosphate residue to it. This is precisely what has happened, what has evolved. A large number of metabolites, including several coenzymes, are phosphate esters.

In particular, the genetic tape consists of phosphate diesters. The tape that has evolved is built of units — the nucleoside phosphates — that can be assembled and then disassembled, and reused. This seems an efficient plan; the nucleosides are complex structures that require considerable metabolic energy to build. If one is to use nucleosides at all, they ought not be wasted. This argument assumes that the genetic tape will be constructed of nucleosides. The argument, and the interests of this symposium, are concerned, among other things, with the question of why Nature chose phosphates, but not why it chose nucleosides. That is an interesting question, too, but separate from the question of phosphates, and deserves a symposium of its own.

If one makes the genetic tape from nucleosides, one must link them together to make the tape. The best way to do this is to take advantage of the hydroxyl groups in deoxyribose — or in ribose to make RNA — and to esterify these hydroxyl groups. In order to link the groups together, one will need a divalent linker, and to make esters, one will then need an acid that is at least dibasic. But if in addition, each unit must carry a charge, so as to retain the material within the cell membrane, then the acid must be tribasic. At least, this is true if the charge must be negative. The reason why the charge must be negative, and why a positive charge positively won't do, is the next topic of discussion. For the present, we note that, if one needs a tribasic acid, then the first compound that comes to mind is phosphoric acid. In a fuller discussion, one can show that no other tribasic acid, or at least none that readily comes to mind, will do.

STABILITY   The next question with respect to the choice of phosphates for the genetic tape concerns stability, and considerations of stability dictate that the charge on the tape must be negative. All metabolites must be sufficiently stable to survive long enough to carry out their biological function. Many compounds are formed and used in brief periods of time, but this is not, and cannot be so for the genetic material; it must be chemically stable for a time comparable to the length of life of the organism in question. This may be less that an hour for bacteria and viruses, but must be many years — even decades, for mammals such as man. Furthermore, the genetic tape has many millions of ester bonds, and although the cleavage of some of them may be tolerated, still the cleavage of only one bond could be fatal. The genetic tape must then be extraordinarily stable against hydrolysis. DNA meets this requirement, and not many other esters do. Furthermore, the reason for its stability lies in the multiple negative charges it carries.

Esters of ordinary carboxylic acids are insufficiently stable to be considered for the genetic material. Even at neutral pH, the half-life of a typical carboxylic acid ester in water at room temperature is only about a month, and one ester bond in a million — enough to break the tape in a hundred places — will hydrolyze every couple of seconds. The rate will be greater — much greater — in alkaline or acid solutions, and although biochemical environments tend toward neutrality, many of them differ significantly from that ideal. Triesters of phosphoric acid are somewhat more stable than esters of carboxylic acids, but nowhere near stable enough to serve as a genetic tape. (Figure 3.)

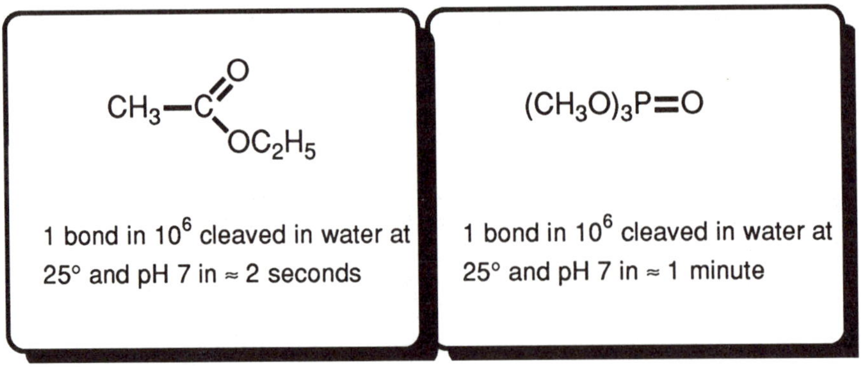

$$CH_3-C{\overset{O}{\underset{OC_2H_5}{\diagup}}} \qquad (CH_3O)_3P{=}O$$

1 bond in 10^6 cleaved in water at 25° and pH 7 in ≈ 2 seconds

1 bond in 10^6 cleaved in water at 25° and pH 7 in ≈ 1 minute

Fig. 3. Rates of hydrolysis of ethyl acetate and trimethyl phosphate.

But fortunately the tape is composed of diester monoanions of phosphoric acid. The tape is multiply charged, and all of these charges are negative. The charges not only perform the function of retaining the genetic material within the cell membrane, but stabilize it against hydrolysis. Similarly, the negative charges in RNA — ribonucleic acid — stabilize it against hydrolysis. Since ribonucleic acid is inherently much less stable than deoxyribonucleic acid, the stabilizing effect of negative charge is even more important here than with DNA. Negative charges repel nucleophiles, such as hydroxide ion and water, and slow the rate of hydrolysis enormously. The negative charges make possible the use of the diester monoanions of phosphoric acid for the genetic tape.

ELECTROSTATICS Quantitatively, how great is the effect of negative charge on the rate of hydrolysis? The effect of a single nearby negative charge on the rate of attack of a negatively charged nucleophile, such as hydroxide ion, is enormous; the rate factor for phosphates is of the order of 10^5. In general, the rate factor depends upon the closeness of approach of the nucleophile to the charge; the electrostatic effect can be small, or it can be enormous, even compared to 10^5.

An estimate of the effect of a negative charge on a negatively charged nucleophile can be obtained by considering the ratio of the first to the second ionization constants of polybasic acids. This ratio is controlled by the electrostatic effect of a negative charge on a proton. Of course, the electrostatic free energy produced by the interaction of a proton and a negative charge will be of opposite sign to that produced by the interaction of two negative charges, but the absolute magnitude should be about the same. (Figure 4.)

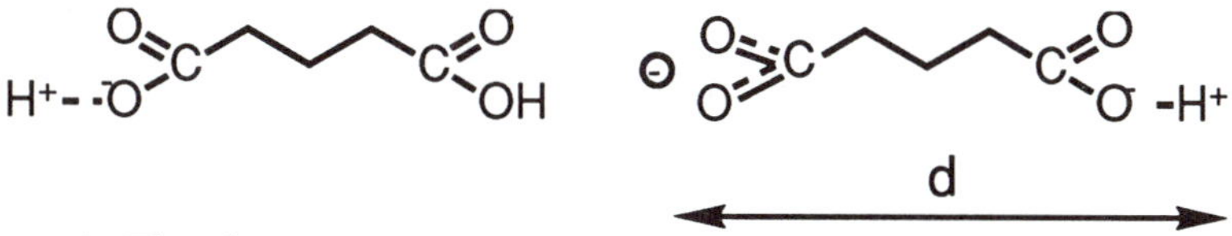

Fig. 4. The electrostatic interaction of a proton with a negative charge in the ionization of a dibasic carboxylic acid.

If we are dealing with a polybasic acid where all the ionizations occur from protons attached to oxygen, then the first ionizable proton is removed against the incipient negative charge of the nascent oxygen anion. The second ionizable proton is removed against a similar force from the nascent anionic charge on the oxygen atom to which it is originally attached, plus an additional electrostatic force arising from the interaction between the proton and the negative charge left behind by the first ionization. If one assumes that the free energy of ionization of a proton from the oxygen atom to which it is attached is roughly the same for the first and for the second ionizations, then the difference in free energy for the two ionizations is simply the electrostatic free energy that arises from removing a proton in the field of the charge left over by the first ionization. The smaller the distance between the second proton and the residual negative charge, the greater the effect. Thus the ratio of the first to the second ionization constants for phosphoric acid is 10^5, whereas that for citric acid is only 50. These numbers need to be corrected for small statistical factors, but the statistical factors don't affect the general conclusion at all, i.e. that the electrostatic effect depends inversely upon the distance between the ionizing groups (Table 1).

TABLE 1
Ionization Constants

	K_1	K_2	K_3, molar
Phosphoric	7.5×10^{-3}	6.2×10^{-8}	2.2×10^{-13}
Citric	7.1×10^{-4}	1.7×10^{-5}	6.4×10^{-6}

A similar effect controls the rate of attack of hydroxide ion on an ester. The attack of hydroxide ion on an electrically neutral ester, such as trimethyl phosphate, involves many interactions. The attack of hydroxide ion on an ester-anion, such as dimethyl phosphate anion, presumably involves the same interactions, plus an additional electrostatic interaction between the hydroxide ion and the residual negative charge on the phosphate anion. (Figure 5.)

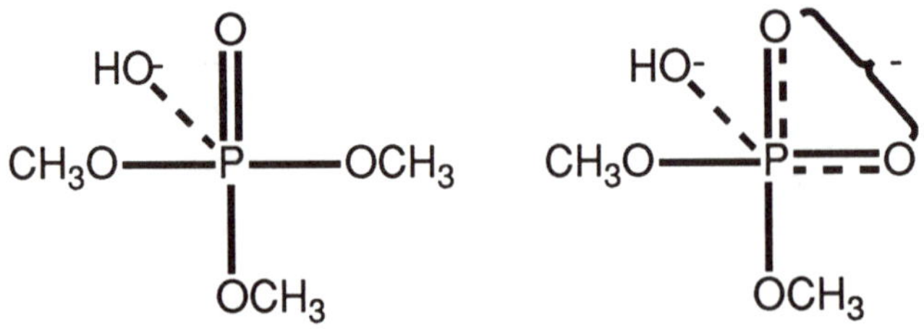

Fig. 5. Attack of hydroxide ion on dimethyl phosphate anion

Since the ratio of the first to the second ionization constants of phosphoric acid is 10^5, we would expect — and find — that the ratio of the rate constants for attack by hydroxide ion on trimethyl phosphate to that for attack on dimethyl phosphate anion is also about 10^5. Uncharged nucleophiles, such as water, are also repelled by negative charges, although of course less strongly. Nevertheless, the rate of attack of water on a negatively charged substrate is much less than that on an electrically neutral one. Any nucleophile, almost by definition, attacks an electrophile with a free electron pair, and this electron pair is repelled by negative charge.

A negative charge, therefore, protects an ester against nucleophilic attack. A positive charge, by contrast, will attract nucleophiles, and a positive charge near the site of reaction makes an ester more susceptible to hydrolysis, and thus less, rather than more stable against hydrolysis. An example of this effect will be presented later in this article. These facts and principles help explain why phosphoric acid is ideal for its role in DNA.

Let me insert a personal footnote here. These electrostatic effects were first discussed in the chemical literature by Niels Bjerrum (5) in 1923, and in 1938 John Kirkwood and I developed a crude model for such electrostatic systems, and worked out an approximate mathematical theory to put these effects on a quantitative base.(6,7,8)

Isn't it astonishing that work I did in 1938, before I knew anything about the role of phosphates in biochemistry, should be pertinent to my own interests more than fifty years later.? When I was a graduate student at Harvard in the early 1930's, I read a paper by Arthur Michael — the Arthur Michael who invented the Michael condensation reaction — one of the early heroes of American organic chemistry. He was, at that time, a Professor Emeritus of Chemistry at Harvard. To the best of my recollection, Michael wrote that "Fifty years ago, my coworkers and I . . . . " (I have not yet found this exact quote, but Michael *(9)* refers to work he did *(10)* fifty-four years earlier.) At the time, this phrase seemed almost hysterically funny; imagine someone maintaining his interest in chemistry for over half a century! At present, I am a Professor Emeritus of Chemistry at Harvard, writing that, more than fifty years ago, Jack Kirkwood and I developed a quantitative theory of electrostatic effects. Perhaps some readers will find this at least mildly amusing. My guess, however, is that most chemists maintain an active interest in chemistry as long as they live, and that my reaction, back in the 1930's, just tells you how young I was then.

In any event, the genetic material is stable against hydrolysis because of multiple negative charges. The negative charges do double duty; they both stabilize the genetic tape against hydrolysis and keep it safe within the lipid cell membrane.

ATP   The principal compound for the storage of chemical energy is adenosine triphosphate; like the genetic tape, it is stabilized against hydrolysis by its negative charges. The free energy for the hydrolysis *(11)* of ATP is -7.3 kcal/mole, as compared to only -2.2 kcal/mole for the hydrolysis of an ordinary phosphate ester. This excess free energy can be used in accomplishing the phosphorylation of various metabolites, or in supplying the needed driving force for many other reactions of biochemistry; the hydrolysis of ATP can be coupled with other metabolic processes that otherwise would be energetically unfavorable. A phosphorylation process involves the transfer of a monomeric metaphosphate unit, PO_3^-, from ATP to the substrate in question. (Figures 6 and 7.)

PO_3^- is a strong electrophile, and a strong phosphorylating agent. In research in my laboratory, my collaborators have shown that, in hydrophobic environments methyl metaphosphate will even phosphorylate the ring in aromatic amines *(12, 13)* Furthermore, we have shown that monomeric metaphosphate ion, presumably but not necessarily free, attacks carbonyl groups, and will convert ketones to the corresponding enol phosphates.*(14, 15)* W.P. Jencks *(16)* has shown that monomeric metaphosphate is never free in aqueous solution; in this regard, it resembles the proton, which is an even stronger electrophile, and similarly is never free in aqueous solution. In acid catalyzed reactions, a proton is transferred from one molecule to

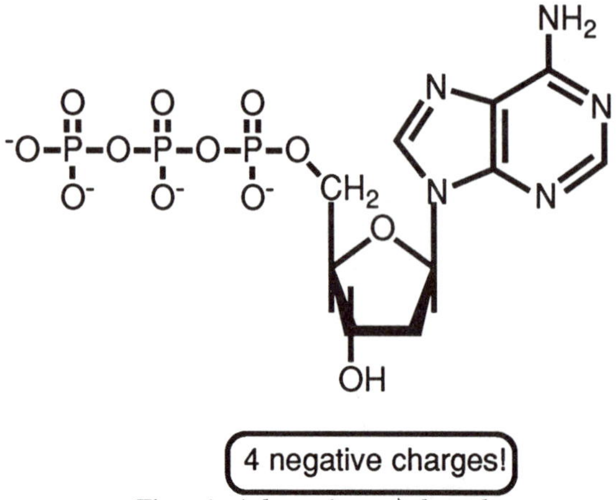

4 negative charges!

Fig . 6 Adenosine triphosphate.

$$[ATP]^{4-} + H_2O \rightarrow [ADP]^{3-} + H_2PO_4^{-}$$

$$\Delta F = -7.3 \text{ kcal/mole}$$

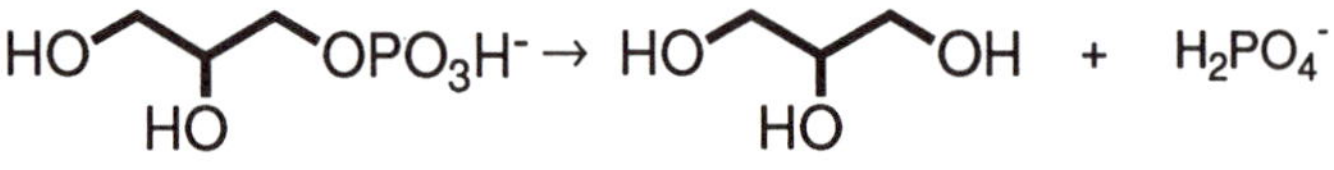

$$\Delta F = -2.2 \text{ kcal/mole}$$

Fig.7. Thermochemistry of hydrolysis of ATP

another; in fact, protons are probably completely free only in the gas phase. Monomeric metaphosphate is nowhere near as electrophilic as protons; nevertheless, Professor Jencks has shown that, at least in aqueous solution, monomeric metaphosphate, like the proton, is transferred from one molecule to another without ever being free.

This is not necessarily the case in less nucleophilic media, and we — the scientific community — will try our best to determine whether the environment of enzymes resembles aqueous solution or a much less polar medium. Such investigations are of real scientific interest, but must not obscure the important fact that the role of monomeric metaphosphate in biochemistry is as an electrophile. Like the proton, its electrophilicity is sharply diminished by water and, at least in aqueous solution, it is transferred from one molecule to another without becoming free. Unlike the proton, which is never free in any liquid medium, monomeric metaphosphate is probably free in apolar solution; like the proton, its role is that of an electrophile.

In any event, water is the enemy of monomeric metaphosphate. ATP, which transfers or generates PO_3^-, depending on the medium, is thermodynamically quite unstable in aqueous solution, but nevertheless kinetically stable. In 1951, Fritz Lipmann *(17)* noted that ATP is stabilized against hydrolysis by its negative charges. The compound occupies a unique place in biochemistry because it is thermodynamically unstable yet kinetically nearly inert — a marvelous and apparently incompatible set of properties that is related to the polybasic nature of phosphoric acid, and the resulting negative charges on ATP.

SYNTHESIS OF DNA   The last section of this report concerns the Letsinger-Caruthers synthesis of DNA. It seems to me that an introductory paper should concentrate on the most important topics, even if my own contribution to the topic is pretty small. Biochemists today isolate and replicate the genes for many enzymes, and then express and isolate proteins related to those genes. They can work up a piece of nucleic acid from a tiny sample — a few billion molecules — to make milligrams of nucleic acid and eventually express many milligrams of enzyme.

Biochemists and molecular biologists perform these miracles routinely, in months or sometimes even in weeks. Fortunately for the progress of science, some chemists have joined in — in the Chemistry Department at Harvard alone, Stuart Schreiber, Gregory Verdine, George Whitesides, and — until he became the Dean of the Faculty of Arts and Sciences — Jeremy Knowles — together with their students — are using these techniques, and of course many more chemists are doing the same sort of thing elsewhere. The techniques allow site directed mutagenesis of enzymes, among other research directions, and this allows a much more exacting test of the detailed chemical mechanism of enzyme action. In this connection let me mention the use of site-directed mutagenesis by my former graduate students, John Gerlt and George Kenyon *(18,19,20)* in determining the mechanism of action of mandelate racemase. There are, of course, many more examples.

Essential to these exercises in molecular biology is the ability to synthesize pieces of nucleic acid of specified sequence and reasonable length, that is, to make polynucleotide sequences thirty to a hundred nucleotides long. These pieces of synthetic nucleic acid are needed, among other uses, to fish out specific genes for which only a bit of sequence is known, and to fish them out from a biochemical soup of almost unbelievable complexity. Chemists now have automatic machines into which you can type the needed sequence, turn on the machine and walk away — though in fact you had best not walk away — and get a fifty-step synthesis in a day. The product can be purified by chromatography, and the sequence can readily be checked by the techniques for which Frederick Sanger and Walter Gilbert received the Nobel prize in 1980.*(21, 22)*

Molecular biologists who use this material seem to take the synthesis for granted. If you need a shirt, you go to a clothing store and buy one in your size. If you need a specific polynucleotide, you go to a chemist, and he synthesizes it for you. There can't be much to it; a machine does it. But really, the machine is a miracle of chemical inventiveness. How many molecular biologists even recognize the names of Robert Letsinger and Marvin Caruthers, their benfactors? One famous practitioner, who generally acknowledges the importance of chemistry, accepted without question that chemists synthesize things, just as carpenters make tables; apparently there's nothing to it.

LETSINGER-CARUTHERS SYNTHESIS Letsinger's original synthesis *(23)* condensed phosphites with the hydroxyl groups of deoxyribosides to build a growing chain that resembles the genetic material except that the last residue is a phosphite, instead of a phosphate. The phosphite is then gently oxidized to a phosphate. The earlier syntheses of polynucleotides, pioneered by Alex Todd *(24)* and by H. Gobind Khorana *(25)* , lighted the way to modern methods, but although these earlier syntheses were absolutely essential to the development of the field, they are much too time-consuming to be really practical. One can understand the advantage of the phosphite method when one realizes how much more reactive phosphites are than phosphates.

Specifically, the rate of acid hydrolysis of an ester of phosphorous acid — that is to say, a phosphite — is about 10^{12} times as great as that for the corresponding phosphate *(26)* - a thousand billion times as great.   This advantage corresponds to the best that enzymes can manage; enzymes usually increase reaction rates by a factor of 10^9 - 10^{12}.   Smart chemistry makes this synthesis the equivalent of an enzymatic one.

The Letsinger-Caruthers synthesis is unrelated to the biochemical one except in the area of efficiency. Because phosphites are so reactive, they would be totally unsuitable for the genetic tape itself, but because they are so reactive, they are ideally suited as intermediates in the chemical synthesis of polynucleotides. One can accomplish the needed condensation reactions with phosphites under strictly anhydrous conditions with very weakly acidic catalysts, and then stabilize the product by a mild oxidation of the phosphite to the desired phosphate with iodine.

The essential feature of the phosphites — the reason why the phosphites are so susceptible to acid hydrolysis or alcoholysis, is that the phosphorus atom of the phosphite is blessed with an unshared electron pair. A proton can then add to the unshared electron pair of a phosphite, or of a phosphoramidite. (Phosphoramidites are discussed below.)   The positive charge that results from attaching a proton directly on phosphorus atom sensitizes the phosphorus   to nucleophilic attack. This is the same sort of electrostatic effect that is discussed above in connection with the retarding effect of a negative charge on the hydrolysis of phosphates, but the effect of a positive charge increases the rate of nucleophilic attack, and the effect is of course enormously greater because the positive charge is directly on phosphorus, whereas the negative charge in a phosphate anion is on an oxygen atom, one bond length removed from the phosphorus atom. (Figure 8.)

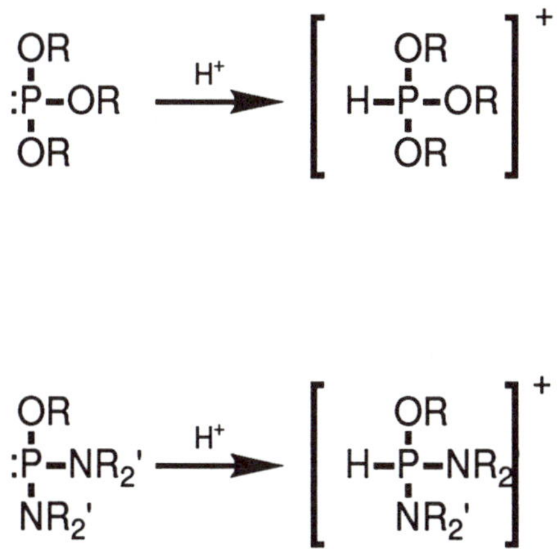

Fig. 8. Protonation of phosphites and phosphoramidites.

With courage and an equation for electrostatic potential, one could have predicted the approximate magnitude of this electrostatic effect, but in fact no one did.

Although Letsinger's original phosphite synthesis is an efficient process, it was still too difficult for general use, as it required dry ice temperatures and strictly anhydrous conditions. The method, however, was modified by Marvin Caruthers (27) who substituted monoester diamidites for phosphites. The reactions proceed much more smoothly with the amidites, which are, relatively, resistant to hydrolysis and to autoxidation. A beginning of the physical organic chemistry of the phosphites has been published (22), but to the best of my knowledge none has been published for the phosphoramidites. Perhaps this review will stimulate someone to take up this problem, just as Todd's lectures stimulated me to work in this area.

In some detail, the Letsinger-Caruthers synthesis involves attaching a nucleotide to a solid support, such as silica gel, and preparing protected phosphite reagents for the four nucleotides. A 2,4-dimethoxytrityl group, which can be easily removed when desired, is usually used for temporary protections of the 5'-hydroxyl group of the nucleoside. Methoxyl groups, or more frequently cyanoethyl groups, are used to protect the phosphorus-atom during the synthesis. One such protected reagent is then allowed to react with the 5'-hydroxyl group of the nucleotide tethered to a solid support. The resulting phosphate-phosphite compound is then oxidized with iodine to produce a protected dinucleotide.

The dimethoxytrityl group that protects the 5'-end of the tethered dinucleotide can then be removed, by mild acid treatment, and the chain lengthened by repeating the process with the desired phosphoramidite. Finally, at the end of the synthesis, the protecting methoxyl groups, or whatever groups have been used to protect the phosphates, must be removed, protecting groups must be removed from the bases, and the completed molecule must be liberated from the solid support. The individual steps can be carried out with 99% yields, and long polynucleotides, with fifty or more units, can be prepared in this way. Some details of the chemical synthesis are not fully understood; in particular, the physical-organic chemistry is still incomplete or missing. But even without a full understanding of all the details, the synthetic procedure works, and works wonderfully. The chemical community can be proud of this contribution of chemistry to molecular biology. (Figure 9.)

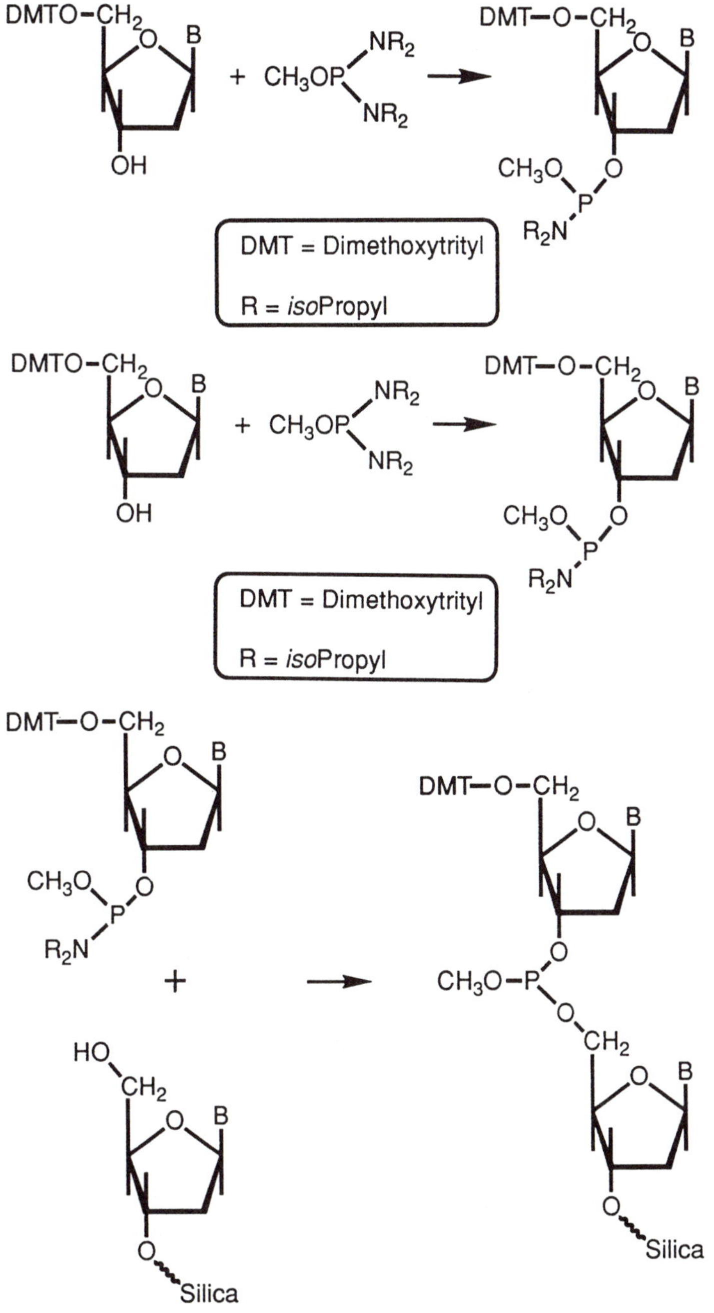

Fig. 9. The Letsinger - Caruthers synthesis of a protected form of nucleotides.

Continued on next page

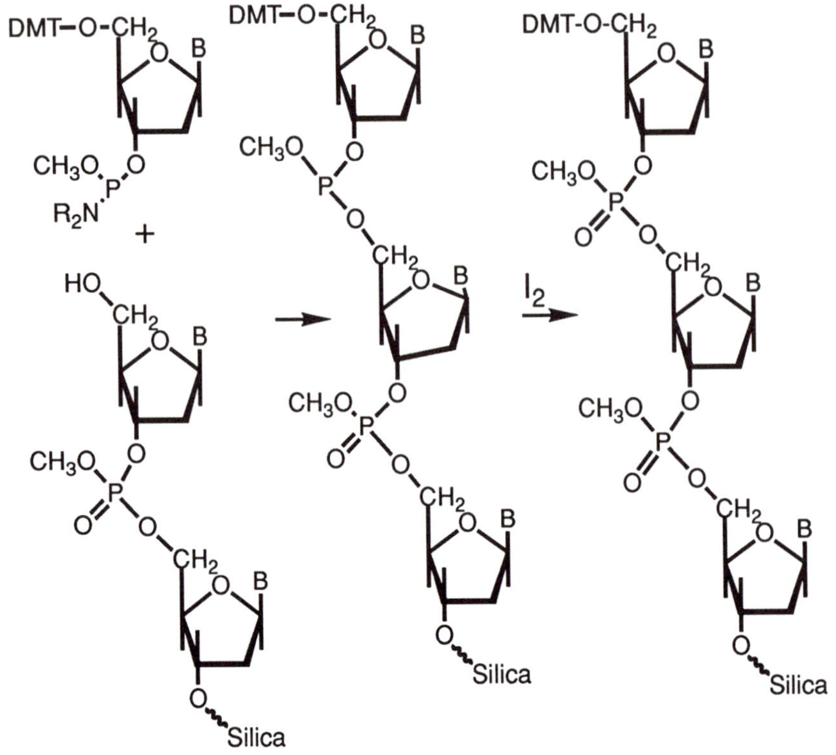

Fig. 9. *Continued*

Phosphorus chemistry and biochemistry are enormous topics, and obviously this essay touches on only a few aspects of the subject; many other aspects are discussed in detail in this volume by others. This introduction to the symposium is built   around the commanding importance of electrostatics in the chemistry and biochemistry of phosphates and phosphites, but that emphasis necessitated omitting a discussion of many other topics.   These include, for example, pseudorotation in the hydrolysis of cyclic esters of phosphorus, a fascinating topic to which Edward Dennis and David Gorenstein and others in my laboratory have made significant contributions *(28)* . The brilliant work of Usher, Richardson and Eckstein *(29)* which (if you will pardon a pun) ushered in the determinations of the stereochemistry at phosphorus in the reactions of phosphate esters, has been omitted, as has the work of Jeremy Knowles *(30)* and his coworkers, who demonstrated how to use O^{16}, O^{17}, and O^{18}, to produce chirality at phosphorus. I have completely ignored the role of phosphorylation in the control of enzymic processes. But perhaps the work cited demonstrates the importance and conveys the excitement of phosphorus chemistry, and so provides an introduction to this symposium.

LITERATURE CITED

(1) J. W. Mellor *Treatise on Inorganic Chemistry,* VIII, 736 ff Longmans-Green:(1928).

(2) L. Stryer *Biochemistry* (3rd Ed.) W. W. Freeman, N,Y. (1988).

(3) F.H. Westheimer, *SCIENCE* **1987,** *235* 173

(4) B. Davis *Archives Biochem. Biophys.* **1958,** *78* 497 .

(5) N. Bjerrum *Z. physik. Chem.* **1923,** *106* 219 .

(6) J. G. Kirkwood; F. H. Westheimer, *J. Chem Phys* **1938,** *6,* 506

(7) F. H. Westheimer; J. G. Kirkwood *ibid* **1938,** 513.

(8) F. H. Westheimer; M. W. Shookhoff *J. Am. Chem. Soc.* **1940,** *62* 269.

(9) A. Michael; J. Ross *J. Am. Chem. Soc.* **1933,** *55* , 3684.

(10) A. Michael *Am. Chem. J.* **1879,** *1* , 312.

(11) L. Stryer, op. cit. p 317.

(12) C. H. Clapp; A. Satterthwait; F. H. Westheimer *J. Am. Chem. Soc.* **1975,** *97,* 6873.

(13) A. Satterthwait; F. H. Westheimer *ibid.* **1978,** *100,* 3197; *102* 4464 (1980).

(14) A. Satterthwait; F. H. Westheimer *J. Am. Chem. Soc.* **1981,** *103* 1177.

(15) K. C. Calvo; F. H. Westheimer *ibid.* **1983,** *105* , 2827.

(16) W. P. Jencks *Acc. Chem. Res.* **1980,** *13,* 161

(17) F. Lipmann in *Phosphorus Metabolism,* W. D. McElroy and H. B. Glass, Eds.; Johns Hopkins Press: Baltimore, 1951, Vol I, p 521.

(18) V. M. Powers et al. *Biochemistry,* **1991,** *30* 9255

(19) D. J. Neidhart et al. *ibid* **1991,** *30* 9264

(20) J. A. Landro et al. *ibid* **1991,** *30* 9274

(21) A. M. Maxam; W. Gilbert *Proc. Natl. Acad. Sci.* **1977,** *74,* 560

(22) F. Sanger; S. Nicklen; A. R. Coulson *ibid* , 5463 (1977).

(23) R. L. Letsinger; W.B. Lunsford *J. Am. Chem. Soc.* **1976,** *98* 3655

(24) A. Todd *Prospectives in Organic Chemistry,* (A. Todd, Ed.), Interscience, N.Y. 1956, p 245.

(25) H. G. Khorana *Pure Appl. Chem,* **1968** *,* 17, 349 (1968).

(26) F. H. Westheimer; S. Huang; F. Covitz *J. Am. Chem. Soc.* **1988,** *110,* 181. Errata, *Ibid,* 2993.

(27) M. H. Caruthers *SCIENCE,* **1985,** *230* 281.

(28) F. H. Westheimer *Acc. Chem. Res.* **1968,** *1* 70 .

(29) D. A. Usher; D. I. Richardson, Jr.; F. Eckstein, *NATURE* (London), **1970,** *228* 663

(30) S. J. Abbott et al., *J. Am. Chem. Soc.* **1978,** *100* 2558.

RECEIVED December 17, 1991

Chapter 2

Cyclic Oxyphosphoranes

Models for Reaction Intermediates

Robert R. Holmes, Roberta O. Day, Joan A. Deiters,
K. C. Kumara Swamy, Joan M. Holmes, Johannes Hans, Sarah D. Burton,
and T. K. Prakasha

Department of Chemistry, University of Massachusetts,
Amherst, MA 01003

Recently studied cyclic oxyphosphoranes contain ring sizes varying from five- to eight-membered and include oxygen, sulfur, and nitrogen heteroatoms. These phosphoranes have proven their usefulness as models for intermediates in enzymatic reactions of cyclic adenosine monophosphate (c-AMP) and for nonenzymatic reactions of tetracoordinated phosphates. All X-ray studies of pentaoxyphosphoranes show that six-membered rings are located in axial-equatorial (a-e) positions of trigonal bipyramidal (TBP) geometries with the boat form most common. *Ab-initio* calculations support activation energies from NMR studies indicating that the diequatorial (e-e) orientation is less favorable. With eight-membered rings, X-ray studies have shown the existence of the (e-e) orientation in pentaoxyphosphoranes. These studies suggest that if enzymes favor oxyphosphoranes as intermediates to reach proposed activated states with six-membered rings in (e-e) orientations, a combination of hydrogen bonding, steric factors, and ring constraints may be required.

A considerable number of reactions of cyclic phosphorus compounds have been proposed to proceed via trigonal bipyramidal activated states or intermediates (1-4) usually following the Westheimer (5) model that has the entering nucleophile and departing group do so from axial positions. Of concern in nonenzymatic and enzymatic reactions is the location of ring substituents in these intermediate stages of a mechanistic route for they determine the stereochemical outcome of the reaction. For phosphorus compounds with five-membered rings, e.g., cyclic phosphate esters (5), it is well established that ring strain relief occurs when the

0097–6156/92/0486–0018$06.75/0
© 1992 American Chemical Society

rings occupy axial-equatorial (a-e) sites of a trigonal bipyramid. Model cyclic oxyphosphoranes invariably exhibit this structural arrangement (6).

Until recently such model phosphoranes have been lacking for reactions of phosphorus compounds containing larger ring systems. In fact, the first structural study of a pentaoxyphosphorane having a six-membered ring was reported in 1988 by D. Schomburg et al. (7). The compound was a bicyclic derivative with the rings located in (a-e) sites.

Since then we have carried out X-ray and NMR studies (8-13) of a variety of cyclic oxyphosphoranes containing ring sizes from five- to eight-membered in an attempt to learn structural and conformational preferences as ring size varies and to understand what the important factors are that may induce structural and conformational changes. The results should prove useful in modeling proposed intermediates and activated states in a host of reactions that have been studied and for ones that may be developed in future work.

To illustrate potential complications, this review will address mechanisms of action of cyclic adenosine 3',5'-monophosphate (c-AMP) with phosphodiesterases and protein kinases. Proposals exist which depict reactions proceeding via intermediates and activated states that involve pentacoordinated phosphorus entities. Although the underlying basis for proposing such states has not been studied in any adequate way, it is useful to gauge the likelihood of these pentacoordinated states in terms of what is learned of structural preferences and energetics on model oxyphosphoranes. cAMP is an important effector molecule which regulates a wide variety of biochemical processes (14). It contains a six-membered ring with an attached trans-fused ribose component. Hence, it provides an interesting challenge for mechanistic interpretation.

cAMP Action

Mechanisms. cAMP plays a central role as a second messenger in the regulation of cell metabolism (14). Binding of cAMP with the regulatory subunit of a protein kinase initiates a cascade of enzymatic reactions that ultimately lead to the breakdown of glycogen and release of glucose to the blood stream. The intracellular concentration of cAMP represents a balance between the action of adenylate cyclase (which produces cAMP from adenosine triphosphate in response to a hormone signal), and 3'-5'-cyclic nucleotide phosphodiesterase (which catalyzes the hydrolysis of cAMP into 5'-adenosine monophosphate (5'-AMP)). The structural requirements for the binding of cAMP to the regulatory subunit of protein kinases, as well as to phosphodiesterases have been a subject of numerous investigations (15-18).

More recently increased attention has been devoted to the type of activated states that may form in these enzyme reactions. Several proposals invoke the formation of a pentacoordinated phosphorus (P^V) state having a trigonal bipyramidal (TBP) geometry. These TBP activated states or intermediates, as the case may be, can be generated by way of nucleophilic attack at phosphorus with the result that the six-membered ring resides either in a diequatorial or an axial-equatorial orientation. Although the phosphate ring of cAMP is known in the chair form from X-ray (19) and ^{1}H NMR work (20), little is known about the structures or ring conformations of pentacovalent intermediates that may form. As will be outlined in this section, a knowledge of these structures and ring conformations provides an important key to the understanding of the ensuing enzymatic reactions. Ring conformational changes themselves within pentacoordinated phosphorus structures may exert a controlling feature in the binding process at an enzyme active site and in the subsequent catalytic reaction.

Specifically, the phosphodiesterase-catalyzed hydrolysis of cAMP which leads to its 5'-monophosphate has been proposed (21-26) to proceed via a pentacovalent phosphorus adduct formed at the enzyme active site that involves cAMP and a water molecule and/or a nucleophilic moiety from the peptide chain of the enzyme itself. In one such proposal, val Ool and Buck (26) suggest the formation of a pentacoordinated phosphorus intermediate with the phosphorinane ring occupying axial-equatorial sites as the controlling feature for phosphodiesterase action (see Route A, Scheme 1).

The stereochemistry is known to proceed by inversion of configuration (22, 27-29). In order to accomplish this by Route A, the initial enzyme bound-P^V intermediate (a) forms c by in-line nucleophilic displacement, an inversion route. The adduct c is subsequently hydrolyzed by an adjacent attack of the water molecule to give 5'-AMP with retention. The latter step is complicated requiring axial entry of water adjacent to the departing enzyme residue. Departure of this residue from an axial position occurs after pseudorotation (intramolecular positional rearrangement of ligands) of the P^V complex. Although the process leads to overall inversion of configuration as required by the known stereochemistry, to date there is no precedent for the occurrence of pseudorotation at an enzyme active site (30-31). A simpler mechanism for phosphodiesterase hydrolysis suggested by Bentrude (32-33) would take place with in-line attack by a water molecule if enzyme constraints are able to position the water molecule for axial attack opposite the 3'-oxygen atom.

Scheme 1

Phosphodiesterase Activity on c-AMP

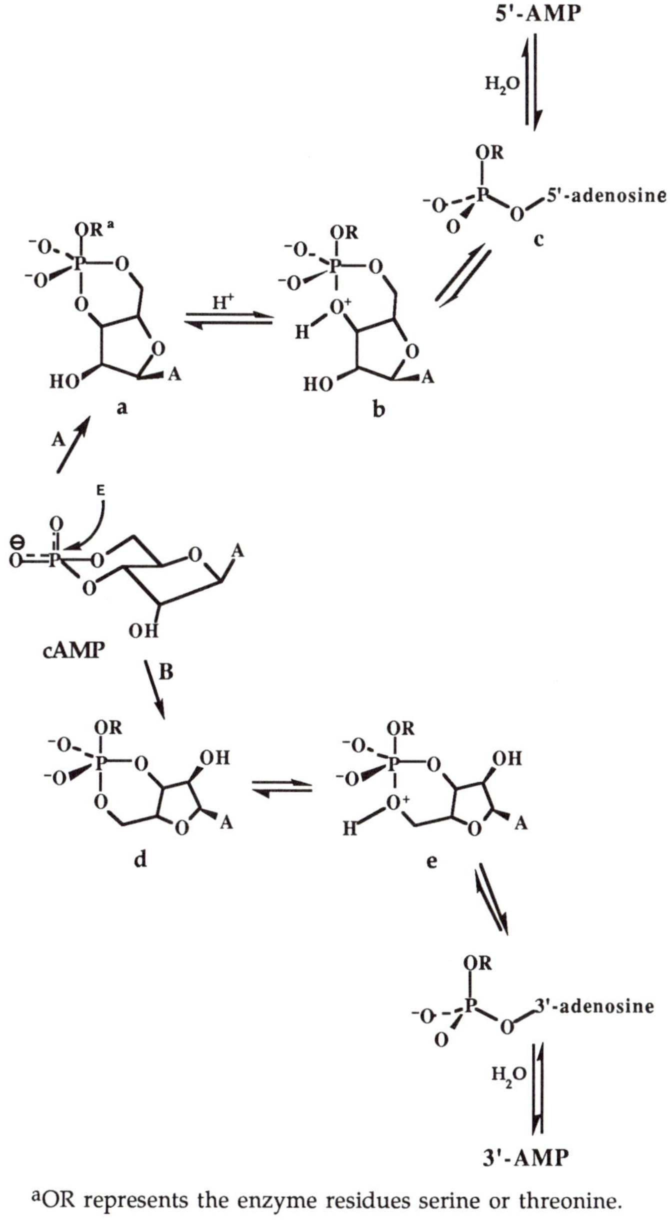

aOR represents the enzyme residues serine or threonine.

Formation of 5'-AMP with overall inversion of configuration favors Route A. (Adapted from ref. 26)

Scheme 2

Enzymatic Hydrolysis of c-AMP

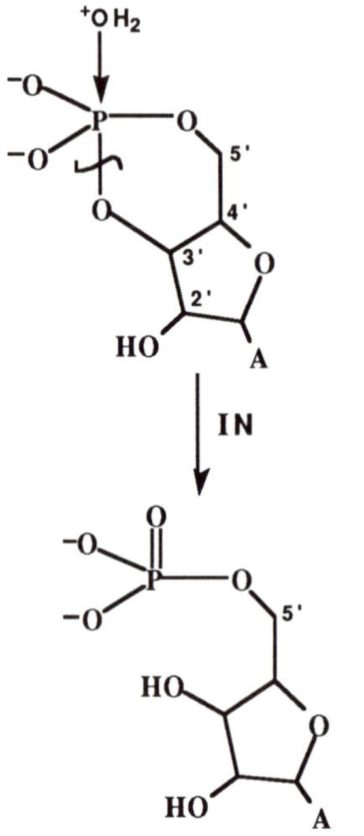

5'-AMP

Formation of 5'-AMP with inversion favors in-line attack opposite the
3'-oxygen atom with phosphodiesterases.

In nonenzymatic hydrolysis of cAMP, it is known that 3'-AMP forms instead of 5'-AMP but does so also with inversion of configuration (23). A simple in-line displacement is envisioned.

Scheme 3
Nonenzymatic Hydrolysis of c-AMP

3'-AMP

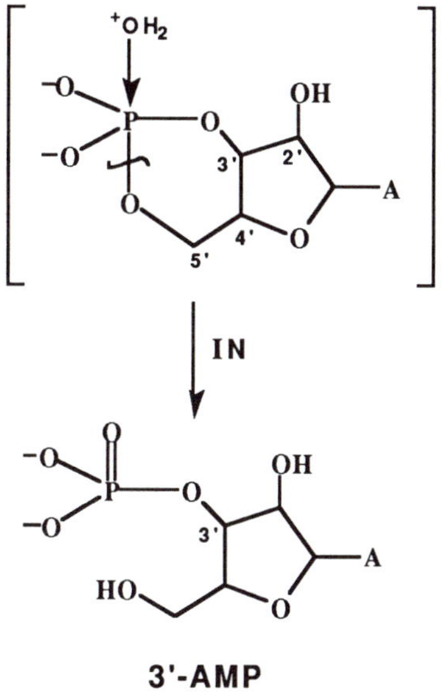

Formation of 3'-AMP with inversion of configuration favors in-line attack of H_2O opposite the 5'-oxygen atom (23).

The latter scheme requires the 5'-oxygen atom to be located in an axial position in contrast to Schemes 1 and 2 which require the 3'-oxygen atom to be so situated. Thus, the enzyme appears effective in bringing about this structural change.

In the triggering of protein kinases by cAMP, enzyme-cAMP adduct formation via a P^V-TBP having a diequatorial ring orientation has been proposed (24-26, 34) as an essential step in the dissociation of the haloenzyme to the free catalytic subunit, Scheme 4.

Since the phosphorinane ring is not cleaved in the process, diequatorial ring placement seems favored, i.e., out of the way of the reaction sites which conventionally involve axial entry and axial departure for activated states of pentacoordinated TBP phosphorus (5, 6).

A further mechanism involving an enzyme-bound P^V activated state with a diequatorially placed phosphorinane ring of cAMP has been advanced (21-26) for phosphodiesterase hydrolysis, Scheme 5. Pseudorotation is necessary. Subsequent ring cleavage yields a covalent enzyme-nucleotide intermediate. Displacement of the enzyme attachment must proceed in-line by attack of a water molecule to give the required overall inversion of configuration.

Scheme 4

Proposed Intermediate in the Activation of Protein Kinases by c-AMP.
(Adapted from ref. 26)

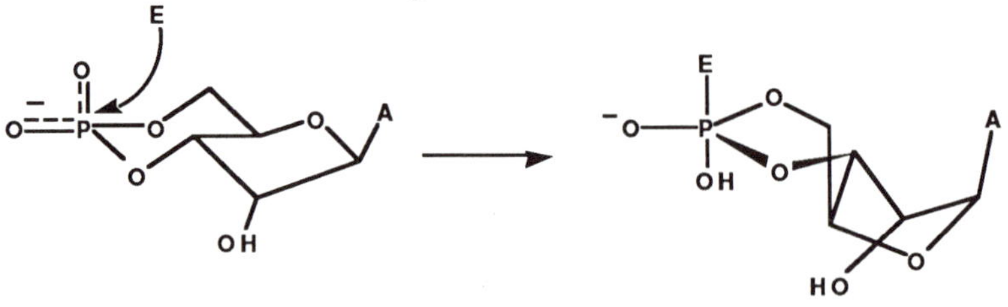

g

Scheme 5

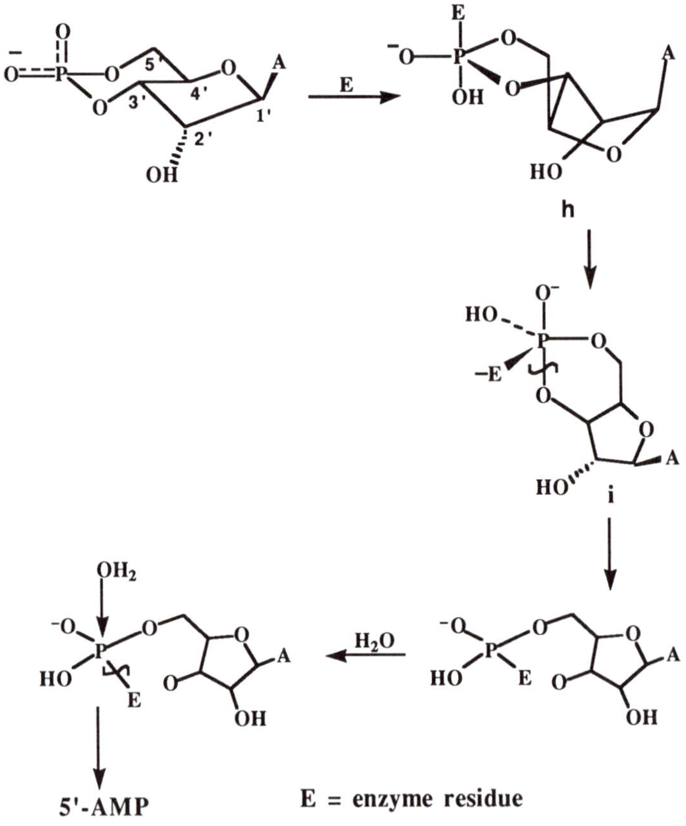

h

i

5'-AMP   E = enzyme residue

Cyclic Oxyphosphorane Structures

Let us review some of the recent structural work on cyclic oxyphosphoranes relative to the proposed activated states in the above schemes.   While preferred structures and conformational properties of cyclic oxyphosphoranes containing five-membered rings have been well established (6), corresponding work on cyclic oxyphosphoranes with six-membered rings that could serve as models for P^V activated states and intermediates in cAMP reactions, is practically nonexistent until very recently.   In fact, from structural studies (8-12, 32, 33) of a range of cyclic oxyphosphoranes containing six-membered ring assemblies, we have dispelled the notion that six-membered rings prefer diequatorial rather than axial-equatorial positions of a TBP oxyphosphorane (6). Single crystal X-ray analyses and complimentary variable temperature NMR studies (8-12, 32, 33) have shown that, like five-membered ring compositions in oxyphosphoranes, six-membered rings uniformly are positioned at axial-equatorial sites.   A sampling of some of these are shown in Table I.

An interesting observation is found in the first X-ray structures (12, 32, 33) of oxyphosphoranes containing *trans*-fused five-membered rings. Trans annelation of the ribose component is a property of cAMP.

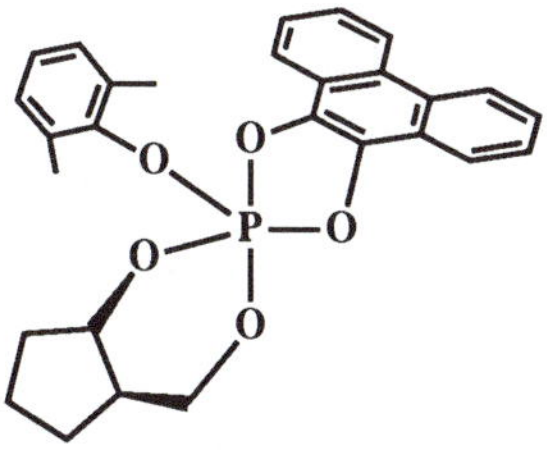

E, Holmes et al. (*12*) 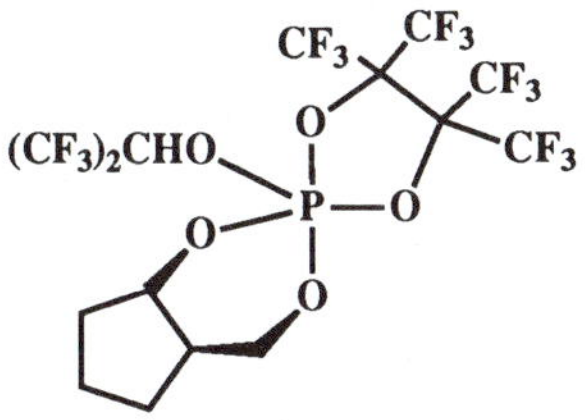   F, Bentrude et al. (*32, 33*)

In both of the cyclic P^V derivatives, the 5'-oxygen atom (what would be the 5'-oxygen atom if the five-membered ring were ribose) is located in an axial position, similar to that proposed in Scheme 3 for the nonenzymatic hydrolysis of cAMP. It would be interesting to learn what features influence the positioning of a *trans*-fused ring such that the 3'-oxygen atom resides in the axial position, as proposed in Scheme 2 for the enzymatic hydrolysis of cAMP with phosphodiesterases.

Table I. Structures of Cyclic Oxyphosphoranes and Their
          Intramolecular Exhange Processes

No.	Compound	^1H NMR behavior	Ref.
A	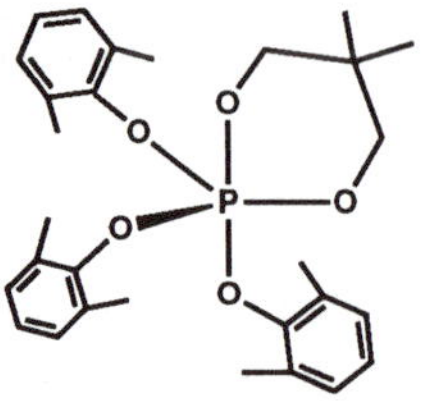	a-e ⇌ e-a slowed below -50°C.	(11)
B	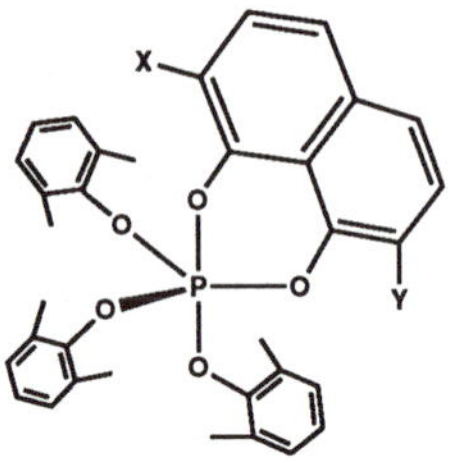 X = Y = H X = Y = Cl X = H, Y = Cl	a-e ⇌ e-a and C-O bond rotation slowed below -60°C.	(11)
C	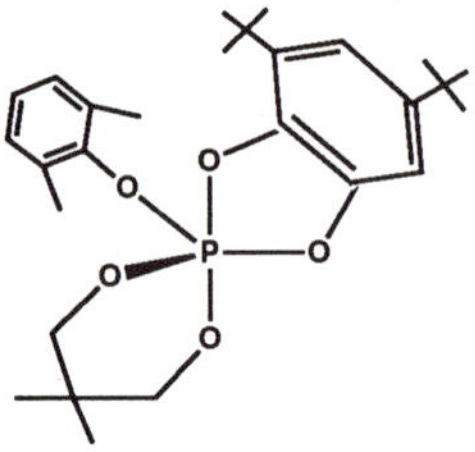	a-e ⇌ e-a and C-O bond rotation not stopped down to -90°C.	(8)
D	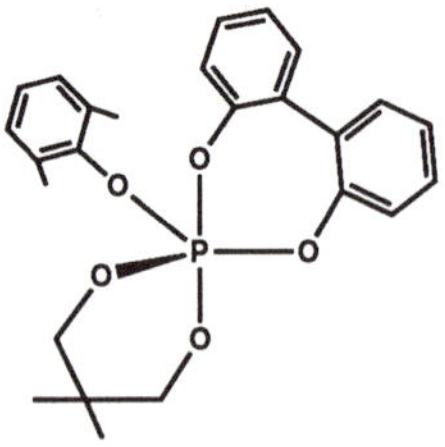	a-e ⇌ e-a slowed below -70°C. C-O bond rotation stopped below -60°C. a-e ⇌ e-e above 0°C via seven-membered ring.	(10)

An additional property of cyclic oxyphosphoranes of interest for mechanistic considerations is that they belong to a class of nonrigid molecules and undergo fast intramolecular ligand permutation as shown by our VT ^{1}H and ^{13}C NMR studies (*8-12*). Two processes are encountered, a so-called low temperature and high temperature process, illustrated here with five-membered ring containing phosphoranes.

Ligand Exchange (Pseudorotation)

Low Temperature Process

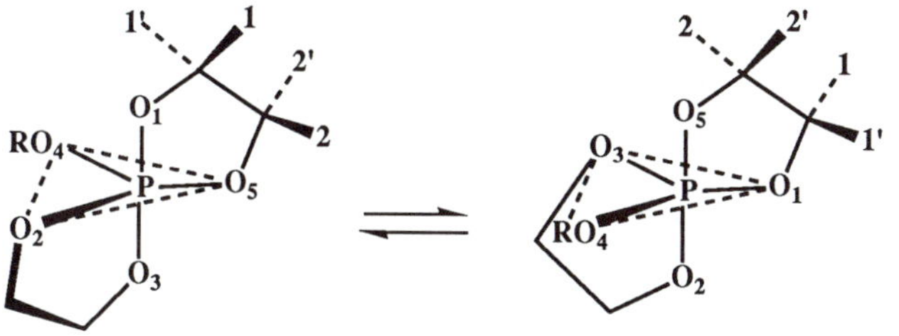

High Temperature Process

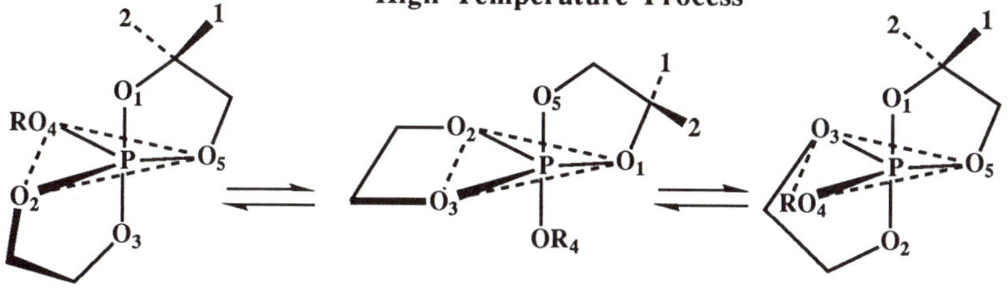

Since the high temperature process takes the ground state axial-equatorial (a-e) ring orientation to a diequatorial (e-e) configuration, the activation energy for this process represents a reasonable estimate of the energy needed to reach the normally less stable diequatorial orientation. As seen in the above tabulation, structure **D** exhibits this process, as do others not listed (*8-12*). Barrier energies ranging from 6 kcal/mol for simply structured cyclic oxyphosphranes with six-membered rings (*35*) to 11 kcal/mol for more encumbered formulations (*10-12*) suggest that active site environments are capable of supplying this relatively low amount of stabilization energy assuming a number of enzyme interactions are present. Due to variations from one enzyme system to another, a P^V cyclic oxyphosphorane activated state may have its phosphorinane ring stabilized in either of the two basic structural arrangements, i.e., (a-e) or (e-e).

Theoretical calculations by van Ool and Buck (1981) (26) at the semi-empirical level (CNDO/2) have been carried out on the model cAMP P^V intermediate (j).

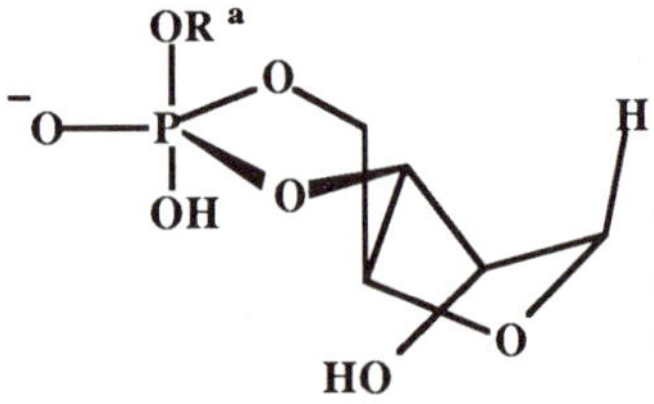

[a] OR simulates the enzyme residue responsible for the enzyme-nucleotide interaction.

j

The results show a preference for the diequatorial ring arrangement over the corresponding axial-equatorial configuration (k) by 28.2 kcal/mol. In a much more recent paper by Broeders, Koole, and Buck (1990) (36), similar semi-empirical calculations (MNDO) on a very analogous sytem, where the hydrogen atom at the adenine position was replaced by an NH_2 group, favored an axial-equatorial ring orientation by 2.8 kcal/mol. The results of the first mentioned calculations favoring (e-e) ring occupancy were used by van Ool and Buck in 1984 (24) to conclude that this was the ring arrangement in compound (H). Compound (G) was shown to have (a-e) ring occupancy by low temperature ^{13}C NMR where pseudorotation was halted.

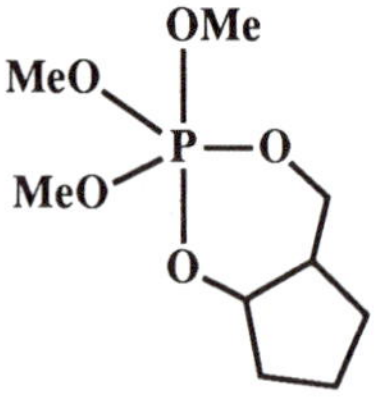

G 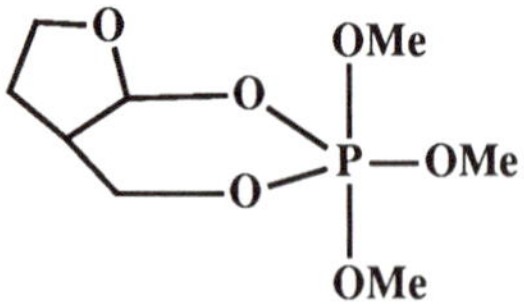   H

Afterwards, Bentrude et al. (33) using a 500 MHz NMR instrument stopped ligand exchange in compound (H) and obtained a ^{13}C pattern consistent with (a-e) ring occupancy, the same as that found for (G). The difficulties inherent in relying on semi-empirical methods as an aid to experimental work is evident here and has prompted us to employ the much more accurate *ab-initio* molecular orbital method to evaluate energy differences. As described in the next section, the calculations favor (a-e) occupancy for simply constituted P^V model intermediates which agree with activation energies from VT NMR on related compounds.

Ring Conformations

The above discussion emphasized phosphorinane ring orientation in isolated and structurally characterized cyclic oxyphosphoranes and their relation to proposed P^V activated states in enzyme reactions of cAMP. Here, we concentrate on ring conformation and its projected role in cAMP interactions based largely on our recent structural work and preliminary investigations of new systems.

In 1974 Trippett (35) made the interesting observation from molecular models that the boat conformation for a six-membered ring located (a-e) in an oxyphosphorane is the most stable since it is the only one that positions the lone pair of electrons on the equatorial ring oxygen atom in the favored equatorial plane for π back-bonding to empty phosphorus d orbitals. For this requirement to be met, the dihedral angle

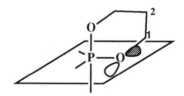

formed by the plane of the $P-O_{eq}-C_1$ ring atoms and the equatorial plane must be 90°. Measurement of this angle for oxyphosphoranes containing six-membered rings that we studied by X-ray diffraction largely follows this assertion when the boat form is present. The boat form has the axial oxygen atom and opposite ring carbon atom C_1 at the prow and the stern of the boat. The average of this angle is 81° for six such structures (11, 12), whereas phosphorane **B** in Table I contains a near planar phosphorinane ring and has a value of 56° for the dihedral angle of interest. However, the ring is still oriented (a-e) in the trigonal bipyramidal geometry. In fact, the boat form is the predominant form found in a-e sites in the absence of additional constraints. No one has yet structurally characterized a six-membered ring in (e-e) sites despite attempts to do so.

In an effort to induce diequatorial ring formation, hydrogen bonding was introduced as a constraint on the system in the preparation of a series of tetraoxyphosphoranes containing an imino function (13, 37). Some representative examples are displayed here.

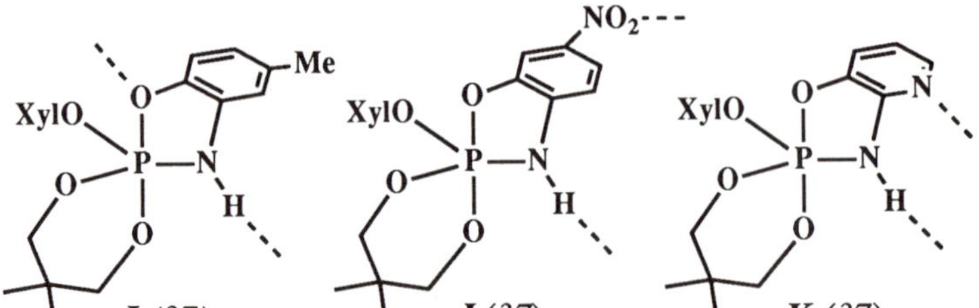

What occurred as found from X-ray studies was the formation of chair conformations for the first time for phosphorinane rings situated (a-e). For **I** (37) and **K** (37), hydrogen bonded dimers formed. For **I**, one six-membered ring was in a chair form and the other was in a boat conformation. The boat is shown schematically in Figure 1(a). A chair form also was present in **J** (37). Here the hydrogen bonding occurred intermolecularly and gave a chain arrangement of phosphorane units. The dimer for **K** (37) had both six-membered rings in twisted chair conformations represented in Figure 1(b). The dihedral angle between the endocyclic P-O_{eq}-C plane and equatorial plane for the phosphorine rings in these hydrogen-bonded phosphoranes average 66° independent of whether the boat or chair form is considered. This angle is much lower than that found for oxyphosphoranes having (a-e) phosphorinane rings in boat conformations that lack hydrogen bonding interactions.

For the latter class, Figure 1(c) represents the type of boat conformation in the dimer for all members studied (8, 10) and shows the four atom base of the boat tipped in an opposite direction from that found for the boat conformation in hydrogen bonded derivatives, Figure 1(a).

The torsion angle X-P-O_A-C_A defined in Figure 1 provides a measure of which ring form is present. Hydrogen bonded phosphoranes like **I-K**, which have chair and boat ring conformations, have values of this angle in the range of 151-169°, whereas phosphoranes lacking hydrogen bonding interactions show boat forms with torsion angles less than 69° (13, 37).

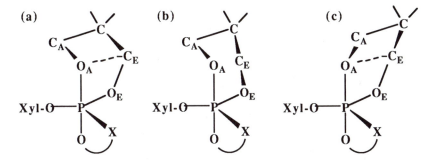

Figure 1. (a) Twist-boat as found in **I** (*13*) (X = NH). The fold is along a line between "prow" atoms OA and CE such that CA is away from X. The torsion angle, X-P-OA-CA > 90°. (b) Chair as in **J** (*37*) and twist-chair as found in **K** (*37*) and **I** (*37*). This conformation can be converted to a twist-boat by moving only atom CE into a prow position. The torsion angle, X-P-O$_A$-C$_A$ is essentially unchanged (> 90°). (c) Boat as found in spirocyclic oxyphosphoranes (*8, 10, 12*) in the absence of hydrogen bonding (X = O). The fold along the line between "prow" atoms OA and CE is such that CA is towards X. The torsion angle, X-P-O$_A$-C$_A$ < 90°.

Whether these chair ring conformations are retained in solution or revert to boat forms is a matter of some speculation. Detailed examination of neighboring group interactions in the dimer formulations suggests the reason the particular chair and boat conformations orient the way they do, Figure 1(b) and (a), respectively, is due to a best steric fit (13). If the dimer dissociates in solution, the chair conformation present in the dimer may revert to a boat. However, VT ^{1}H NMR work indicates retention of the hydrogen bonding interaction which decreases as the temperature increases (13, 37). In an enzyme active site with a preponderance of hydrogen bonding interactions, retention of the chair conformation of the six-membered ring is more likely. One can envision that changes in interactions during the course of a reaction at the active site involving a cAMP P^V activated state might induce a monomer to dimer formation with accompanying changes in ring conformation from (c) to (a) or (b) of Figure 1 and provide a trigger for enzyme action. The possibility of this dimer formation has not been considered before but it is one that may have some merit, especially in mammalian cAMP dependent protein kinase action for which the tetrameric holoenzyme consists of two catalytic and two regulatory subunits and each regulatory subunit binds two molecules of cAMP to distinct sites (34, 38).

Variable temperature ^{1}H and ^{13}C NMR investigations support retention of the basic ground state TBP structures of I-K and other cyclic oxyphosphoranes (39a) in solution having (a-e) phosphorinane ring orientations. Bentrude and coworkers (32, 39b) recently reported detailed ^{1}H NMR assignments for the phosphorinane ring in cyclic oxyphosphoranes (L-N) (32) designed as models of P^V H$_2$O-cAMP or enzyme-cAMP adducts. These derivatives did not incorporate any special constraints such as hydrogen bonding or steric effects. They concluded (32) that the structures are retained in solution with the phosphorinane rings situated (a-e) in nonchair, boat or twist conformations and that these conformations are intrinsically more stable than the chair form. These results are encouraging and nicely support the principal features of our investigations on related oxyphosphoranes lacking special constraints.

We have explored other factors in a preliminary way to ascertain their influence on P^V geometry and ring conformation and have isolated the monocyclic phosphorane (O) for study. X-ray analysis (40) shows the usual (a-e) ring orientation but the presence of a chair conformation. This represents the first example of the appearance of the chair conformation stabilized in the absence of hydrogen bonding interactions. Comparison with the related dithiaphosphorinane P^V analog (P) shows the same (a-e) ring orientation but in the normally observed boat conformation (40).

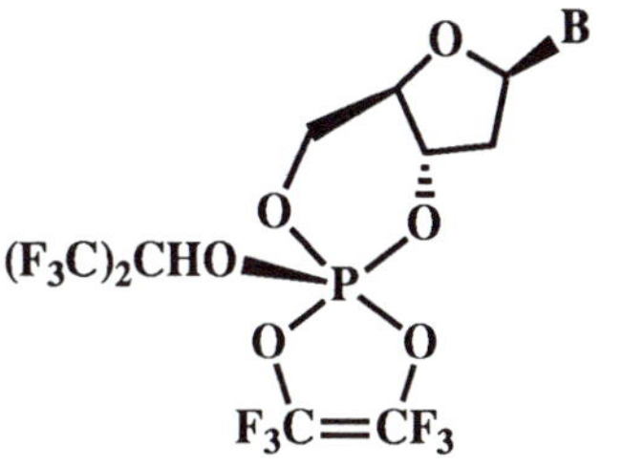

L (B = H)
M (B = thymin-1-yl)

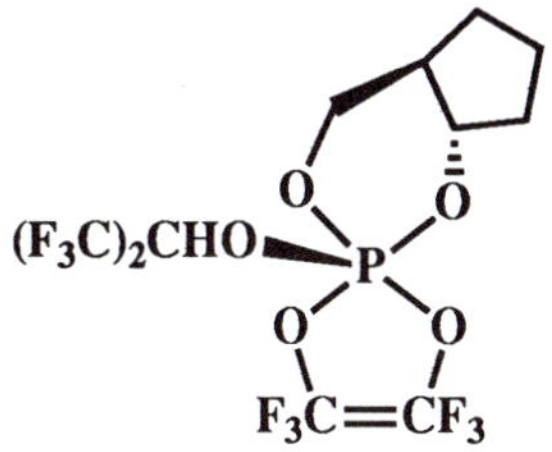

N

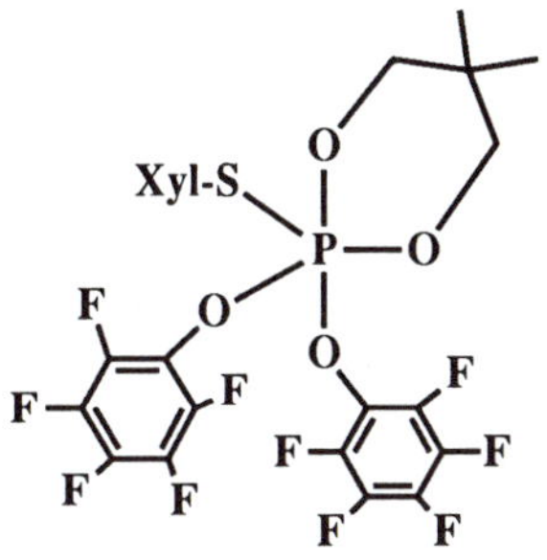

O (*40*)

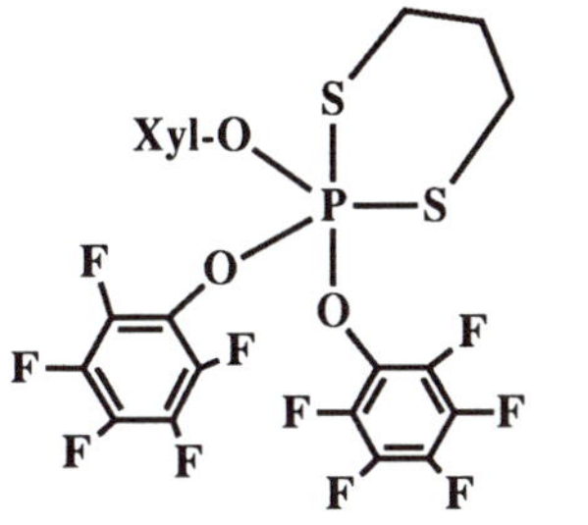

P (*40*)

We reason that the special electronegativity and steric requirements of the pentafluorophenyl ligands contribute to the stabilization of the chair form over the boat form. Derivative **O** is also unique in having equal axial and equatorial P–O bond lengths for the six-membered ring, a feature we surmise (*40*) results from the high electron withdrawing capacity of the opposite axial OC$_6$F$_5$ group. As a consequence, the balance is tipped toward stabilization of the more symmetrical chair arrangement over the boat form. Here again, the formation of this type of electronegativity interaction at an enzyme active site combined with hydrogen bonding may be effective in stabilizing the chair form as the normal conformation, whereas in a nonenzyme environment, as we have found in the absence of these effects, the reverse is true.

By way of molecular orbital calculations, we have examined various phosphorinane ring conformations in both (a-e) and (e-e) orientations in a TBP geometry and have obtained final geometries and energies for saturated five- and six-membered rings in the oxyphosphorane molecules **l** - **o**.

Ab-Initio Calculation - Five-Membered Ring

3-21G* Basis Set

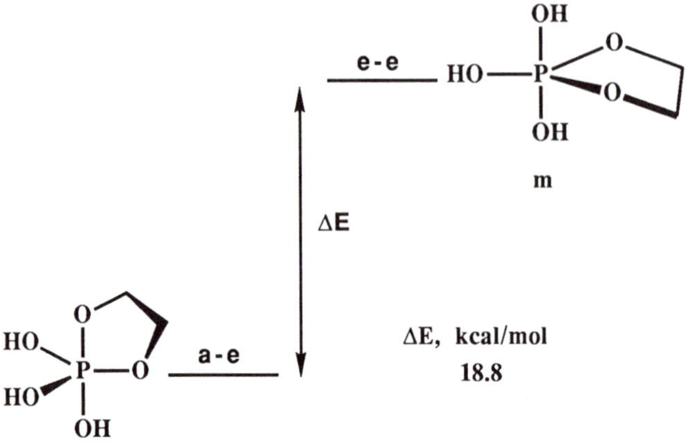

ΔE, kcal/mol
18.8

Ab-Initio Calculation - Six-Membered Ring

3-21G* Basis Set

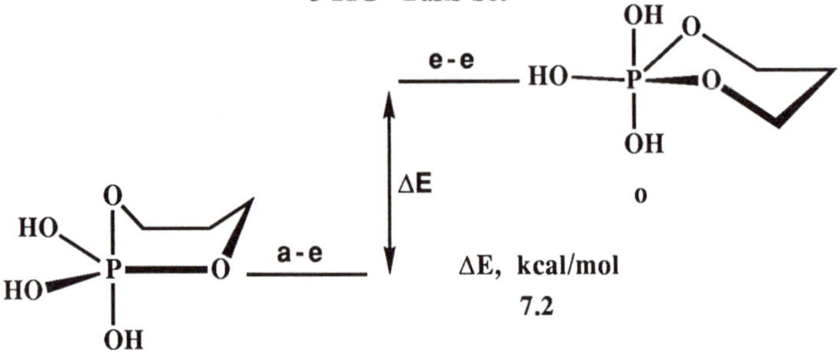

ΔE, kcal/mol
7.2

(Deiters, J. A. and Holmes, R. R., unpublished work)

Ligand Exchange Barriers from ^{19}F NMR

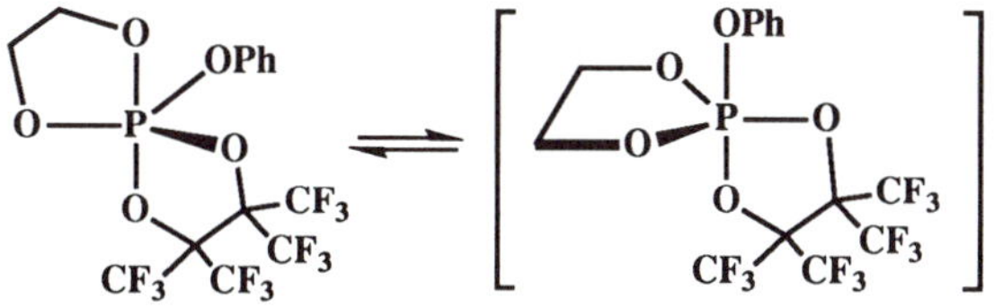

17.4

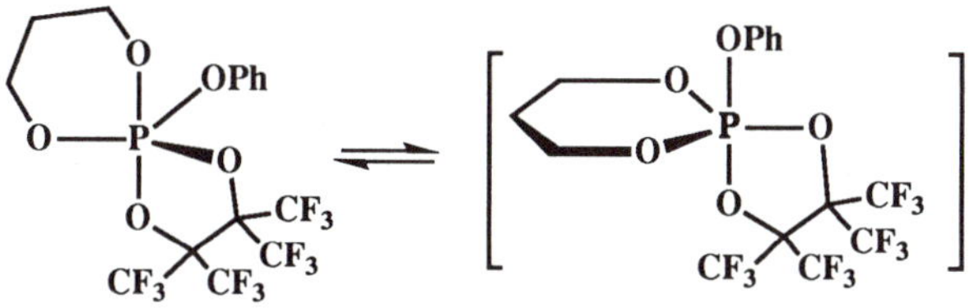

6.1

These are *ab initio* calculations carried out with Gaussian programs on the Cray computer at the San Diego Supercomputer Center either on line or via the UMass Engineering Computer Services VAX cluster. A sophisticated basis set was used, termed 3-21G* (split-level valence with polarization functions on second- and third-row atoms) (*41*). For the phosphorane with the six-membered ring, a typical boat conformation for the (a-e) orientation (**n**) was compared with a chair conformation expected for an (e-e) ring orientation (**o**). There is considerably less energy difference between the two structures in the case of phosphoranes with six-membered rings, 7.2 kcal/mol, compared to that when five-membered rings are introduced, 18.8 kcal/mol. This partly reflects the lower strain in placing a six-membered ring diequatorial in a TBP compared to that encountered with a five-membered ring. These results agree nicely with activation energies from VT ^{19}F NMR (*35*) for pseudorotation of phosphoranes **Q** and **R** containing these rings. The process occurring causes ring reorientation from a ground state phosphorane with a ring occupying (a-e) sites to an activated state having the ring in (e-e) sites. The close energy fit provides confidence in the comparison.

The *ab initio* calculations of energy on fully minimized structures give high accuracy with the basis sets employed. However, one must simplify the system in order to conserve computer time in a judicious manner so as to preserve the main features of interest. It would be prohibitive in time to compute the structure of **Q** and **R** and their exchange intermediates. It took about 35-40 hours of actual supercomputer time to obtain the results for the saturated five- and six-membered ring systems 1 and o. The close agreement between theory and experiment encountered here indicates that the energy differences obtained are controlled by ring strain effects of the five- or six-membered ring going from (a-e) to (e-e) orientations and not by the other phosphorane substituents in any significant way.

Recent Work

In view of the relatively small energy difference estimated between (a-e) and (e-e) orientations of a phosphorinane ring in an oxyphosphorane, it is somewhat surprising that work so far has not stabilized the higher energy (e-e) ring orientation. Working with oxyphosphoranes (**S-V**) containing larger ring systems, seven- and eight-membered, Denney and coworkers (42) conducted a solution state NMR study and concluded that the rings were positioned at (e-e) sites of a TBP in all of these derivatives.

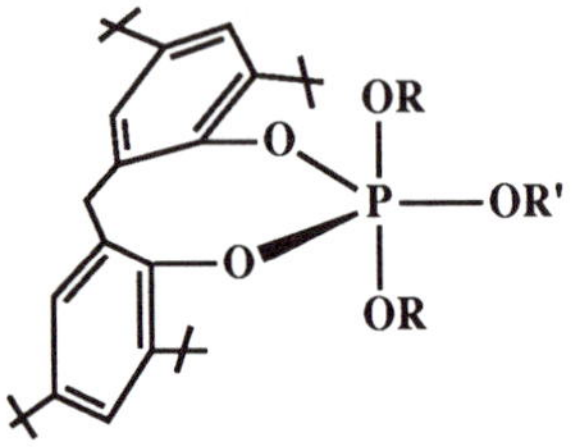

S, R = R' = CH_2CF_3 $\Delta G^{\ddagger}$ = 15.4 kcal/mol
T, R = CH_2CF_3; R' = C_2H_5 $\Delta G^{\ddagger}$ = 15.7 kcal/mol

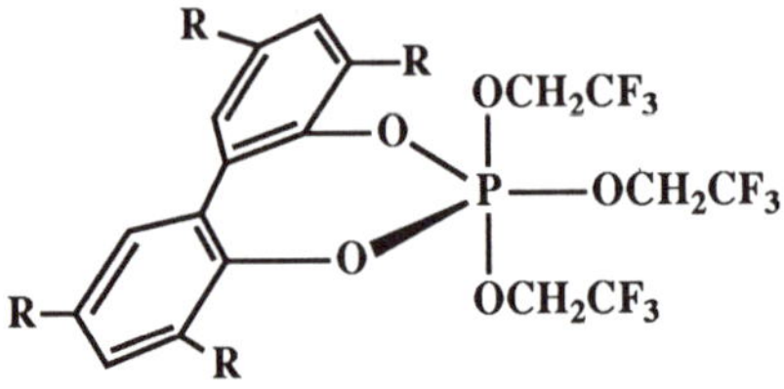

U, R = t-Bu $\Delta G^{\ddagger}$ = 14.6 kcal/mol
V, R = H   Exchange not stopped to -55°C

Recently, we synthesized the related phosphorane **W** and showed the formation of the (e-e) ring orientation by X-ray analysis (Prakasha, T. K.; Day, R. O.; Holmes, R. R., unpublished work). This represents the first solid state structural evidence of a diequatorial ring orientation in any cyclic oxyphosphorane system. However, when the phenyl ring substituents are replaced by protons, X-ray analysis of **X** (Prakasha, T. K.;

Day, R. O.; Holmes, R. R., unpublished work) reveals the (a-e) ring orientation found in all other X-ray studies of cyclic oxyphosphoranes. Apparently the energy difference for the two orientations of this ring system in a TBP is very small such that steric effects are influential in controlling ring geometry. Inspection of the structure reveals a high degree of ring flexibility due to the presence of the methylene unit connecting the phenyl portions of the cyclic system.

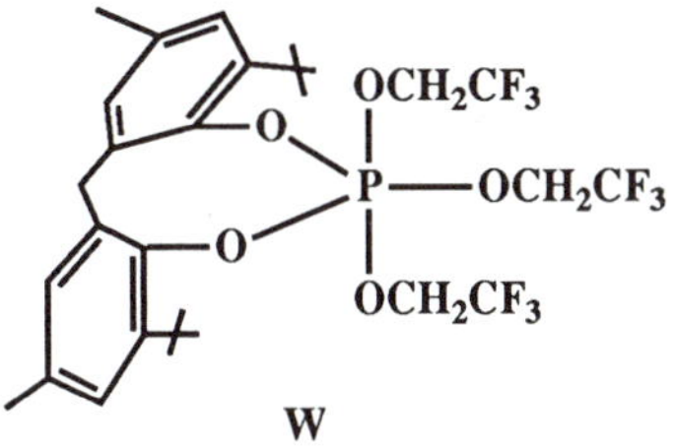

An X-ray study of oxyphosphorane **U** showed that the more constrained seven-membered ring resides in (a-e) sites of a TBP (Prakasha, T. K.; Day, R. O.; Holmes, R. R., unpublished work). This contrasts with Denney's conclusion (*42*) of an (e-e) orientation in solution for **U**. The implication is that packing effects are controlling the ring arrangement in the solid state compared to that indicated in solution. If so, this would be the first case where a phosphorane underwent a significant structural change on going from the solid to the solution state.

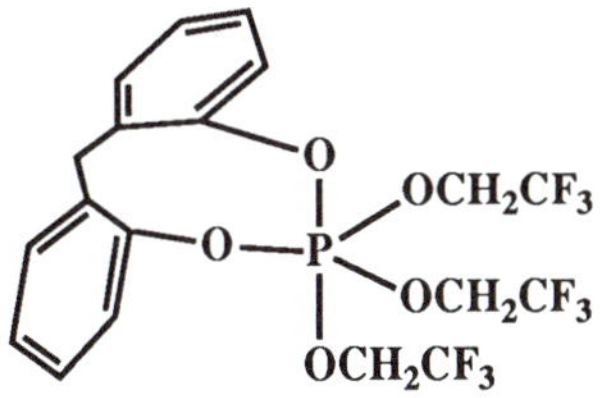

Only one other study exists suggesting (e-e) ring orientation, a recent NMR study by Broeders et al. (*36*) on the interesting bicyclic oxyphosphorane **Y** which has a ribose component trans-annelated to a phosphorinane ring, as in cAMP. Indirect evidence from an orientational effect indicated to be present in the axial acyclic attachment suggests that isomers of **Y** with an (e-e) and (a-e) ring orientations coexist in solution. It would prove worthwhile if this or related derivatives can be isolated and structurally identified to verify this structural indication.

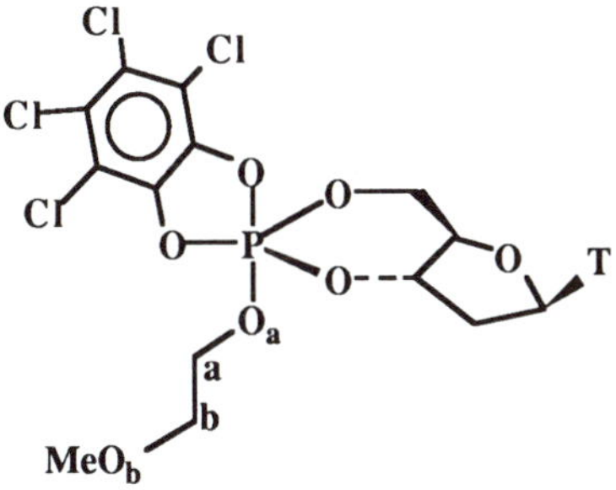

Relative to **Y**, we have prepared Z_1 and Z_2 and have obtained preliminary NMR data. Derivative Z_1 has a cis-fused ring system and Z_2 a trans-fused one.

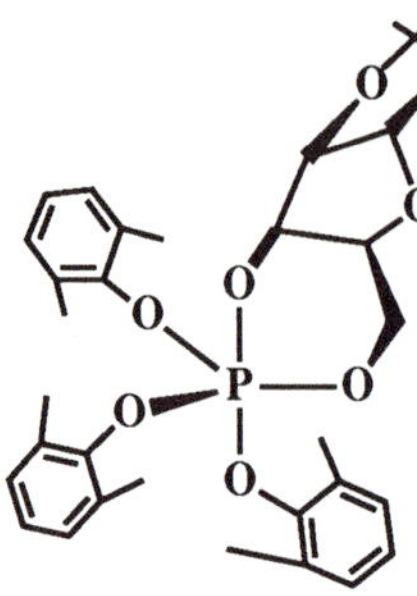

Z_1 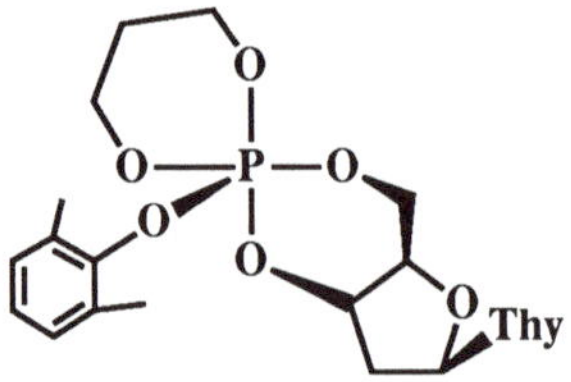 Z_2

In conclusion, we may expect further advances in understanding factors controlling structural and conformational preferences of cyclic oxyphosphoranes, both from an experimental as well as a theoretical point of view. As the area continues to develop, the use of oxyphosphoranes as models in mechanistic interpretation should enhance our understanding of pathways followed in nucleophilic displacement reactions of tetracoordinated phosphorus compounds.

Acknowledgment

The support of this research by the National Science Foundation (Grant CHE88-19152 and the Army Research Office is gratefully acknowledged.

Literature Cited

1. Corriu, R. J. P. *Phosphorus Sulfur* **1986**, *27*, 1, and references cited therein.
2. Mikolajczyk, M.; Krzywanski, J.; Ziemnicka, B. *Tetrahedron Lett.* **1975**, 1607.
3. Hall, C. R.; Inch, T. D. *Tetrahedron* **1980**, *36*, 2059.
4. Holmes, R. R. *Pentacoordinated Phosphorus, Reaction Mechanisms*, ACS Monograph 176; American Chemical Society: Washington, DC, 1980; Vol. II, Chapter 2 and references cited therein.
5. Westheimer, F. H. *Acc. Chem. Res.* **1968**, *1*, 70, and references cited therein.
6. Holmes, R. R. *Pentacoordinated Phosphorus--Structure and Spectroscopy*, American Chemical Society: Washington, DC, 1980; Vol. I, ACS Monograph No. 175, and references cited therein.
7. Schomburg, D.; Hacklin, H.; Röschenthaler, G.-V. *Phosphorus and Sulfur* **1988**, *35*, 241.

8. Kumara Swamy, K. C.; Burton, S. D.; Holmes, J. M.; Day, R. O.; Holmes, R. R., *Phosphorus, Sulfur, and Silicon*, **1990**, *53*, 437.
9. Kumara Swamy, K. C.; Holmes, J. M.; Day, R. O.; Holmes, R. R., *J. Am. Chem. Soc.*, **1990**, *112*, 6092.
10. Kumara Swamy, K. C.; Day, R. O.; Holmes, J. M.; Holmes, R. R., *J. Am. Chem. Soc.*, **1990**, *112*, 6095.
11. Burton, S. D.; Kumara Swamy, K. C.; Holmes, J. M.; Day, R. O.; Holmes, R. R., *J. Am. Chem. Soc.*, **1990**, *112*, 6104.
12. Holmes, R. R.; Kumara Swamy, K. C.; Holmes, J. M.; Day, R. O., *Inorg. Chem.*, **1991**, *30*, 1052.
13. Day, R. O.; Kumara Swamy, K. C.; Fairchild, L.; Holmes, J. M.; Holmes, R. R. *J. Am. Chem. Soc.* **1991**, *113*, 1627-1635.
14. (a) Stryer, L., in *Biochemistry*, Freeman and Co., San Francisco (1980); (b) Review Series, *Advances in Cyclic Nucleotide Research*: Greengard, P.; Robinson, G. A., Sr., Eds.; Raven Press: New York, 1980-1988, Vols. 11-18.
15. Revenkar, G. R.; Robins, R. K., in *Handbook of Experimental Pharmacology*; Nathanson, J. A.; Kebabian, J. W., Eds.; Springer Verlag: Berlin and Heidelberg, West Germany, **1982**; Vol. 58/I, Chapter 2.
16. Miller, J. P., *Adv. Cyclic Nucl. Res.*, **1981**, *14*, 335.
17. Meyer, R. B., Jr., in *Burger's Medicinal Chemistry*, 4th ed., Wolff, M. E., Ed.; Wiley Interscience: New York, **1979**, Chapter 34, Part II.
18. Miller, J. P., in *Cyclic Nucleotides*: Mechanisms of Action; Cramer, H.; Schultz, J., Eds.; Wiley: London, **1977**, pp. 77-105.
19. Varughese, K. I.; Lu, C. T.; Kartha, O., *J. Am. Chem. Soc.*, **1982**, *104*, 3398.
20. Blackburn, B. J.; Lapper, R. D.; Smith, I. C. P., *J. Am. Chem. Soc.*, **1973**, *95*, 2873.
21. van Haastert, P. J. M.; Dijkgraaf, P. A. M.; Konijn, T. M.; Abbad, E. G.; Petridis, G.; Jastorff, B. *Eur. J. Biochem.*, **1983**, *131*, 659
22. Burgers, P. M. J.; Eckstein, F.; Hunneman, D. H.; Baraniak, J.; Kinas, R. W.; Lesiak, K.; Stec, W. J., *J. Biol. Chem.*, **1979**, *254*, 9959.
23. Mehdi, S.; Coderre, J. A.; Gerlt, J. A., *Tetrahedron*, **1983**, *39*, 3483.
24. van Ool, P. J. J. M.; Buck, H. M., *Recl. Trav. Chim. Pays Bas*, **1984**, *103*, 119.
25. van Ool, P. J. J. M.; Buck, H. M., *Eur. J. Biochem.*, **1982**, *121*, 329.
26. van Ool, P. J. J. M.; Buck, H. M., *Recl. Trav. Chim. Pays Bas*, **1981**, *100*, 79.
27. Coderre, J. A.; Mehdi, S.; Gerlt, J. A., *J. Am. Chem. Soc.*, **1981**, *103*, 1872.
28. Cullis, P. M.; Jarvest, R. L.; Lowe, G.; Potter, B. V. L., *J. Chem. Soc., Chem. Commun.*, **1981**, 245.
29. Jarvest, R. L.; Lowe, G.; Baraniak, J.; Stec, W. J., *Biochem. J.*, **1982**, *203*, 461.
30. Holmes, R. R.; Deiters, J. A.; Gallucci, J. C., *J. Am. Chem. Soc.*, **1978**, *100*, 7393.

31. Holmes, R. R., *Pentacoordinated Phosphorus--Reaction Mechanisms*, ACS Monograph No. 176, American Chemical Society: Washington, DC, 1980; Vol. II, pp. 180-222.
32. Yu, J. H.; Arif, A. M.; Bentrude, W. G., *J. Am. Chem. Soc.*, **1990**, *112*, 7451.
33. Yu, J. H.; Sopchik, A. E.; Arif, A. M.; Bentrude, W. G., *J. Org. Chem.*, **1990**, *55*, 3444.
34. de Wit, R. J. W.; Hekstra, D.; Jastorff, B.; Stec, W. J.; Baraniak, J.; van Driel, R.; van Haastert, P. J. M., *Eur. J. Biochem.*, **1984**, *142*, 255.
35. Trippett, S., *Pure Appl. Chem.*, **1974**, *40*, 595.
36. Broeders, N. L. H. L.; Koole, L. H.; Buck, H. M., *J. Am. Chem. Soc.*, **1990**, *112*, 7475.
37. Hans, J.; Day, R. O.; Holmes, R. R., *Inorg. Chem.*, **1991**, *30*, in press.
38. Lincoln, T. M.; Corbin, J. D., *J. Cycl. Nucl. Res.*, **1978**, *4*, 3.
39. (a)   Yu, J.; Sopchik, A. E.; Arif, A. M.; Bentrude, W. G.; Roeschenthaler, G.-V. *Heteroatom Chem.* **1991**, *2*, 177.   (b) Yu, J. H.; Bentrude, W. G., *Tetrahedron Lett.*, **1989**, *30*, 2195.
40. Hans, J.; Day, R. O.; Howe, L.; Holmes, R. R., *Inorg. Chem.*, **1991**, *30*, in press.
41. Frisch, M. J.; Head-Gordon, M.; Trucks, G. W.; Foresman, J. B.; Schlegel, H. B.; Raghavachari, K.; Robb, M. A.; Binkley, J. S.; Gonzalez, C.; Defrees, D. J.; Fox, D. J.; Whiteside, R. A.; Seeger, R.; Melius, C. F.; Baker, J.; Martin, R. L.; Kahn, L. R.; Stewart, J. J. P.; Topiol, S.; Pople, J. A., *Gaussian 90*, Gaussian, Inc., Pittsburgh, PA, 1990.
42. Abdou, W. M.; Denney, D. B.; Denney, D. Z.; Pastor, S. D. *Phosphorus and Sulfur* **1985**, *22*, 99.

RECEIVED December 10, 1991

Chapter 3

Quantitative Analysis of Inorganic Phosphates Using ^{31}P NMR Spectroscopy

Janice K. Gard, David R. Gard, and Clayton F. Callis[1]

Monsanto Company, 800 North Lindbergh Boulevard, St. Louis, MO 63167

Phosphorus-31 nuclear magnetic resonance (^{31}P NMR) is optimized with respect to accuracy, precision, and analysis time for the characterization of inorganic phosphates. Species determinations in commercial sodium tripolyphosphate are routinely achieved with an accuracy and precision (0.1-0.5%) comparable to that obtained by chromatographic methods as determined in interlaboratory analyses. The method has been completely automated using a robotic sample changer and an algorithm for data analysis. In a demonstration of the precision attainable, the hydrolysis kinetics of tripolyphosphate is obtained with a correlation coefficient of 0.998. Extension of ^{31}P NMR to the analysis of higher oligophosphate mixtures (i.e. sodium phosphate glass) has recently been examined using homonuclear 2DJ-resolved spectroscopy to separate the coupling constant and chemical shift information. Semi-quantitative analyses are achieved using curve deconvolution.

The ubiquitous nature and broad importance of phosphates demands exacting analytical methods for their characterization. Phosphorus-31 nuclear magnetic resonance (^{31}P NMR) has been used as a method for the quantitative analysis of small inorganic phosphates (1-4). Several potential advantages are offered by ^{31}P NMR including observation of only the phosphorus-containing species, structural information which may complement or aid

[1]Retired

0097–6156/92/0486–0041$06.00/0
© 1992 American Chemical Society

quantitation, and quantitation with an elemental as opposed
to a molecular standard. Limitations of phosphorus-31 NMR
include   sensitivity,   complexity   of   spectra   for
oligophosphates   higher   than   tripolyphosphate,   and,
occasionally, slow relaxation times. Little appears in the
literature, however, concerning the precision or accuracy
of phosphate analysis by ^{31}P NMR. Recent advances in
instrumentation have revolutionized NMR spectroscopy,
particularly with respect to sensitivity and resolution,
and has prompted this reexamination of the quantitative
capabilities of ^{31}P NMR.

Further enhancements of the NMR method have been
carried out in our laboratory by optimization of
experimental   parameters   with   respect   to   accuracy,
precision, and analysis time using commercial sodium
tripolyphosphate (STP). The NMR technique was then
directly compared with chromatographic methods in a
controlled interlaboratory study with the use of Lorentzian
lineshape analysis to improve precision yet further. The
accuracy and precision of the NMR method is illustrated by
the   high   value   of   the   correlation   coefficients   in
monitoring   the   kinetics   of   hydrolysis   of   sodium
tripolyphosphate. With the installation of a robotic
sample changer and customized robotic software the
efficiency of NMR quantitation is greatly enhanced.
Demonstration   of   the   complete   automation   of   NMR
quantitative analysis is herein described, including full
data reduction and generation of the analytical report.

Efforts are currently underway to expand the utility
of ^{31}P NMR to qualitative and quantitative analysis of much
more complex oligophosphate mixtures. A novel application
of   homonuclear   two-dimensional   J-resolved   (2DJ)
spectroscopy of sodium polyphosphate glass is shown to
effectively yield ^{31}P-^{31}P decoupled spectra. Used in
conjunction with Lorentzian lineshape analysis and curve
deconvolution, semiquantitative analyses of these mixtures
has been achieved.

Analytical Methods

One-Dimensional Quantitative ^{31}P NMR Spectroscopy. Spectra
were collected on automated Varian XL-200 or VXR300S
Fourier   transform   NMR   spectrometers,   operating   at
phosphorus frequencies of 80.98 and 121.42 MHz. The
phosphates were typically prepared as 2-5 weight per cent
in D_2O with the pH maintained near 9 in order to optimize
the signal separation and the longitudinal relaxation times
(5,6). T_1 values were determined using the Fast Inversion
Recovery Fourier Transform (FIRFT) method (7) and
acquisition parameters were optimized to maximize the
observable magnetization with respect to analysis time (8).
Spectra were accumulated using a 20-25 degree pulse width,
0.50 sec acquisition time, a 3 to 5 second repetition

delay, gated Waltz decoupling, collection of 16K data points prior to zero-filling to 32K points, and application of a 1.0 Hz exponential line broadening. In the interlaboratory analyses 1024 transients were generally collected requiring about 1 hr. of total scan time. A spectrum of commercial sodium tripolyphosphate acquired under these conditions at 80.98 MHz is shown in Figure 1. Non-linear least squares Lorenzian lineshape analysis was also examined for curve fitting and integrations in the interlaboratory analysis (9).

Robotic ^{31}P NMR Assays. Spectra were collected on a Varian VXR300S spectrometer equipped with a Varian automatic sample management system (robot) with a 50 sample tray using the parameters described above. Customized software was developed for robot control providing full automated work-up of data including the calculation of weight percents and generation of analytical reports.

Two-Dimensional Semiquantitative ^{31}P NMR. Spectra were collected on a Varian Unity 400 Fourier transform NMR spectrometer at 161.90 MHz. The phosphates were prepared at 3-5 weight per cent in water and a D_2O insert used for locking purposes. Homonuclear 2DJ-resolved spectra were accumulated using an 8K X 0.2K data set with an acquisition time in the F_2 dimension of 0.946 sec, four steady state pulses, 128 transients, and 200 increments in the F_1 domain. Spectra were analyzed with zero-filling to 16K X 0.5K and application of a sine bell or shifted sine bell weighting function on a Sun Microsystems Sparc 1+ computer.

Chromatography. Separations by ion exchange column chromatography were performed according to ASTM methods (10,11). The Technicon AutoAnalyzer (Bran and Luebbe) (12,13) was employed for analysis of the eluent from ASTM method D 2761 (11-13). ASTM method D 501 was followed as written, using the ammonium molybdate colorimetric analysis of the hydrolyzed fractions.

Interlaboratory Analyses of Commercial Sodium Tripolyphosphate

An interlaboratory study was conducted to assess the capabilities of standard ion-exchange chromatographic methods and NMR at 80.9 MHz for the analysis of sodium tripolyphosphate. Three well-mixed, but separate, samples of commercial sodium tripolyphosphate were submitted to each laboratory in triplicate, representing three different levels of tripolyphosphate. Protocol was maintained consistent among methods/sites.

The results comparing the methods are presented in Table I (Gard, D.R.; Burquin, J.C.; Gard, J.K.; submitted for publication.). The AutoAnalyzer and the NMR analyses were each performed by two different analysts; results by

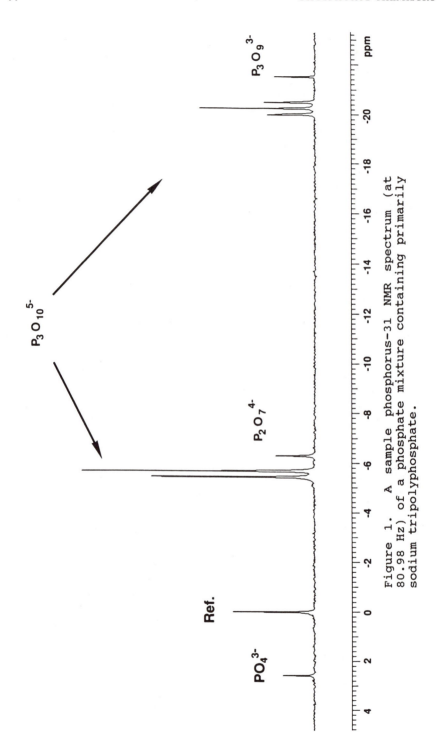

Figure 1. A sample phosphorus-31 NMR spectrum (at 80.98 Hz) of a phosphate mixture containing primarily sodium tripolyphosphate.

each analyst are listed separately in Table I. The line shape analysis and curve fitting routine was employed on one set of NMR data (row 4 of Table I); these results are presented in row 6 of Table I. IECC designates ASTM method D 501.

Table I
Interlaboratory Assay of Sodium Tripolyphosphate

% Total P_2O_5 as-

Mean/Standard Deviation of 9 Samples

Method	Ortho-phosphate	Pyro-phosphate	Tripoly-phosphate	Trimeta-phosphate
AutoAn.	0.31/0.09	7.36/0.73	92.19/0.78	0.14/0.12
AutoAn.	0.25/0.11	6.60/0.38	92.84/0.34	0.30/0.03
IECC	0.55/0.16	7.66/0.17	91.03/0.23	0.73/0.10
^{31}P NMR*	0.35/0.11	7.35/0.48	91.86/0.51	0.44/0.14
^{31}P NMR*	0.36/0.15	7.20/0.56	92.09/0.61	0.37/0.18
^{31}P NMR**	0.51/0.28	6.68/0.05	92.33/0.35	0.51/0.17

* Separate integration of the tripolyphosphate doublet and the pyrophosphate singlet

** Using line shape analysis and curve fitting

Two-tailed tests of the null hypothesis are used to identify significant differences among the methods (14). The IEC determinations are consistently to one end of the range of results. The determination of tripolyphosphate by IEC is lower than for the other methods while that for the other species is higher. Hydrolytic degradation during the IEC analysis would account for this difference since tripolyphosphate is the species most susceptible to hydrolysis. This hypothesis is independently supported by the fact that little or no orthophosphate (a hydrolysis product) could be detected by NMR in separate analyses of tetrasodium pyrophosphate (≤ 0.1%, detection limit ~0.01%), while significant concentrations (0.5%) were observed by IEC. Differences in the accuracy between direct absorbance measurements for ASTM D 501 vs peak integration for the AutoAnalyzer may also be important.

Significant differences in the AutoAnalyzer results are ~1% difference in the determination of the major species between the two sets of data. This is mainly thought to reflect error in peak integration; although the means of the AutoAnalyzer determinations for pyro- and tripolyphosphate are closer in the first interlaboratory analysis, the standard deviation is larger. On the other

hand, the two sets of results from ^{31}P NMR using simple electronic integration are in close agreement with each other, especially considering different analysts performed the analyses. Results by ^{31}P NMR agree well with one set of the AutoAnalyzer results, the only significant difference being in the determination of trimetaphosphate.

The precision for the determination of the major phosphate species was significantly increased with the application of line shape analysis. The tripolyphosphate contribution is, however, weighted more at the expense of pyrophosphate using line shape analysis and curve fitting for the ^{31}P NMR, compared to manual integrations. More accurate determinations are achieved since curve fitting better accounts for signal overlap, baseline noise, and contribution of the wings of the Lorentzian peaks, especially for the lower concentration species (15,16).

Application of ^{31}P NMR to Kinetics

The high precision and accuracy of ^{31}P NMR for the analysis of sodium tripolyphosphate indicated the technique might also be suitable for application to precise kinetic studies involving oligophosphates. The hydrolytic degradation of 0.035 M sodium tripolyphosphate was examined in the presence of various concentrations of Me_4NCl at 60.3°C. The hydrolysis was followed over approximately two half-lives. Even with large changes in the relative concentrations of phosphate species over time, excellent correlation for pseudo-first order kinetic plots are obtained allowing distinctions to be made among the rates under different Me_4NCl concentrations (Figure 2). The high correlation for these plots illustrates the combined high accuracy and precision of the NMR method (Table II). Values for the rate constants are consistent with those of Watanabe et al (17) and Crowther and Westman (18).

Table II
Hydrolysis of Sodium Tripolyphosphate with Added
Tetramethylammonium Chloride as Followed by ^{31}P NMR

(35mM STP, pH 9.0, 60.3°C)

Me_4NCl (M)	10^3 k (hr^{-1})	Correlation Coeff. (R)
0	7.56	0.998
0.43	7.29	0.998
1.38	6.83	0.998

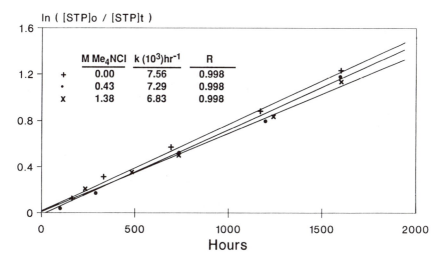

Figure 2. Pseudo-first order kinetics plot of the hydrolysis of 35 mM sodium tripolyphosphate (STP) as followed by ^{31}P NMR (pH 9.0, 60.3°C) in the presence of various concentrations of Me₄NCl. Me₄NCl concentrations, observed rate constants, and correlation coefficients (R) are given in the inset.

Robotic ^{31}P NMR Assays of Sodium Tripolyphosphate

Although ^{31}P NMR yields rapid, reproducible quantification of oligophosphates, quantitative spectra require longer accumulations than simple survey spectra. Manual data reduction of the spectra is a subjective process, and can be time-consuming, especially when large numbers of samples are involved. In order to improve spectrometer efficiency, the phosphate assays are now being performed on a VXR300S equipped with a robotic sample changer. Utilizing easy-to-use menu-driven software, laboratory personnel designate the preferred solvent, duration and type of shimming (magnetic field optimization), automation operating parameter set (environment), and plotter choice. Initiation of the automation run is simply begun by clicking a single menu button. The customized software then automatically accumulates the data, stores it in a common area for robotic runs, and performs the Fourier transform and integration, and calculates the relative distribution of phosphate species. Spectra comparable to that in Figure 1 are generated without further operator intervention. Improvements in the sensitivity are observed at the higher field (121.4 MHz) of the VXR300S, resulting in shorter overall accumulations.

The data reduction software Fourier transforms and phases the data after zero-filling and application of the designated exponential filter function. The usual chemical shift range of the peaks is noted, as the system searches for the primary phosphorus-containing species. A signal-to-noise check is performed over that region, and peaks generating a value less than three are designated as not being present (or approximately 0.2 weight percent, the lower concentration limit for the routine ^{31}P assay). Once the peaks are found, the spectrum is automatically integrated. The integral values and molecular weight of each of the species are then used in the computation of the relative weight percent distribution of the phosphate species. The computer-derived data reduction compares quite favorably to that performed by human analysts as seen in Table III. The two methods commonly differ by a relative error of less than 0.2 weight percent for the major species. The mean for each species is shown to illustrate the lack of bias between the two techniques. The main advantages of the computer method are its speed and lack of subjectivity. The automated quantitative analyses are normally performed at night, allowing access to the spectrometer during the day, when demand for the instrument is high.

Table III
Manual vs. Robotic (ASM) Data Reduction
Comparison of STP Assays

Relative Wt % as Sodium Salts

Sample No.	Ortho-phosphate manual	ASM	Pyro-phosphate manual	ASM	Tripoly-phosphate manual	ASM	Trimeta-phosphate manual	ASM
A	0.35	0.39	8.63	8.62	89.39	89.48	1.63	1.51
B	0.17	0.15	6.62	6.61	90.99	91.18	2.22	2.05
C	0.24	0.21	4.29	4.35	93.56	93.68	1.91	1.76
D	0.11	0.08	6.82	6.72	91.05	91.31	2.03	1.90
E	0.14	0.12	6.31	6.16	91.60	91.84	1.95	1.87
F	0.23	0.18	5.41	5.36	92.76	93.01	1.60	1.45
Mean	0.21	0.19	6.35	6.30	91.56	91.75	1.89	1.76

Homonuclear Two-Dimensional J-Resolved Spectroscopy

Due to the complexity of the one-dimensional spectra of oligophosphate glasses, species determination by curve-fitting has been successfully applied only to relatively pure samples of individual oligophosphate species (19). In an effort to resolve the overlapping multiplets in complex spectra, two-dimensional NMR is being examined. A one-dimensional ^{31}P spectrum of a sodium phosphate glass with an average chain length of 4.11 is shown in Figure 3. (The average chain length, $\bar{n}$, is determined by elemental analysis.) As can be observed, a large number of closely spaced resonances are obtained. An expansion of the "ends" region is also seen in the inset of Figure 3. Although some of the peaks are separated at high field, assignments, and therefore quantitation, of each of the varying chain length resonances, is complicated by the presence of ^{31}P-^{31}P coupling. Removing this coupling would greatly simplify the spectrum. Simple homonuclear decoupling would be insufficient, as the ability to simultaneously observe all phosphorus-containing species would then be lost.

^{31}P homonuclear 2DJ experiments were performed on the glass to effectively decouple the spectrum. A 2DJ spectrum of the phosphate ends region is shown in Figure 4. The ends region was used for the phosphate species determination because of its relative simplicity and first-order nature in comparison with the middles region. The phosphorus chemical shifts are seen on one axis, and the ^{31}P-^{31}P coupling constants on the other. Pairs (or more) of contours at one chemical shift are the peaks for one (chain length) species, separated by the coupling constant. A vertical projection of the spectrum onto the chemical shift axis clearly delineates the newly-separated resonances,

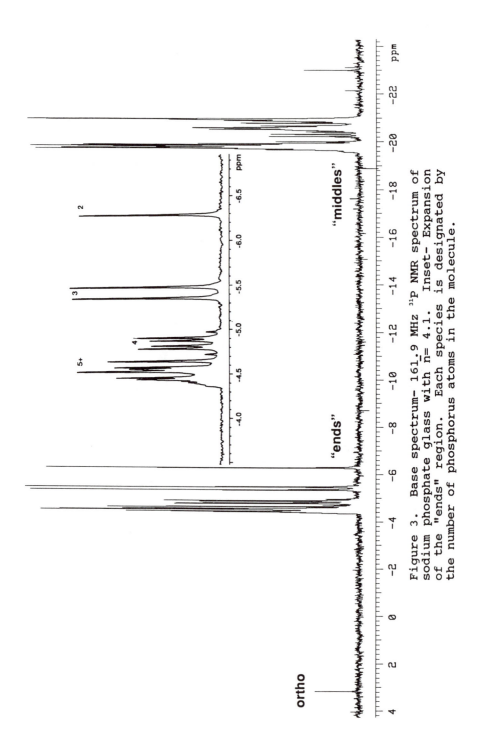

Figure 3. Base spectrum- 161.9 MHz ^{31}P NMR spectrum of sodium phosphate glass with $\underline{n}$= 4.1. Inset- Expansion of the "ends" region. Each species is designated by the number of phosphorus atoms in the molecule.

resulting in a phosphorus-phosphorus decoupled spectrum as shown at the top of Figure 4. A monotonic increase in chemical shift is seen with increasing phosphorus chain length species (20).

In an effort to quantify the amounts of the various chain lengths, this projection was subjected to spectral curve-fitting. In Figure 5 is seen the results of this analysis, with the experimental spectrum at the bottom, and the computer-generated fit above it. The determination of the phosphate species by integration of the signal areas of the fitted spectrum is listed in Table IV. The average chain length determined by 2DJ NMR is 4.13, well within experimental error of the 4.11 value determined by elemental analysis. A comparison is also made in Table IV of the 2DJ results with that previously found for a sodium phosphate glass determined by paper chromatography to be n= 4.0 (21). The results are currently considered to be semi-quantitative, as the 2DJ experiment can give rise to artifacts; true intensities do not always result when projections onto the chemical shift axis are made because of poor lineshapes. More accurate quantification could possibly be achieved by using volume integrals or by utilizing the methods of Xu et al (22) to help eliminate these shortcomings.

Table IV
Composition of Sodium Phosphate Glass

% Total P_2O_5 as-

Phosphate Species	2DJ ^{31}P NMR	_Glass n = 4.0*
Ortho	1.2	0
Pyro	9.4	6.6
Tripoly	13.8	28.2
Tetrapoly	24.9	27.4
Pentapoly	17.5	16.9
Hexapoly	19.7	9.4
Heptapoly	7.8	5.7
Octapoly	-	2.7
Nonapoly	2.8	1.8
Higher	2.9	1.2

Average Chain Length, $\bar{n}$ =

Elemental Analysis	4.11
2DJ ^{31}P NMR	4.13

* Ref. 21

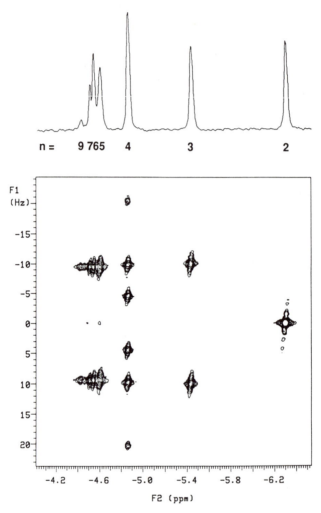

Figure 4. ^{31}P-^{31}P homonuclear 2DJ-resolved NMR spectrum of the same glass whose one-dimensional spectrum is shown in Figure 3. The characteristic J coupling pattern is observed for each species. A projection onto the chemical shift axis is shown, above. The projection for each species is designated by the number of phosphorus atoms in the molecule. The lineshape of the octaphosphate was not distinct enough to allow for independent fitting.

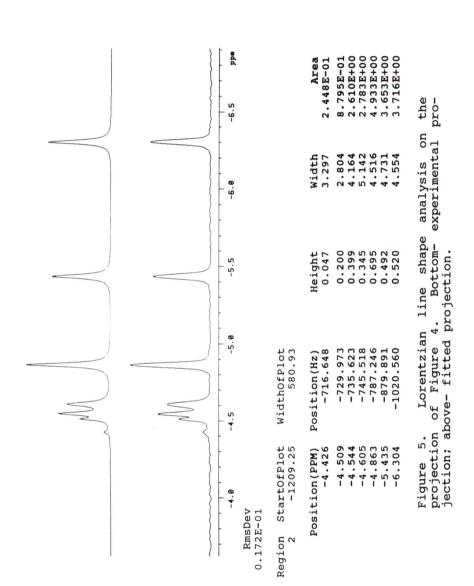

RmsDev
0.172E-01

Region StartofPlot WidthofPlot
2   -1209.25   580.93

Position(PPM)	Position(Hz)	Height	Width	Area
-4.426	-716.648	0.047	3.297	2.448E-01
-4.509	-729.973	0.200	2.804	8.795E-01
-4.544	-735.623	0.399	4.164	2.610E+00
-4.605	-745.518	0.345	5.142	2.783E+00
-4.863	-787.246	0.695	4.516	4.933E+00
-5.435	-879.891	0.492	4.731	3.653E+00
-6.304	-1020.560	0.520	4.554	3.716E+00

Figure 5. Lorentzian line shape analysis on the projection of Figure 4. Bottom— experimental projection; above— fitted projection.

Conclusions

Phosphorus-31 NMR provides an accurate and precise method for the analysis of soluble phosphates and their mixtures. The accuracy and precision are comparable to accepted chromatographic methods. The results signify that the NMR method might also be successfully applied to less straightforward systems with high accuracy and precision and with the additional advantages characteristic of NMR which were enumerated earlier in the article. The most significant source of error in the NMR method is in integration of the signal areas, especially for resonances with chemical shifts lying close to one another. This source of error is alleviated in large measure by employing curve deconvolution and lineshape analysis.

Application of quantitative robotic analysis greatly improves the efficiency, while reliably producing results equivalent to those obtained by manual selection of the integral regions. Little more than clicking a single menu button is required to initiate an automation run. By removing much of the analyst subjectivity, more reproducible and reliable assays are obtained.

By extending the application of the homonuclear 2DJ experiment to the realm of oligophosphate assays, two-dimensional NMR is seen to be a promising extension for unravelling complicated oligophosphate spectra and in quantitative analysis. Coupled with the use of lineshape analysis and curve fitting, reasonable semiquantitative results are obtained that compare well with other established analytical techniques.

Acknowledgments

NMR line shape analysis software developed for use at Monsanto was kindly supplied by Mr. N.G. Hoffman, Research Computing Consortium, Monsanto Co., and Prof. A.J. Duben, Southeast Missouri State University. The authors would also like to thank Mr. Brad Herman of the Research Computing Consortium, Monsanto Co., for his development of the automatic sample management software, Dr. William Wise of the Physical Sciences Center (PSC), Monsanto Co., for his many valuable discussions on homonuclear 2DJ spectroscopy, and Mr. John Burquin (Monsanto PSC) for his assistance in the development of the robotic assay. Appreciation is also expressed to the many analysts who participated in the study.

Literature Cited

1. Colson, J.G.; Marr, D.H. Anal. Chem. 1973, 45, 370.
2. Gurley, T.W.; Ritchey, W.H. Anal. Chem., 1975, 47, 1444.
3. Sojka, S.A.; Wolfe, R.A. Anal. Chem., 1978, 50, 585.
4. Stanislawski, D.A.; Van Wazer, J.R. Anal. Chem., 1980, 52, 96.
5. Ruben, I.B. Anal. Lett., 1984, 17, 1259.
6. Glonek, T.; Wang, P.J.; Van Wazer, J.R. J. Amer. Chem. Soc., 1976, 98, 7968.
7. Hanssum, H. J. Magn. Reson., 1981, 45, 461.
8. Becker, E.D.; Ferretti, J.A.; Gambhir, P.N. Anal. Chem., 1979, 51, 1413.
9. Press, W.H.; Flannery, B.P.; Teukolski, S.A.; Vetterling, W.T. Numerical Recipes: The Art of Scientific Computing, Cambridge Univ. Press, Cambridge, MA, 1986; Chapter 14.
10. ASTM Method D 501, Standard Test Methods of Sampling and Chemical Analysis of Alkaline Detergents, Annual Book of ASTM Standards, American Society for Testing and Materials, Philadelphia, 1989.
11. ASTM Method D 2761, Standard Test Method for Analysis of Sodium Triphosphate by the Simplified Ion Exchange Method, Annual Book of ASTM Standards, American Society for Testing and Materials, Philadelphia, 1989.
12. Lundgren, D.P.; Loeb, N.P. Anal. Chem., 1961, 33, 366.
13. Condensed Phosphates by Column Separation, Industrial Method No.317-74I. Bran and Luebbe, Technicon Industrial Systems, Elmsford, NY.
14. Longley-Cook, L.H. Statistical Problems and How to Solve Them, Barnes & Noble, New York, N.Y., 1970; Chapter 14.
15. Sotak, C.H.; Dumoulin, C.L.; Levy, G.C. Anal. Chem., 1983, 55, 782.
16. Weiss, G.H.; Ferretti, J.A. J. Magn. Reson., 1983, 55, 397.
17. Watanabe, M.; Matsuura, M.; Yamada, T. Bull. Chem. Soc. Japan, 1981, 54, 738.
18. Crowther, J.P.; Westman, A.E.R. Can. J. Chem., 1954, 32, 42.
19. MacDonald, J.C.; Mazurek, M. J. Magn. Reson., 1987, 72, 48.
20. Glonek, T.; Costello, A.J.R.; Myers, T.C.; Van Wazer, J.R. J. Phys. Chem. 1975, 79, 1214.
21. Van Wazer, J.R. Phosphorus and Its Compounds, Interscience Publishers, New York, 1958, Vol. 1, Chap. 12.
22. Xu, P.; Wu, X-L.; Freeman, R. J. Magn. Reson., 1991, 95, 132.

RECEIVED December 10, 1991

Chapter 4

Phosphorus Atoms in Unusual Environments

Alan H. Cowley, Richard A. Jones, and Miguel A. Mardones

Department of Chemistry, University of Texas, Austin, TX 78712

Phosphorus-containing cubanes of the type (RMPR')$_4$ (M=Al, Ga, In) have been prepared by alkane elimination or salt elimination methods. The new compounds have been characterized by X-ray crystallography, NMR, and mass spectroscopy. Conceptually, the group 13/15 cubanes can be regarded as tetramers of RM=PR'. If the steric bulk of the substituents R and R' is increased it is possible to isolate dimers, (RMPR')$_2$.

For several years, our laboratory has been interested in the chemistry of main group elements in low coordination number environments (*1*). Some low coordinate species from Group 15 are shown below. They include two-coordinate cations (phosphenium ions), two-coordinate radicals (phosphinyl radicals), as well as stable double-bonded and even triple-bonded neutral molecules. For many years it was thought that

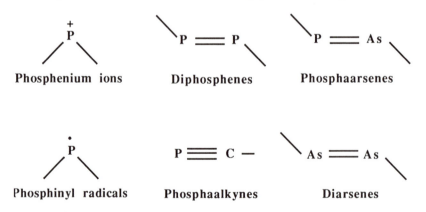

| Phosphenium ions | Diphosphenes | Phosphaarsenes |
| Phosphinyl radicals | Phosphaalkynes | Diarsenes |

compounds with multiple bonds between heavier main group elements would not be isolable because of the inherent weakness of π-bonds when the principal quantum number exceeds 2. However, despite the unfavorable thermodynamic features, it is possible to effect kinetic stabilization of these low-coordinate species.

0097–6156/92/0486–0056$06.00/0

Kinetic stabilization is accomplished principally by the use of bulky ligands. As shown below, the bulky groups can be alkyl (e.g. **1** and **2**) or or aryl (e.g. **3**). In some cases, such as the electrophilic phosphenium ions, steric bulk plus conjugative stabilization is necessary. Amido groups (**4**) have proved to be very useful in this context (*2*).

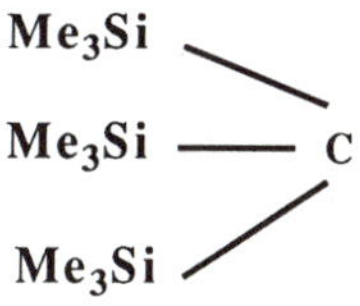

Me$_3$Si, Me$_3$Si — C, Me$_3$Si

1

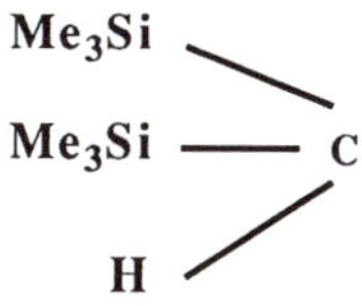

Me$_3$Si, Me$_3$Si — C, H

2

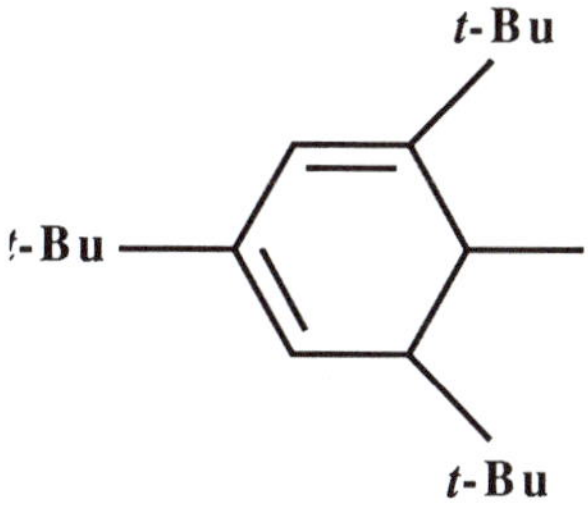

t-Bu, *t*-Bu, *t*-Bu

3

R, N, R

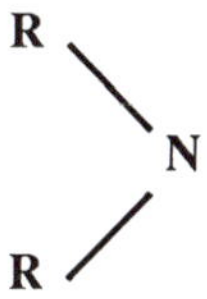

4

Special interest is associated with compounds that feature multiple bonding between heavier group 13 and 15 elements. As shown below, isolobal analogues of alkenes and alkynes can be envisioned. In addition to possessing intrinsically

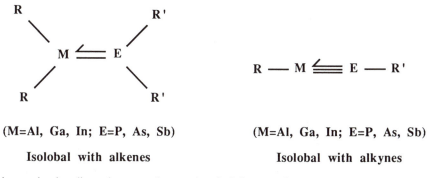

(M=Al, Ga, In; E=P, As, Sb)

Isolobal with alkenes

(M=Al, Ga, In; E=P, As, Sb)

Isolobal with alkynes

interesting bonding schemes and stereochemical features, these two classes of compounds and their oligomers are potentially important as precursors to compound semiconductors. Devices fabricated from gallium arsenide, indium phosphide, and related ternary compounds have a wide range of uses including field effect transistors, light emitting diodes, and photodetectors. However, the chemistry which underlies the organometallic chemical vapor deposition synthesis of GaAs and InP is approximately 30 years old (3).

$$Me_3M + EH_3 \rightarrow ME + 3CH_4$$
$$M=Ga, \; In; \; E=P, \; As$$

Apart from toxic nature of PH_3 and AsH_3 and the pyrophoric character of Me_3Ga, there are also problems with stoichiometry control and the relatively high reaction temperatures. In an effort to overcome some of these problems we have devised the concept of single source precursors (4). The single source precursors are typically rings or clusters in which the 13-15 bonds are already formed. The precursors also feature hydrocarbyl groups such as *t*-butyl, isopropyl, or ethyl that undergo facile alkene elimination via a ß-hydride shift mechanism. So far, most of the GaAs film work has been carried out with **5** (5). Mass spectral studies on this compound indicate that the Ga:As stoichiometry is retained (6) and that hydrocarbyl ligand elimination starts as low as 375 °C. Mass spectroscopic and isotopic labelling studies indicate that monomer formation takes place in the vapor phase.

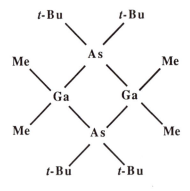

5

However, from the point of view of a synthetic chemical challenge it is the alkyne-analogous 13/15 compounds that are the most interesting. As shown below, there is a potentially rich variety of structural forms and bonding modes that includes alkyne-analogous (**6**), cyclobutadiene-analogous (**7**), and benzene-analogous (**8**) species, as well as higher oligomers. Previous compounds of empirical formula RMER' were confined to the light atoms B, Al, and N (7). There were no examples of compounds where both partners are heavier main group elements. Our initial attempts to prepare the desired compounds were frustrated by the high Lewis acidities of the products. Our next attempt, which involved intramolecular base stabilization, proved to be successful. For this experiment we decided to use the 2,6-bis(dimethylaminomethyl) benzene ligand which has been developed by van Koten and coworkers (8) in the Netherlands for the stabilization of unusual transition metal oxidation states and intermediates. A metathetical reaction of the lithium salt of this ligand with $GaCl_3$ affords the pentacoordinate gallium derivative (**9**). In turn, treatment of this gallium dichloride derivative with the dilithium salt of Ph_3SiPH_2 affords the first base-stabilized diphosphagalletane (**10**) (9). As revealed by an X-ray crystallographic study, one amino group of each ligand is uncoordinated, hence the geometry at gallium is approximately tetrahedral. The central Ga_2P_2 core is planar and resides on a center of symmetry. The Ga-P bond distance is relatively short (2.338(1)Å) *(10)*. However, the phosphorus geometry is pyramidal, thus if there is any p_π–p_π delocalization in the ring, it is slight.

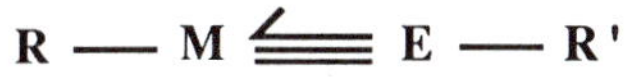

R —— M $\Longequiv$ E —— R'

6

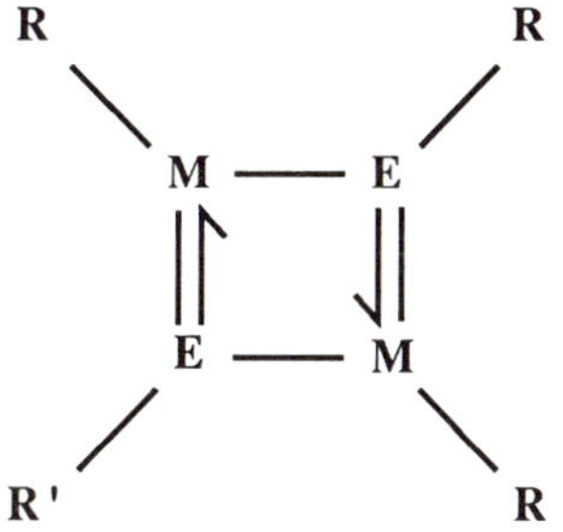

7

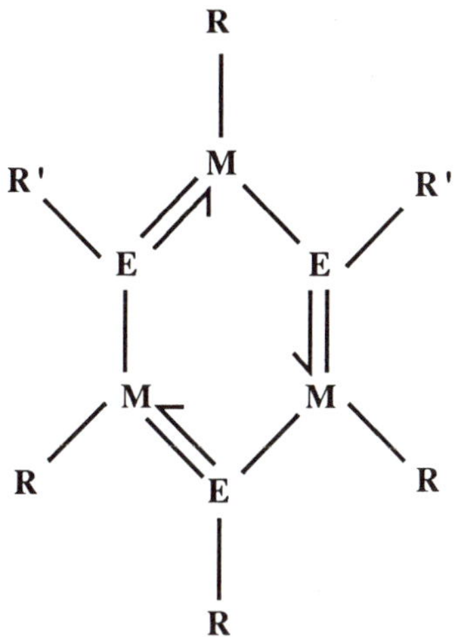

8

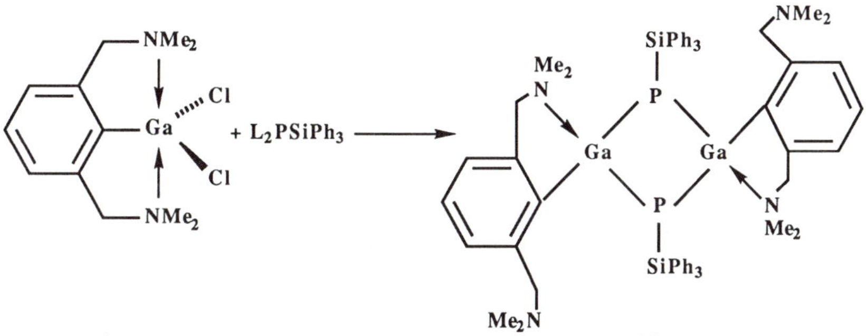

9

10

Our next objective was to prepare a base-free example of a diphosphadigalletane. To achieve this goal, it was necessary to develop a new synthetic approach. The reaction of $(t\text{-BuGaCl}_2)_2$ with the monolithium salt of $(2,4,6\text{-}t\text{-Bu}_3C_6H_2)PH_2$ affords the bis(phosphido)gallane (**11**) *(11)*. Sublimation of compound (**11**) or heating in toluene solution results in elimination of the primary phosphine and conversion to the first base-free diphosphagalletane (**12**). Compounds (**11**) and (**12**) have been characterized by X-ray crystallography. The yellow bis(phosphido)gallane adopts a trigonal planar geometry at gallium and the gallium-phosphorus bond lengths, which average 2.324(5)Å, are quite short *(10)*. Although the P–H hydrogens were not detected *(12)*, it is clear that the geometry at each phosphorus atom is pyramidal. The observed conformation is the one which minimizes the steric interactions between the bulky aryl groups. The base-free diphosphadigalletane possesses a planar Ga_2P_2 ring and a Ga-P bond distance that is extremely short (2.274(4)Å) *(10)*. There is some flattening of the phosphorus pyramid – the sum of angles being 314.7°. However, the structure is not indicative of extensive p_π–p_π bonding.

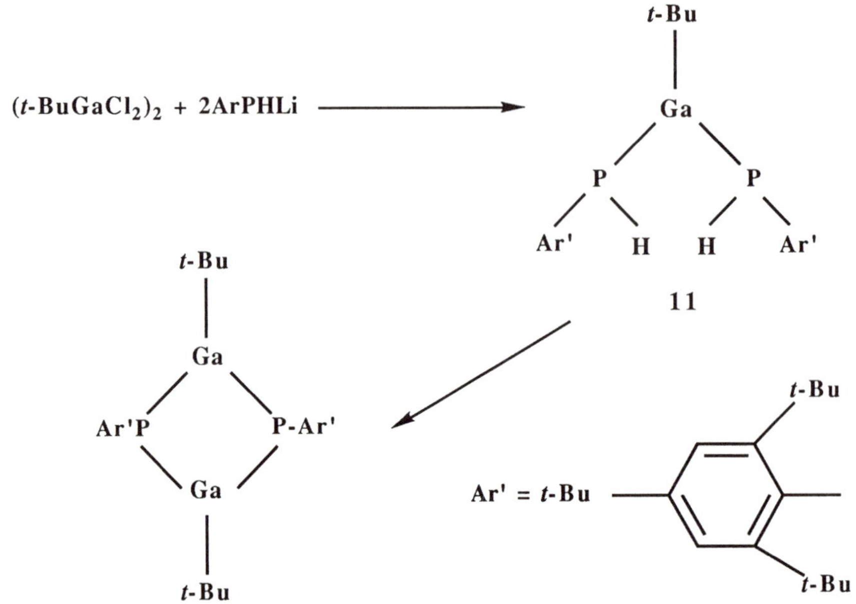

The reaction of i-Bu$_2$AlH with Ph$_3$SiPH$_2$ proved to be very interesting *(13)*. The initially isolated product (**13**) resulted from the elimination of molecular hydrogen at room temperature. However, refluxing (**13**) in toluene for 12 hours caused isobutane elimination and conversion to the first aluminum-phosphorus cubane (**14**). The cubane structure was confirmed by X-ray crystallography which also revealed that the cube is slightly distorted in the sense that the internal bond angles at phosphorus are < 90°, while those at aluminum slightly exceed 90°. The average aluminum-phosphorus distance of 2.414(4)Å is slightly less than that observed in aluminum-phosphorus dimers which span the range 2.433(4) to 2.475(1) *(14)*. The aluminum-phosphorus cubane is reactive to both electrophiles and nucleophiles. Preliminary thermolysis and X-ray photoelectron spectroscopic studies reveal that the aluminum-phosphorus cubane is also a potentially useful, low temperature precursor to the wide band gap semiconductor, aluminum phosphide. Very recently, we have discovered that other cubanes can be formed via metathesis reactions. For example, the reaction of i-PrInI$_2$ with the dilithium salt of Ph$_3$SiPH$_2$ affords the first indium-phosphorus cubane. Like gallium arsenide, indium phosphide crystallizes in the zinc blend (ZnS) structure. However, under high pressures indium phosphide undergoes a phase change to another, denser cubic modification. The new indium-phosphorus cubane might therefore serve as a model for the high pressure phase of indium phosphide. One final comment about the structure – like the aluminum-phosphorus cubane, the indium-phosphorus cubane is distorted. However, the average bond angle at phosphorus exceeds 90° in the indium compound.

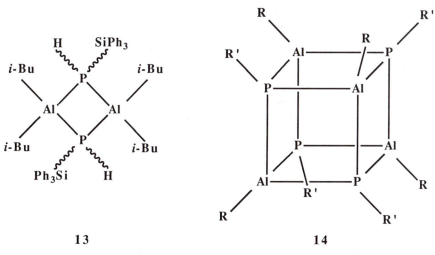

13   14

A characteristic feature of synthetic chemistry is the element of surprise. One such surprise was the formation of a novel gallium-phosphorus cluster in which a molecule of *t*-BuGaCl₂ behaves as both a Lewis acid and a Lewis base *(15)*. The reaction of the bis(tertiarybutyl)diphosphide anion with *t*-BuGaCl₂ was expected to produce a three- or six-membered gallium-phosphorus ring, *t*-BuGa(P-*t*-Bu)₂ or [*t*-BuGa(P-*t*-Bu)₂]₂. However, the mass spectrum indicated that the product contained an extra molecule of *t*-BuGaCl₂. Moreover the ^{31}P NMR spectrum showed two resonances of equal abundance. *t*-Bu-Ga resonances in 2 : 1 abundance were evident in the ^{1}H spectra. Elucidation of the structure needed an X-ray analysis. A convenient way to visualize the structure of (**15**) is to regard it as a complex of *t*-BuGaCl₂ with the

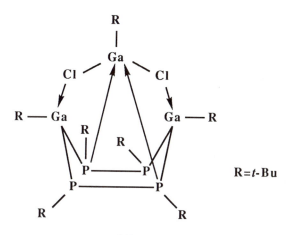

15

six-membered gallium-phosphorus ring. Two of the phosphorus atoms are involved in electron pair donation to the t-BuGaCl$_2$ molecule and the bonding in the cluster is completed by the formation of Cl bridges between t-BuGaCl$_2$ and the ring gallium atoms. In a formal sense, therefore, the t-BuGaCl$_2$ molecule plays the role of both Lewis acid and Lewis base. In this context, the geometry around the Ga of the t-BuGaCl$_2$ moiety is best described as approximately trigonal bipyramidal. It should be noted however that the Ga-Cl–Ga bridges are very unsymmetrical and that the Ga-Cl distances are very long (2.681(2)Å). The boat conformation of the Ga$_2$P$_4$ ring is undoubtedly caused by the interations with the t-BuGaCl$_2$ molecule.

So far, much of the discussion has focussed on the bonding of phosphinidene units to aluminum, gallium, or indium. However, we are also interested in the coordination of phosphinidenes to transition metals – particularly in cases where the phosphinidene unit coordinates in a terminal fashion. As indicated below, terminal phosphinidene complexes, like imido complexes, can exist in angular or linear forms.

| Angular | Linear |

Much elegant work has been done with terminal phosphinidenes as reactive intermediates – particularly by Mathey and coworkers (*16*). However, it was only in 1987 that Lappert and coworkers (*17*) reported the first angular terminal phosphinidene complex. We have recently succeeded in preparing the first linear terminal phosphinidene complex WCl$_2$(PMePh$_2$)(CO)($\equiv$PAr') [(**16**), Ar'=2,4,6-t-Bu$_3$C$_6$H$_2$]. (*18*). The tungsten (II) phosphine complex, WCl$_2$(PMePh$_2$)$_4$, undergoes oxidative addition via insertion into the phosphorus-carbon double bond of the phosphaketene (2,4,6-t-Bu$_3$C$_6$H$_2$)P=C=O. The ^{31}P chemical shift of (**16**) (δ+193.0) is relatively upfield, and the phosphorus-tungsten coupling constant is very large (649 Hz). These data suggest the presence of a triple bond between tungsten and phosphorus. This suggestion was confirmed subsequently by an X-ray analysis which revealed that the P-W bond distance is quite short (2.169(1)Å and the C-P-W angle is 168.2(2)°.

Acknowledgment

We are grateful to the National Science Foundation and the Robert A. Welch Foundation for generous financial support.

Literature Cited

(*1*)   For a review, see Cowley, A.H. *J. Organometal. Chem.* **1990**, *400*, 71.
(*2*)   See, for example, Kopp, R.W.; Bond, A.C. Parry, R.W. *Inorg. Chem.* **1976**, *15*, 3042; Schultz, C.W.; Parry, R.W. *Inorg. Chem.* **1976**, *15*, 3046 and references therein.
(*3*)   Manasevit, *Appl. Phys. Lett.* **1968**, *12*, 156.
(*4*)   Cowley, A.H.; Jones. R.A. *Angew. Chem. Int. Ed. Engl.* **1989**, *28*, 1208.
(*5*)   Cowley, A.H.; Benac, B.L.; Ekerdt, J.G.; Jones, R.A.; Kidd, K.B.; Lee, J.Y.; Miller, J.E. *J. Am. Chem. Soc.* **1988**, *110*, 6248.

(6)   It is recognized that miniscule deviations from exact 1:1 stoichiometry can have major effects on the electrical and/or optical properties of the resulting semiconductor material. This aspect is under active investigation.

(7)   For representative examples, see Paetzold, P. *Adv. Inorg. Chem.* **1987**, *31*, 123; Nöth, H. *Angew. Chem., Int. Ed. Engl.* **1988**, *27*, 1603; Waggoner, K.M.; Hope, H.; Power, P.P. *ibid.* **1988**, *27*, 1699.

(8)   Van Koten, G. *Pure Appl. Chem.* **1989**, *61*, 1681.

(9)   Cowley, A.H.; Jones, R.A.; Mardones, M.A.; Ruiz, J.; Atwood, J.L.; Bott, S.G. *Angew. Chem., Int. Ed. Engl.* **1990**, *29*, 1150.

(10)   See reference 4 for a discussion of Ga-P bond distances.

(11)   Atwood, D.A.; Cowley, A.H.; Jones, R.A.; Mardones, M.A. *J. Am. Chem. Soc.* **1991**, *113*, 7050.

(12)   The presence of P-H bonds was confirmed by ^{31}P NMR Spectroscopy: $\delta_P = -110.3$; $^1J_{PH} = 203.5$Hz.

(13)   Cowley, A.H.; Jones, R.A.; Mardones, M.A.; Atwood, J.L.; Bott, S.G. *Angew Chem, Int. Ed. Engl.* **1990**, *29*, 1409.

(14)   Janik, J.F.; Duesler, E.N.; McNamara, W.F.; Westerhausen, M.; Paine, R.T. *Organometallics* **1989**, *8*, 506. For additional structural information on this class of compound, see Sangokoya, S.A.; Pennington, W.T.; Robinson, G.H.; Hrncir, D.C. *J. Organomet. Chem.* **1990**, *385*, 23.

(15)   Cowley, A.H.; Jones, R.A.; Mardones, M.A.; Atwood, J.L.; Bott, S.G. *Angew. Chem., Int. Ed. Engl.* **1991**, **30**, 1141.

(16)   For a review, see Mathey, F. *Angew Chem. Int. Ed. Engl.* **1987**, *26*, 275.

(17)   Hitchcock, P.B.; Lappert, M.F.; Leung, W.P. *J. Chem Soc., Chem. Commun.* **1987**, 1282.

(18)   Cowley, A.H.; Pellerin, B.; Atwood, J.L.; Bott, S.G. *J. Am. Chem. Soc.* **1990**, *112*, 6734.

RECEIVED December 10, 1991

Chapter 5

Five-Coordinate and Quasi-Five-Coordinate Phosphorus

John G. Verkade

Department of Chemistry, Iowa State University, Ames, IA 50011

Some time ago we described the synthesis of two novel related classes of polycyclic phosphorus compounds known as prophosphatranes (**I**) and phosphatranes (**II**) (*1-8*). In more recent years we have extended these classes to include the analogous pro-azaphosphatranes (**III**) and azaphosphatranes (**IV**) (*9-14*).

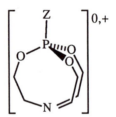

Prophosphatranes
I

Phosphatranes
II

Of interest in systems of type **I-IV** is the relationship of the substituent Z or Z^+ to the stabilization of structure **I** relative to **II** and that of **III** relative to **IV**. Although the former relationship is reasonably well characterized (*4*), the latter one is currently under investigation in our laboratories. By utilizing the proper combination of R and Z, it is possible to imagine the possibility of inducing long-range bridgehead-bridgehead

0097–6156/92/0486–0064$06.00/0
© 1992 American Chemical Society

interactions of varying strength of the sort depicted in **V**. Similarly, certain Z groups may give rise to partial transannular interaction in quasiphosphatranes as shown in **VI**.

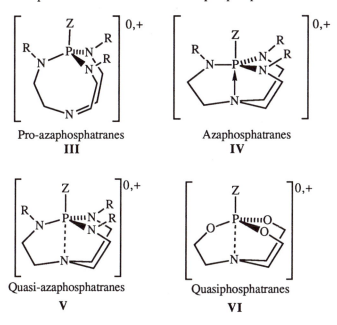

Pro-azaphosphatranes   Azaphosphatranes
III   **IV**

Quasi-azaphosphatranes   Quasiphosphatranes
V   **VI**

Synthesis of Azaphosphatranes

Azaphosphatranes were accidentally synthesized for the first time as a result of a frustrated attempt to synthesize pro-azaphosphatranes **1** and **2** in the direct approach depicted in reaction 1. In the case of R = H, an insoluble polymer is slowly formed (*15*). When R = Me, we observed by ^{31}P NMR spectroscopy the slow formation of **2**

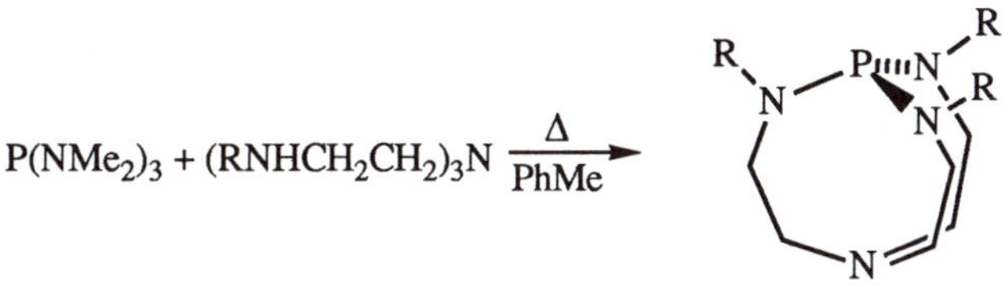

$$P(NMe_2)_3 + (RNHCH_2CH_2)_3N \xrightarrow[\text{PhMe}]{\Delta}$$

(1)

1 R = H
2 R = Me
3 R = CH$_2$Ph

over a period of weeks in an equilibrium mixture also containing mono and disubstituted product. Upon work-up, the reaction mixture gave a mediocre yield (20-50%) of **2** (*10*). This contrasts with the rapid and excellent conversion of (MeNHCH$_2$)$_3$CMe in the presence of P(NMe$_2$)$_3$ to P(MeNCH$_2$)$_3$CMe (*16*).

Rather than attempt to make the more sterically hindered **3** *via* reaction 1, we sought to speed up the formation of pro-azaphosphatranes by employing the more reactive phosphorus reagent indicated in reaction 2. Indeed the reaction was over within an hour at room temperature, but to our surprise, the azaphosphatranes **4-6** formed in virtually quantitative yield (*14, 15*). Thus in spite of the presence of the

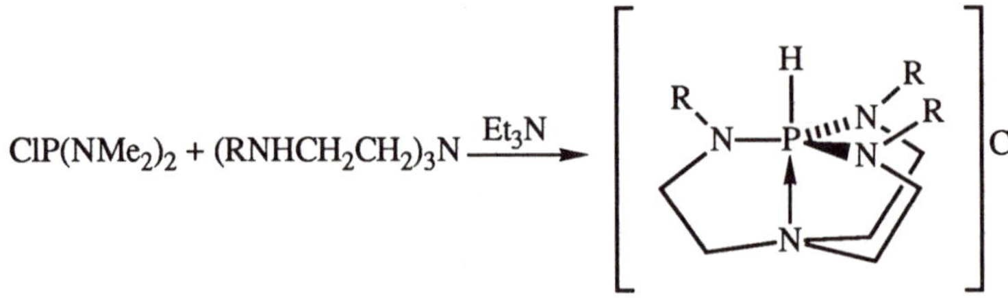

$$\text{ClP(NMe}_2)_2 + \text{(RNHCH}_2\text{CH}_2)_3\text{N} \xrightarrow{\text{Et}_3\text{N}} \left[\text{image} \right] \text{Cl} \quad (2)$$

4(Cl) R = H
5(Cl) R = Me
6(Cl) R = CH$_2$Ph

strong base Et$_3$N for the neutralization of the HCl formed, the latter protonated the pro-azaphosphatranes which (as is discussed later) are considerably more basic than any amine known (*10, 14, 15, 17*). Reaction 2 also occurs quantitatively in CH$_2$Cl$_2$ at 0 °C in the *absence* of Et$_3$N. The rapid formation of **4(Cl)** in reaction 2, in contrast to the polymer formed in reaction 1, suggests that a protonated intermediate such as **4(Cl)** may facilitate rapid further nucleophilic attack on phosphorus of the pendant primary

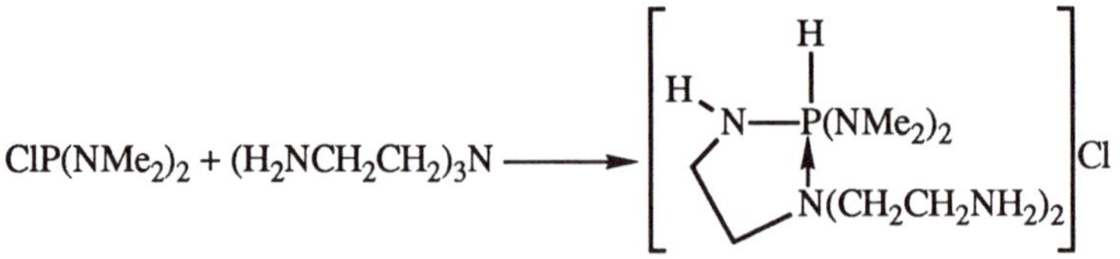

$$\text{ClP(NMe}_2)_2 + \text{(H}_2\text{NCH}_2\text{CH}_2)_3\text{N} \longrightarrow \left[\text{image} \right] \text{Cl} \quad (3)$$

7(Cl)

amine branches. Once **4(Cl)** is formed, polymer formation (presumably by intermolecular transamination via the secondary amines) is inhibited. As might be expected then, **4(OTf)** can be synthesized within an hour in quantitative yield *via* reaction 4 in which a proton source is present from the start in equimolar concentration.

$$\text{P(NMe}_2)_3 + \text{HOTf} + \text{(H}_2\text{NCH}_2\text{CH}_2)_3\text{N} \xrightarrow[\text{CH}_2\text{Cl}_2]{\text{r.t.}} \textbf{4(OTf)} \quad (4)$$

The molecular structures of **4(Cl)** (*14*) and **5(BF$_4$)** (*10*) have been confirmed by X-ray means. Formally, these structures can be considered to contain a P(III) atom which is stabilized through proton-induced chelation by the tertiary nitrogen, rendering the P(III) atom five-coordinate. Although a more sophisticated view of the bonding in these cations involves three-center four-electron MOs along the three-fold axis, the oxidation state of the phosphorus can be viewed as trivalent since electron withdrawal by the proton is compensated by electron donation from the axial nitrogen. The presence of these pentacoordinate structures is signalled by upfield ^{31}P chemical shifts in the -10 to -42 ppm region, with $^1J_{PH}$ coupling values from 453 to 506 Hz (15).

It might be predicted that cationic compounds of type **IV** wherein Z is an alkyl carbocation would be stable. As will be discussed later, a CH_3^+ species does not induce transannulation, but the more electronegative CH_2Br^+ cation does ($9, 18$). Similarly, Cl_2 provides a Cl^+ which induces transannulation in **2** to give azaphosphatrane **9**(Cl) ($9, 18$).

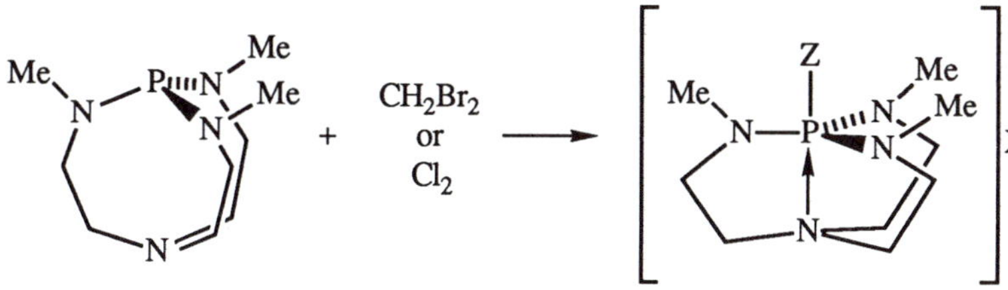

$$\text{2} \qquad\qquad \begin{array}{l}\text{8 (Br) Z = CH}_2\text{Br, X = Br}\\\text{9 (Cl) Z = Cl, X = Cl}\end{array} \qquad (5)$$

The examples of azaphosphatranes cited so far possess five-coordinate P(III) atoms which formally began as three-coordinate P(III) in a pro-azaphosphatrane. It is also possible to induce five-coordination in the P(V) pro-azaphosphatrane **10** with BF_3 (12). The ^{31}P chemical shift of this compound (-2.2 ppm) is upfield as expected, and

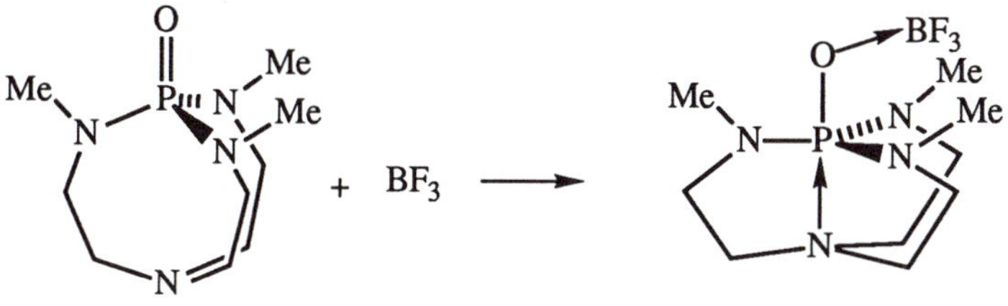

$$\text{10} \qquad\qquad\qquad\qquad \text{11} \qquad (6)$$

no P=O stretching frequency is detected in its IR spectrum. We observed similar results for the reaction of prophosphatrane **12** with BF_3 (6) as well as with the other Lewis acids shown in reactions 7-13 ($6, 7, 19$). Although no structural metrics could be determined for **13-19**, reaction 14 gave a crystalline product **21** which was structured by X-ray means ($7, 19$) whereas reactions 15-17 (7) did not provide single crystals. The robust character of the chelated structure of these P(V) azaphosphatranes and phosphatranes is undoubtedly a strong driving force for the diminution of the very strong P-chalcogen multiple bond to a formally single bond. A factor in the dibasic character of the axial chalcogen in some of the above reactions is undoubtedly the electron induction provided by the axial nitrogen. Evidence has also been adduced for a transannulated H_3PO_4 adduct of **12** in which two acidic hydrogens are hydrogen bonded to the P=O oxygen while the third protonates this atom (6). That chelation plays a key role in the stabilization of azaphosphatranes and phosphatranes is consistent with our failed attempts to find ^{31}P NMR evidence for pentacoordinate phosphorus species in mixtures of protonic acids, P(OR)$_3$ or P(NMe$_2$)$_3$, and NR$_3$. The observation

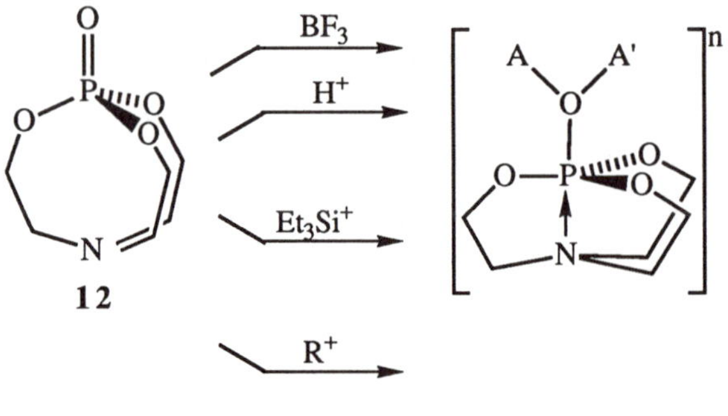

	A	A'	n	
13	BF_3	lp	0	(7)
14	H	lp	1+	(8)
15	H	H	2+	(9)
16	Et_3Si	lp	1+	(10)
17	Et_3Si	Et_3Si	2+	(11)
18	R	lp	1+	(12)
19	R	R	2+	(13)

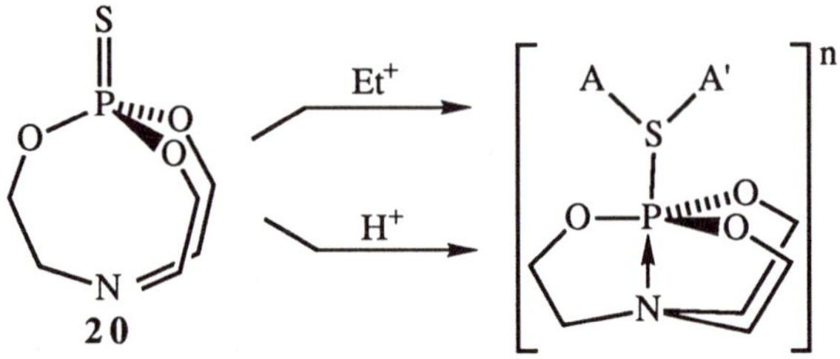

	A	A'	n	
21	Et	lp	1+	(14)
22	Et	Et	2+	(15)
23	H	lp	1+	(16)
24	H	H	2+	(17)

by Müller et al. that cation **25** (detected by ^{31}P NMR in solution) could not be isolated (*20*), however (which contrasts our characterization of **26** by X-ray means (*1*))

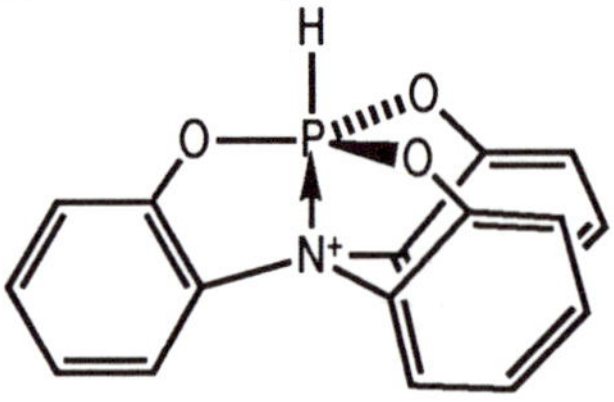

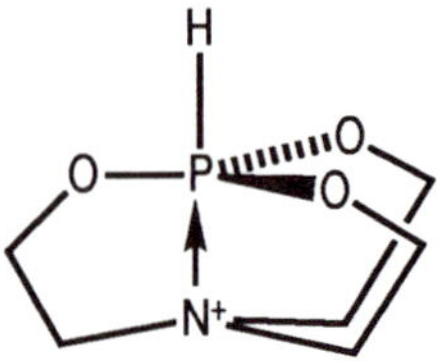

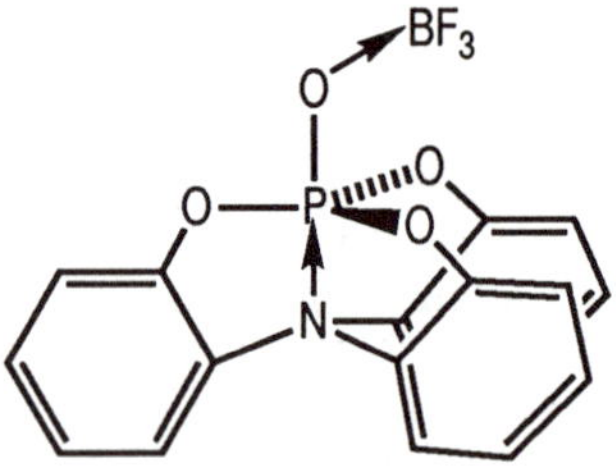

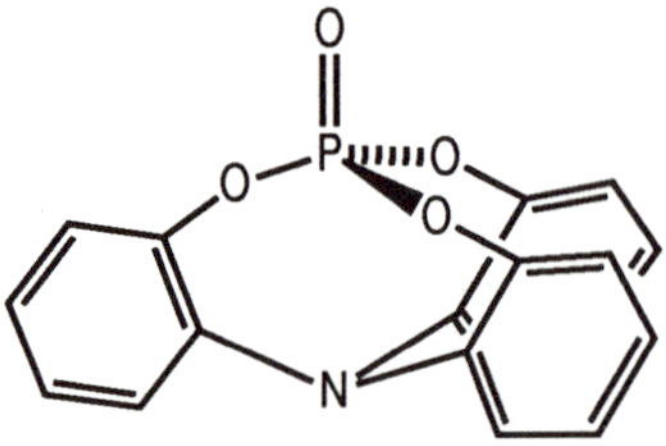

indicates that unsaturation in the ring system of **25** diminishes the stability of chelation, probably predominantly by means of strain imposed by the sp^2 phenylene ortho-substituted carbons on the substituent angles. This is consistent with the considerably smaller upfield ³¹P chemical shift of **27** relative to the parent prophosphatrane **28** compared with that in **13** (*6*) and **12** (*12*).

Synthesis of Pro-azaphosphatranes

Although **2** can be made in low yield after *ca.* two weeks by reaction 1, it is more convenient to synthesize the azaphosphatrane cations **4-6** and then deprotonate them. However, a strong base such as *t*-BuO⁻ is required (*10, 14, 15*). In reaction 18, **4**(OTf) is used instead of the corresponding chloride salt owing to better solubility of the triflate. Pro-azaphosphatranes **2** and **3** are isolable sublimable solids, but **1** like **29** (see reaction 19) rearranges to an intractable polymer upon attempted isolation (*4*).

$$4(\text{OTf}), 5(\text{Cl}) \text{ or } 6(\text{Cl}) \xrightarrow{\ t\text{-BuO}^- \ } 1, 2, \text{ or } 3, \text{ respectively} \tag{18}$$

The P(III) pro-azaphosphatranes **1 - 3** can be converted to P(IV) adducts or oxidized to P(V) compounds. Depending upon the degree of polarization of the phosphorus lone pair by the Lewis acid Z, it might be expected that either a pro-azaphosphatrane or an azaphosphatrane structure will be realized. This is not the case, however, and structures of type V have been observed (see later). These quasi five-coordinate phosphorus systems offer a unique opportunity to study the stepwise formation of a (transannular) bond. In the remainder of this section we focus our attention on systems that are either known from structural metrics to be prophosphatranes and pro-azaphosphatranes, or are known from ³¹P NMR spectroscopy not to be phosphatranes and azaphosphatranes. Thus the latter may or may not be quasi-phosphatranes and quasi-azaphosphatranes. In the next section we discuss structures determined by X-ray means in which there is evidence of a partial transannular P ←--N interaction, even though the phenomenon does not register in a characteristic upfield shift of the ³¹P nucleus. In Table I are summarized the reagents and products in reaction 19 that can on the basis of the aforementioned criteria (i.e., structural metrics or a characteristic ³¹P chemical shift) be classified as pro-azaphosphatranes and prophosphatranes. In *trans*-(**2**)₂PtCl₂ (**36**) (*13*) and *cis*-(**2**)BrRe(CO)₄ (**39**) (*21*), the transannular P-N distances are 3.33 and 3.307 Å, respectively, which lie close to the sum of the van der Waals radii of 3.35 Å (*22*). Not surprisingly, the bridgehead nitrogens in prophosphatranes **42 - 45** are sufficiently basic to allow quaternization by Me⁺ to yield the isolable corresponding ammonium derivatives (*4*).

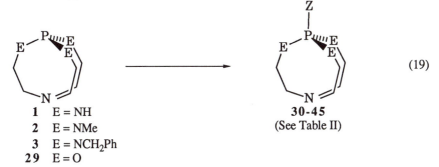

$$\tag{19}$$

1 E = NH
2 E = NMe
3 E = NCH₂Ph
29 E = O

30-45
(See Table II)

The ³¹P chemical shifts of most these compounds are in the normal range expected for four-coordinate phosphorus. Compound **42** deserves some comment, however. The ^{31}P chemical shift for this compound (-6.6 ppm) is considerably above that of O=P(OR)$_3$ (-1.07 ± 1.00 ppm (23)) and this shift moves downfield to +7.2 ppm upon quaternization of the bridgehead nitrogen to give [O=P(OCH$_2$CH$_2$)$_3$NMe]$^+$ (4). While it is tempting in the current absence of a structure determination of 42 to suspect a quasiphosphatrane configuration, it must also be borne in mind that the ^{31}P chemical shift of O=P(OCH$_2$)$_3$CMe is at -7.97 ppm (24).

By reactions analogous to those discussed above, the prophosphatranes 46 (20, 25) and 47 (25) have also been synthesized. The crystal structure of 48 (26) reveals no evidence of transannular interaction.

Table I.   Reagents and Products Associated with Reaction 19

Starting Material	Reagent	No.	E	Z	Reference
			Product		
1	O$_2$	30	NH	O	(14)
1	Se	31	NH	Se	(14)
2	Se	32	NMe	Se	(11)
2	(Me$_3$SiO)$_2$	33	NMe	O	(11)
2	BH$_3$	34	NMe	BH$_3$	(9)
2	CH$_3$$^+$	35	NMe	CH$_3$$^+$	(9)
2	(PhCN)$_2$PtCl$_2$	36	NMe	PtCl$_2$(2)	(11)
2	PhN$_3$	37	NMe	N$_3$Ph	(11)
2	PhN$_3$, Δ	38	NMe	NPh	(11)
2	BrRe(CO)$_4$	39	NMe	BrRe(CO)$_4$	(21)
2	MeHgI	40	NMe	Hg(2)$_2$I$_2$	(21)
3	Se	41	NCH$_2$Ph	Se	(14)
29	KO$_2$	42	O	O	(4)
29	Se	43	O	Se	(4)
29	(OC)$_5$WNCMe	44	O	W(CO)$_5$	(4)
29	Mo(CO)$_6$	45	O	Mo(CO)$_5$	(4)

By reactions analogous to those discussed above, the prophosphatranes **46** (*20*, *25*) and **47** (*25*) have also been synthesized. The crystral structure of **48** (*26*) reveals no evidence of transannular interaction.

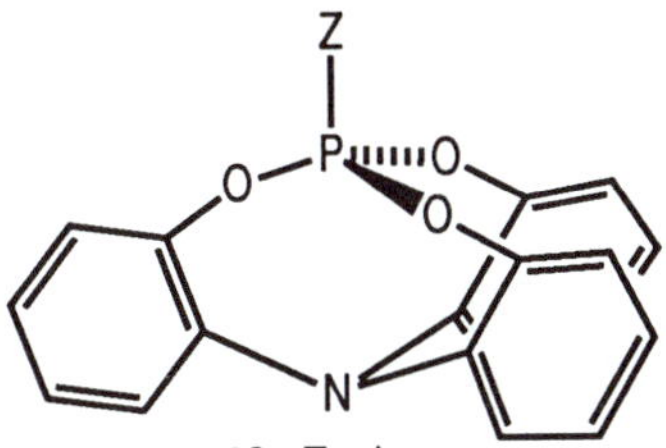

46 Z = lp
47 Z = BH$_3$

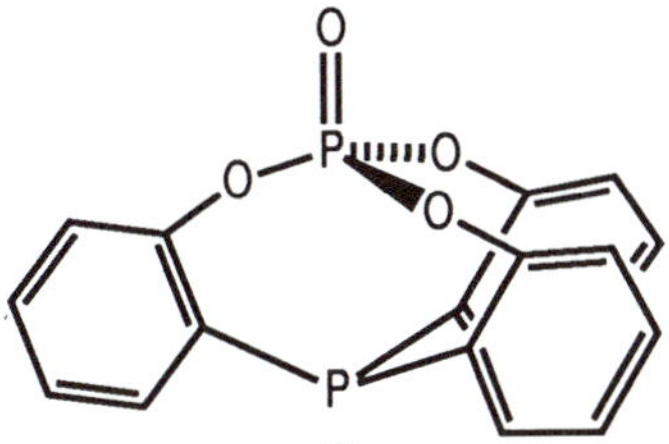

48

Synthesis of Quasi-azaphosphatranes and Quasiphosphatranes

The preparations of the title compounds that give X-ray structural evidence of some degree of transannular interaction are summarized in reactions 20-25. In Table II it is seen that the transannular distances range from 3% to 17% shorter than the sum of the

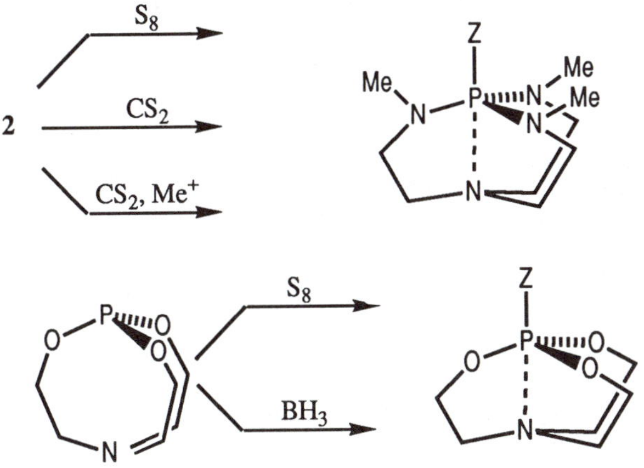

$$2$$

S$_8$ ⟶

CS$_2$ ⟶

CS$_2$, Me$^+$ ⟶

49 (*13*) S $\underline{Z}$ (20)

50 (*13*) CS$_2$ (21)

51 (*27*) S=$\overset{+}{C}$SMe (22)

S$_8$ ⟶

BH$_3$ ⟶

52 (13) $\underline{Z}$ S (23)

53 (13) BH$_3$ (24)

P(NMe$_2$)$_3$ ⟶

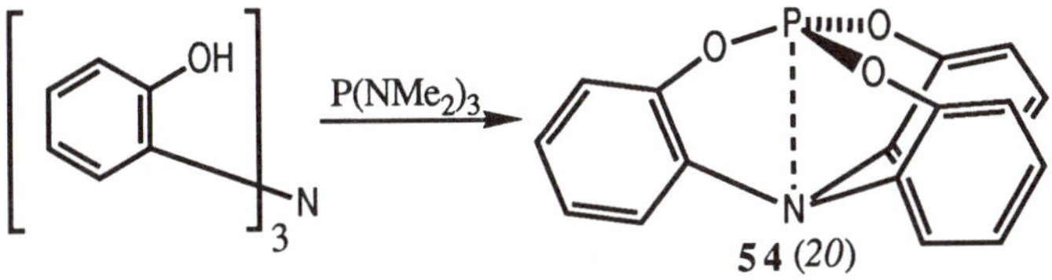

54 (*20*) (25)

van der Waals radii (3.35 Å (*22*)). The contraction of this distance in **51** is considerable, especially when it is recognized that a full transannular bond as in **4$^+$**

PHOSPHORUS CHEMISTRY

(2.0778 Å (*14*)), **5**[+] (1.976 Å (*10*)) **21** (2.0 Å (*7, 19*)) or **26** (1.986 Å (*4*)) is only *ca.* 40% shorter than the van der Waals radii sum and that the pseudo equatorial P-N bonds in **49-51** are 50% shorter than this sum.

Table II.   Selected Distances and Angles in Quasi-azaphosphatranes and Quasiphosphatranes

Compound	P-N, Å[a]	C-N-C, deg.	E-P-E, deg[b]	Ref.
49	3.250	119.6	106.8	(*2*)
40	3.143	120.0	108.3	(*21*)
50	3.008	120.0	110.27	(*27*)
51	2.771	119.4	113.4	(*27*)
52	3.132	119.2	108.1	(*2*)
53	3.098	119.7	107.4	(*3*)
54	3.136	118.5	102.7	(*20*)

[a]Bridgehead-bridgehead distance. [b]E refers to the hetero atom (N or O) in the pseudo equatorial position.

It is interesting to note that whereas the bridgehead nitrogen geometry is quite insensitive to increasing transannulation from **49** to **39** to **50** to **51** (Table II) the E-P-E angle increases linearly with decreasing bridgehead-bridgehead distance. It should also be recognized that hydrogen-hydrogen van der Waals interactions among CH_2 groups of different bridging moieties contribute to flattening of the angles around the bridgehead nitrogen in **49-53**. This phenomenon also rationalizes the 115.5° C-N-C and 113.9° C-C(bridgehead)-C bond angles in **55** (*28, 29*). In spite of the fact that such hydrogen-hydrogen interactions are absent in **54**, there is a small shrinkage of the P-N distance over the van der Waals radii sum. This distance decreases to 2.926 Å in its hydrolysis product **56** wherein hydrogen bonding helps preserve the cage structure.

In view of the structural indications of a transannular interaction in **39, 49-54** and **56**, the determination of the crystal structures of **10, 12** and **28** should be interesting since the phosphoryl oxygen presumably may withdraw electron density better than sulfur or a BH_3 group. In addition, there is calculated to be only a small energy difference between structure **12** and its transannulated form **57** (*5*).

Unfortunately, the position of the ^{31}P NMR chemical shift does not signal the presence of a partially transannulated structure. Thus the chemical shift of **50** (21.3 ppm) actually moves slightly *downfield* compared with **51** (24.2 ppm) despite the decrease in the bridgehead atom interatomic distance (Table II). How close the bridgehead atoms must approach one another before a definitive upfield ^{31}P NMR shift is observed remains to be seen. There also appear to be no firm correlations of ^{1}H or ^{13}C NMR parameters with partial transannulation in these systems.

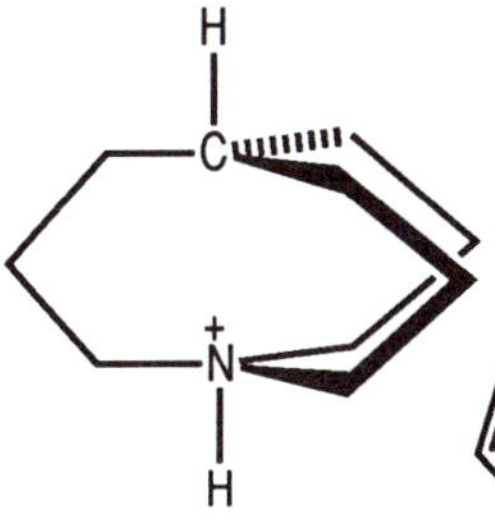

| 55 | 56 | 57 |

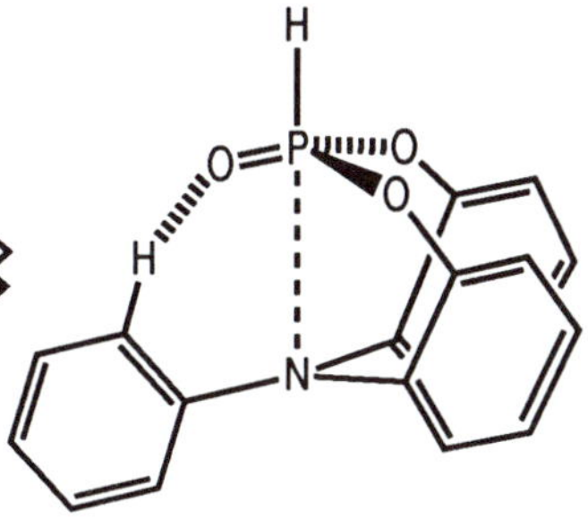

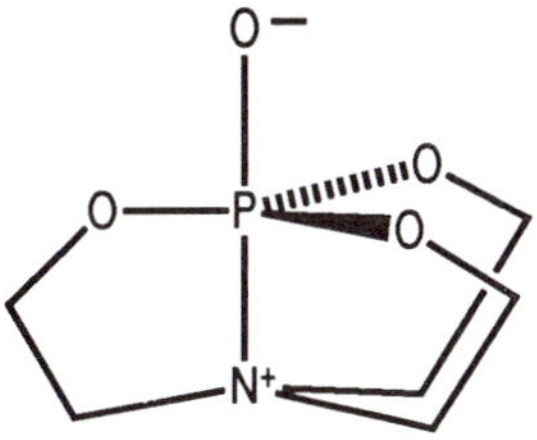

Transannulation and basicity

By using a deficit of base in equilibria 26 and 27, it was determined by ^{31}P NMR spectroscopy that the order of acidities of the azaphosphatrane cations is 6(Cl) > 5(Cl) > 4(Cl), implying the basicity order $3 < 2 < 1$ for the conjugate bases of the

$$5(Cl) \ + \ 6(Cl) \quad \xrightarrow[\substack{- K^+ \\ - t\text{-BuOH}}]{KO\text{-}t\text{-Bu}} \quad 2 \ + \ 3 \quad \quad (26)$$

$$4(Cl) \ + \ 5(Cl) \quad \xrightarrow[\substack{- K^+ \\ - t\text{-BuOH}}]{KO\text{-}t\text{-Bu}} \quad 1 \ + \ 2 \quad \quad (27)$$

azaphosphatranes (*14*). From the NMR data, a pK$_a$ in DMSO of 29.6 was calculated for 4(Cl) and an upper limit of pK$_a$ = 26.8 was estimated for 5(Cl) and 6(Cl). Thus the azaphosphatrane cations 4^+, 5^+ and 6^+ are orders of magnitude less acidic than typical R$_3$PH$^+$ (pK$_a$ = 8-9 (*30*)) or ammonium cations such as protonated "Proton Sponge" (1, 8-(bisdimethylamino)naphthalene•HI, pK$_a$ = 12.3 (*31*)).

The basicity order $1 > 2 > 3$ appears anomalous in that 1 seems out of order on electron induction considerations of the substituent on the nitrogen in the bridging groups. Thus the order expected would be $2 > 1 > 3$. The anomalous basicity of 1 cannot be attributed to a shorter (stabilizing) transannular P-N bond length in 4^+ than in 5^+, because it is in fact longer (*14*). It is tempting to ascribe the higher acidites of 5^+ and 6^+ to sterically assisted proton departure by the Me and CH$_2$Ph groups, respectively. Molecular mechanics calculations on the cations of 4(Cl) and 5(Cl) suggest a higher van der Waals interaction energy between the P-H hydrogen and the equatorial nitrogen substituent by a maximum of about 5 kcal/mole (*32*). Another possibility is that equilibria such as those in Scheme I are involved, which are expected to strengthen the P-H bond. The unusual stability of the P-H bond in 4(Cl) may also be due to relatively little phosphorus orbital imbalance, which is signalled by its strongly negative ^{31}P chemical shift (-42 ppm) compared with that in 5(Cl) (-10.1 ppm) and 6(Cl) (-11.0 ppm) (*14*).

Scheme I

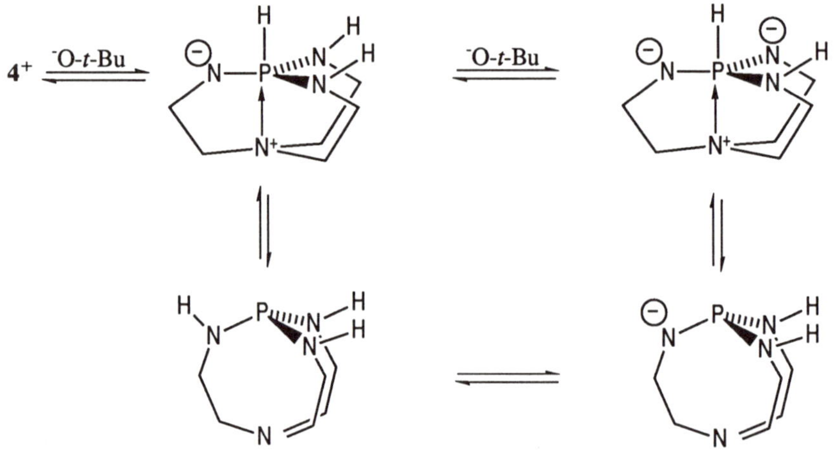

It is interesting that the $^1J_{PH}$ couplings of these azaphosphatrane cations decrease in the same order as the acidity: 6(Cl) (506 Hz) > 5(Cl) (491 Hz) > 4(Cl) (453 Hz) (*14*). Since the structures of 4(Cl) and 5(Cl) are so similar that hybridzational changes are likely to be small (*14*) the coupling trend indicates a buildup of negative charge on the phosphorus, in accord with the decrease in acidity.

The decreasing acidity order 6(Cl) > 5(Cl) > 4(Cl) is paralleled by the apparent acidity order of phosphorus in the selenide derivatives **31, 32** and **41**. This order, is **41 > 32 > 31** as reflected in their $^1J_{PSe}$ couplings (774 Hz, 754 Hz and 590 Hz, respectively) (*14*). The couplings in these latter compounds are distinctly lower than in SeP(NEt$_2$)$_3$ or SeP(NMe$_2$)$_3$ (790 and 794 Hz, respectively) (*33*) suggesting that **1, 2** and **3** have more basic phosphorus atoms than their acyclic analogues. From equilibrium studies it was also demonstrated that **29** is more basic toward BH$_3$ than P(OMe)$_3$ (*4*) and that the adduct **53** according to an IR analysis of its B-H stretching frequencies contains a more basic phosphorus than H$_3$BP(OMe)$_3$ (*4*). To what extent the enhanced negative charge on phosphorus indicated by these experiments reflect partial transannulation and/or perhaps repulsion effects from lone pairs on the NR nitrogens or the oxygens in quasi-azaphosphatrane or quasi-phosphatrane structures, respectively, remains a question to be answered by ongoing research efforts.

Acknowledgments

The author is grateful to each of his coworkers named on the publications cited below for their contributions to the success of this work. Also gratefully acknowledged are the donors of the Petroleum Research Fund administered by the American Chemical Society for support of this research, the W. R. Grace Company for samples of (H$_2$NCH$_2$CH$_2$)$_3$N and Denise Miller for help in preparing this manuscript.

Literature Cited

(1) Clardy, J. C.; Milbrath, D. S.; Springer, J. P.; Verkade, J. G. *J. Am. Chem. Soc.* **1976**, *98*, 623.

(2) Milbrath, D. S.; Clardy, J. C.; Verkade, J. G. *J. Am. Chem. Soc.* **1977**, *99*, 631.

(3) Clardy, J. C.; Milbrath, D. S.; Verkade, J. G. *Inorg. Chem.* **1977**, *16*, 2135.

(4) Milbrath, D. S.; Verkade, J. G. *J. Am. Chem. Soc.* **1977**, *99*, 6607.

(5) van Aken, D.; Castelyns, A. M. C. F.; Verkade, J. G.; Buck, H. M. *Recueil, Journal of the Royal Netherlands Chemical Society* **1979**, *98*, 12.

(6) Carpenter, L. E.; Verkade, J. G. *J. Am. Chem. Soc.* **1985**, *107*, 7084.

(7) Carpenter, L. E.; de Ruiter, B.; van Aken, D.; Buck, H. M.; Verkade, J. G. *J. Am. Chem. Soc.* **1986**, *108*, 4918.

(8) Carpenter, L. E.; Verkade, J. G. *J. Org. Chem.* **1986**, *51*, 4287.

(9) Gudat, D.; Lensink, C.; Schmidt, H.; Xi, S.-K.; Verkade, J. G. *Phosphorus, Sulfur and Silicon* **1989**, *41*, 21.

(10) Lensink, C.; Xi, S.-K.; Daniels, L. M.; Verkade, J. G. *J. Am. Chem. Soc.* **1989**, *111*, 3478.

(11) Schmidt, H.; Lensink, C.; Xi, S.-K.; Verkade, J. G. *Z. anorg. allg. Chem.* **1989**, *578*, 75.

(12) Schmidt, H.; Xi, S.-K.; Lensink, C.; Verkade, J. G. *Phosphorus, Sulfur and Silicon* **1990**, *49/50*, 163.

(13) Xi, S. K.; Schmidt, H.; Lensink, C.; Kim, S.; Wintergrass, D.; Daniels, L. M.; Jacobson, R. A.; Verkade, J. G. *Inorg. Chem.* **1990**, *29*, 2214.

(14) Laramay, M. A. H.; Verkade, J. G. *J. Am. Chem. Soc.* **1990**, *112*, 9421.

(15) Laramay, M. A. H.; Verkade, J. G. *Z. anorg. allg. Chem.*, submitted.

(16) Laube, B. L.; Bertrand R. D.; Casedy, G. A.; Compton, R. D.; Verkade, J. G. *Inorg. Chem.* **1967**, *6*, 173.

(17) Verkade, J. G. U.S. Patent, allowed March 26, 1991.

(18) Tang, J.-S.; Lensink, C.; Laramay, M. A. H.; Verkade, J. G. to be published.

(19) van Aken, D.; Merkelbach, I. I.; Koster, A. S.; Buck, H. M. *J. Chem. Soc., Chem. Commun.* **1980**, 1045.

(20) Müller, E.; Burgi, H.-B. *Helv. Chem. Acta* **1987**, *70*, 1063.

(21) Laramay, M. A. H., to be published.

(22) Bondi, A. *J. Phys. Chem.* **1964**, *68*, 441.

(23) Emsley, J. W.; Feeney, J.; Sutcliffe *High Resolution Nuclear Magnetic Resonance Spectroscopy*; Pergamon Press: New York, NY, 1966; Vol. 2, 1066.

(24) Verkade, J. G.; King, R. W. *Inorg. Chem.* **1962**, *1*, 948.

(25) Paz-Sandoval, M. A.; Fernandez-Vincent, C.; Uribe, G.; Contreras, R.; Klaebe, A. *Polyhedron* **1988**, *7*, 679.

(26) Bohn, C.; Davis, W. M.; Halterman, R. L.; Sharpless, K. B. *Angew. Chem.* **1988**, *100*, 882.

(27) Tang, J.-S.; Verkade, J. G., to be published.

(28) Coli, J. C.; Crist, D. R.; Barrio, M. C.; Leonard, N. J. *J. Am. Chem. Soc.* **1972**, *94*, 7092.

(29) Wang, A. H. J.; Missavage, R. J.; Byrn, J. R.; Paul, I. A. *J. Am. Chem. Soc.* **1972**, *94*, 7100.

(30) Streuli, C. A. *Anal. Chem.* **1960**, *32*, 985.

(31) Alder, R. W.; Bowmann, P. S.; Steele, W. R. S.; Winterman, D. R. *J. Chem. Soc. Chem. Commun.* **1968**, 723.

(32) Jacobson, R. A. private communication.

(33) Stec, W. J; Okrusek, A.; Uznanski, B; Michalski, J. *Phosphorus* **1972**, *2*, 97.

RECEIVED November 27, 1991

Chapter 6

New Derivative Chemistry of Two-Coordinate Phosphines

Robert H. Neilson[1] and Christo M. Angelov[2]

[1]Department of Chemistry, Texas Christian University,
Fort Worth, TX 76129
[2]Department of Chemistry, Higher Pedagogical Institute,
9700 Shoumen, Bulgaria

A variety of novel reactions of two "low-coordinate" phosphines, $(Me_3Si)_2NP=ESiMe_3$ (1: E = CH; 2: E = N) with a diverse series of unsaturated organic substrates are reported. For example, phosphines 1 and 2 react readily with allenes via an *ene* process to afford novel phosphorus-substituted dienes or with 2-butyne to give allenic phosphines. Other derivative chemistry, including the addition of terminal acetylenes and acetylenic alcohols to the P=E bond as well as some related reactions with acetylenic halides and both β- and γ-diketones, is also described.

To a large degree, the extensive development of phosphorus chemistry is due to the great variety of different valence states and coordination numbers that the element phosphorus exhibits. Until about twenty years ago, the chemistry of phosphorus compounds included mainly the well known P^{III} and P^V valence states with their common coordination numbers of three, four, and five. In recent years, however, these ideas have changed dramatically and it is now known that phosphorus can form a variety of "low-coordinate" structures (*1*). These novel systems include phosphorus in one- and two-coordinate P^{III} and three-coordinate P^V environments, with (p-p)π bonds to adjacent elements, including C, N, P, Si, B, etc.

During the past decade, many different examples of one- and two-coordinate, trivalent phosphorus compounds have been synthesized and isolated as thermally stable, but chemically reactive, products. The two-coordinate derivatives, in particular, have been found to undergo a variety of interesting and synthetically useful transformations, a few of which are illustrated below. Although several examples of addition, cycloaddition, oxidation, and metal-coordination reactions of two-coordinate phosphines are readily found in the literature (*1*), there is still a great deal of work to be done before the full potential of these new reagents can be realized.

Part of our research program in phosphorus chemistry deals with the study of new chemical reactions of two such low-coordinate phosphines: the (methylene)phosphine (*2*), $(Me_3Si)_2NP=CHSiMe_3$ (1) and the iminophosphine (*3*), $(Me_3Si)_2NP=NSiMe_3$ (2). It is known that the heteroatomic double bond in these two-coordinate phosphines is highly polar in nature with a partial positive charge on phosphorus, thus making it an electrophilic center. The sp^2 hybridized

0097–6156/92/0486–0076$06.00/0

phosphorus atom also carries a lone pair of electrons and, accordingly, might be expected to exhibit nucleophilic character as well. While the nucleophilic nature of the two-coordinate phosphorus is important in some metal complexation reactions, it is of secondary importance in most additions and cycloadditions involving organic reagents.

In general, therefore, our research results are related to the study of the *electrophilic* properties of the two-coordinate, trivalent phosphorus atom in (methylene)phosphine (**1**) and the isoelectronic phosphinimine (**2**). In particular, the novel reactions of these compounds with allenes, acetylenes, acetylenic halides and alcohols as well as selected diketones will be described in this paper.

Reactions with Allenic Compounds

One of the subjects of our investigations involved the interaction of allenes with the P=E derivatives. This work provided some very interesting, unexpected results that may well be of use in synthetic organophosphorus chemistry. Thus, in attempting to study the (2+2)-cycloadditions of the phosphinimine and the (methylene)phosphine systems with the allenic C=C bond, we established that the reactions take a different and more interesting course -- that of an "ene" reaction (eq 1). In the first step of the reaction, the electrophilic phosphorus center apparently attacks the nucleophilic central carbon of the allene system. Then, instead of undergoing nucleophilic attack on the incipient carbonium ion, the anionic center (E) abstracts a proton from the terminal C-H bond, leading to the formation of a new double bond in the phosphorus-substituted 1,3-butadiene derivatives (**3**).

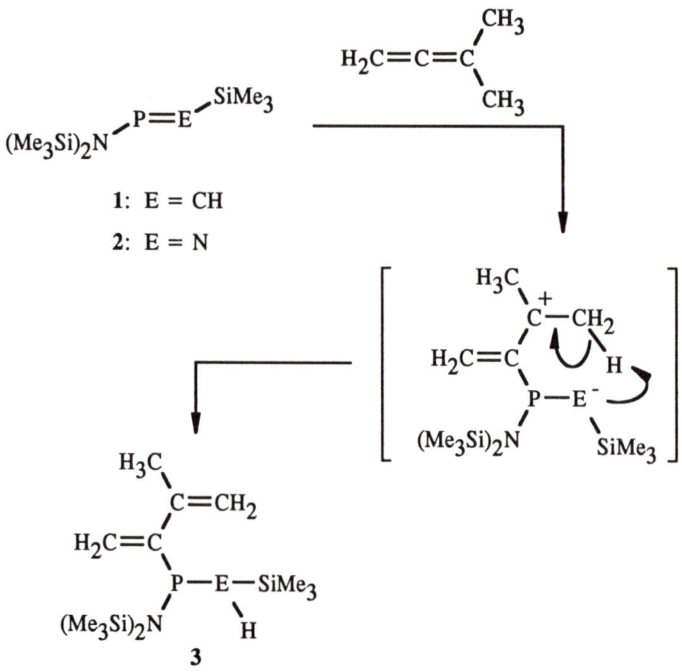

(1)

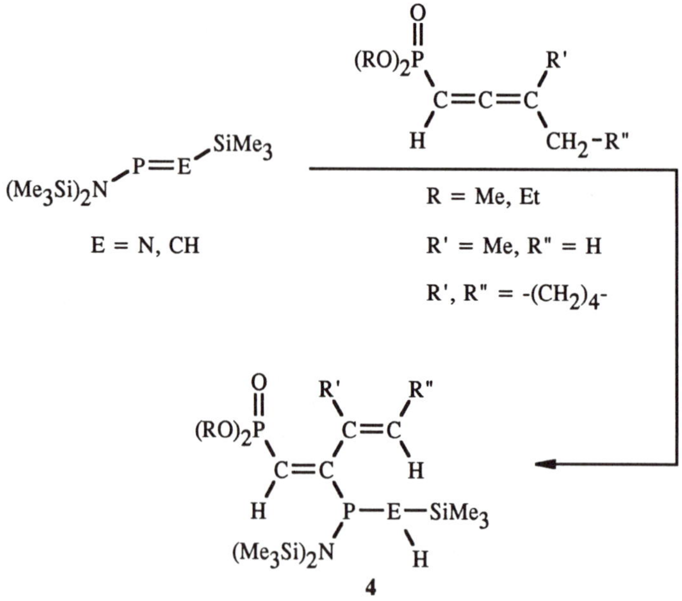

(2)

These reactions are highly chemoselective as indicated by NMR analysis of the crude reaction mixture which showed that only a single product was produced in each case. The reactions with the P=N compound were completed at room temperature in ca. 24 hours, while those involving the (methylene)phosphine required about 40 days to go to completion. This large difference in rate is consistent with the increased polarity of the P=N bond as has been shown in other studies. The same type of addition-elimination process was observed in the reactions of the two-coordinate phosphines with allenic phosphonates (*4*) (eq 2). In this case, however, it is necessary to heat the reaction mixture in benzene (4 - 6 hr) for the phosphinimine and in toluene (6 - 8 hr) for the (methylene)phosphine. Under these conditions, the reactions afforded high yields of the butadienes (*4*) containing two phosphorus substituents in different oxidation states.

Reactions with Acetylenes

Since dialkyl acetylenes contain two *sp* hybridized carbon centers, they have the potential, like allenes, to undergo ene reactions. This prompted us to investigate the reaction of 2-butyne with the two-coordinate phosphines (eq 3). Initially, we found that this reaction was extremely slow at room temperature. Due to the high volatility of 2-butyne, we then carried out the reactions in sealed glass ampoules at 80 °C without solvent. In this case, the reactions were completed after 10 - 15 days and the novel allenic phosphines (**5**) were isolated in good yields. The simplicity of this reaction makes it a very useful method for the synthesis of here-to-fore unknown allenic phosphines.

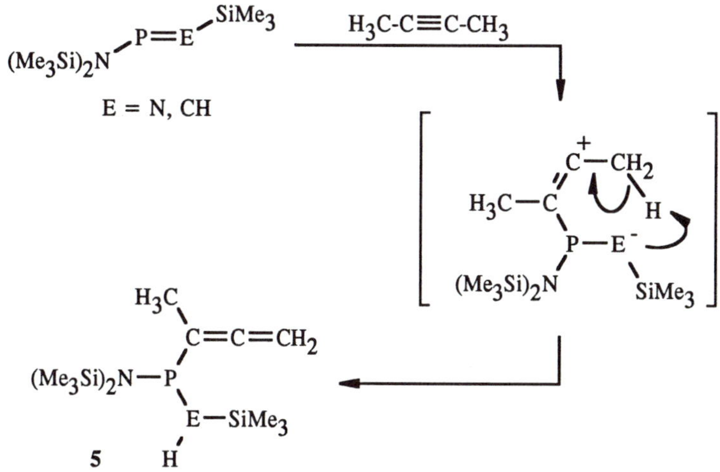

$$(3)$$

Similarly, heating the phosphinimine with 1-hexyne in benzene afforded the analogous allenic phosphine (**6a**: eq 4). It is interesting to note that we observed two diastereomers for this product in the NMR spectra (^{1}H, ^{13}C, and ^{31}P) as a result of the chirality of the allenic moiety in this case. When the same reaction of the phosphinimine with 1-hexyne was carried out at room temperature without a solvent, however, a surprisingly different course was observed (eq 4). Thus, instead of the expected ene reaction, we found that the C-H bond of the acetylene underwent addition to the P=N double bond to yield an acetylenic phosphine (**6b**).

In a similar fashion, we obtained unusual, new acetylenic phosphines (**7**) when the same reaction was carried out with other terminal acetylenes, H-C≡C-R [R = Ph, SiMe$_3$] (eq 5). The (methylene)phosphine, however, did not react with acetylenes in this manner even when the mixtures were heated in benzene for several days. All of these reactions involving addition of the C-H bond to the P=N bond afforded high yields of a single product. The mechanism probably involves electrophilic addition of the relatively acidic C-H proton to the polar P=N double bond.

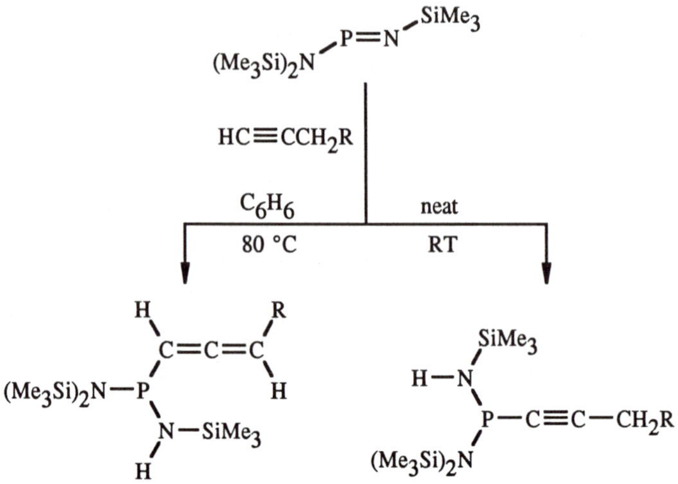

$$(Me_3Si)_2N \diagdown P{=}N \diagup ^{SiMe_3}$$

$$HC{\equiv}CCH_2R$$

$$\begin{array}{cc} C_6H_6 & \text{neat} \\ 80\ ^\circ C & RT \end{array}$$

(4)

6a: R = CH$_2$CH$_2$CH$_3$ 6b: R = CH$_2$CH$_2$CH$_3$

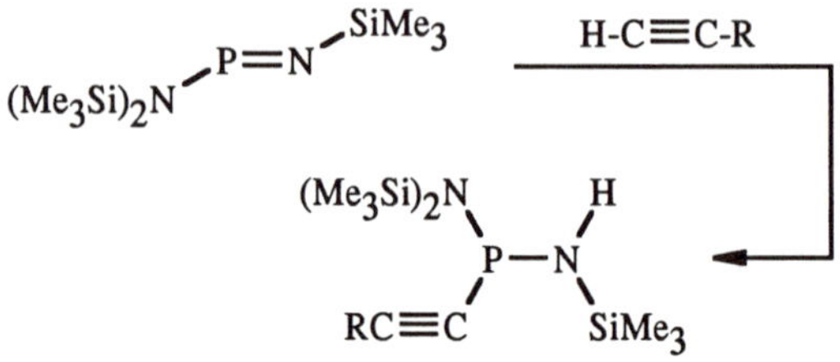

(5)

7: R = Ph, SiMe$_3$

Reactions with Propargyl Halides

The marked difference in reactivity of two- and three-coordinate phosphines was observed in their contrasting chemical behavior toward propargyl halides. It is well known that the reaction of three-coordinate phosphines with alkyl halides proceeds with formation of a phosphonium salt. When a N-trimethylsilyl group is present, however, the elimination of a silyl halide ensues with formation of a phosphoranimine (5). This type of chemical transformation was, in fact, observed when we reacted the (disilylamino)phosphine, (Me$_3$Si)$_2$NPMe$_2$, with propargyl chloride (eq 6). Thus, after addition of propargyl chloride to a CH$_2$Cl$_2$ solution of the phosphine, the initial formation of the propargyl-substituted phosphoranimine (8) was confirmed by ^{31}P NMR spectroscopy. This product could not be isolated, however, due to the occurrence of additional rearrangements. By careful NMR studies, we determined that the Me_3SiCl, eliminated from the phosphonium salt, reacted with the terminal C-H bond of the initial product to yield HCl and the $Me_3Si-C\equiv C-CH_2-P$ derivative (**9**).

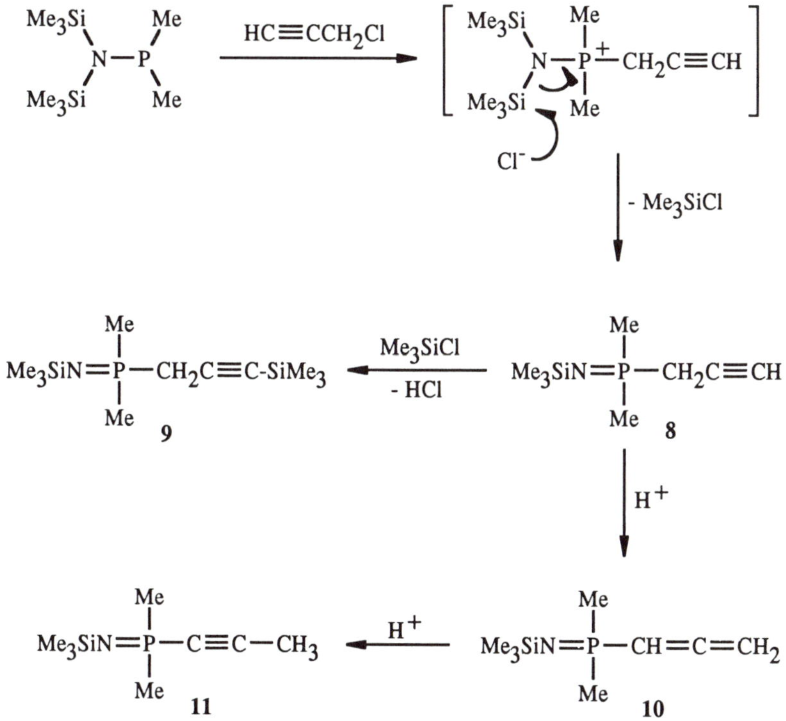

$$- Me_3SiCl \qquad (6)$$

$$(7)$$

In the following steps (eq 7), we observed two HCl-catalyzed processes: an acetylenic-allene rearrangement (to **10**) and an allenic-acetylene rearrangement (to **11**). While both **10** and **11** were easily identified by their characteristic NMR signals and could be distilled as a mixture without decomposition, only the final acetylene derivative (**11**) could be isolated in pure form after fractional distillation. Propargyl bromide reacted similarly with the three-coordinate phosphine, but in this case all of the chemical transformations occurred very rapidly and only the final acetylene phosphoranimine (**11**) could be isolated.

These results prompted us to investigate the reactions of two-coordinate phosphines with propargyl chloride (eq 8). In direct contrast to the results described above, this reaction proceeded not by chloride displacement by the nucleophilic phosphine, but by electrophilic addition of the terminal C-H bond to the P=N linkage. This reaction, therefore, took the same course as was found with other mono-substituted acetylenes (eq 5) and gave the P-C≡C-CH$_2$Cl derivative (12) in high yield after only 24 hours at room temperature. Subsequent treatment of this product with the 3-coordinate phosphine (Me$_3$Si)$_2$NPMe$_2$ gave the novel di-phosphorus substituted acetylene product (13). This last reaction clearly illustrates the differences in the reactivity of the two- and three-coordinate phosphorus centers: electrophilic character of the two-coordinate phosphorus but nucleophilic character of the three-coordinate phosphorus.

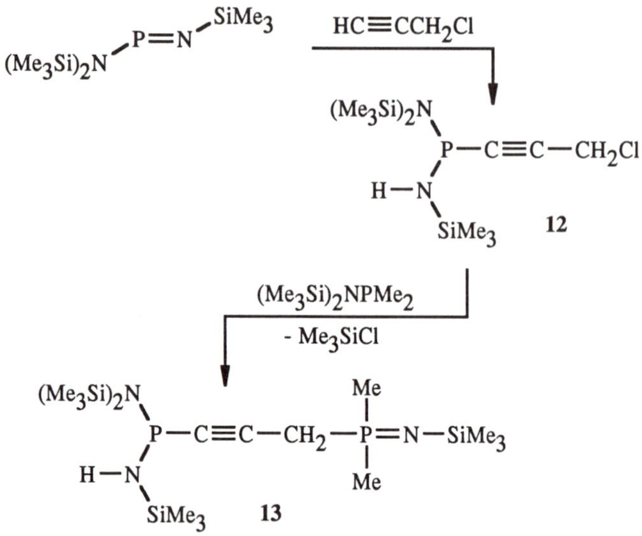

$$(Me_3Si)_2N \diagup P{=}N \diagdown^{SiMe_3} \quad \overset{HC{\equiv}CCH_2Cl}{\underline{\hspace{3cm}}}$$

$$(Me_3Si)_2N \diagdown \atop H{-}N \diagup P{-}C{\equiv}C{-}CH_2Cl \atop \diagdown SiMe_3 \quad \mathbf{12}$$

(8)

$$\overset{(Me_3Si)_2NPMe_2}{\underset{- Me_3SiCl}{\underline{\hspace{3cm}}}}$$

$$(Me_3Si)_2N \diagdown \atop H{-}N \diagup P{-}C{\equiv}C{-}CH_2{-}\overset{Me}{\underset{Me}{P}}{=}N{-}SiMe_3 \atop \diagdown SiMe_3 \quad \mathbf{13}$$

Reactions with Allenic Alcohols

We also obtained interesting results in studying the reactivity of acetylenic alcohols with the two-coordinate phosphines (eq 9). Depending on the position of the OH group relative to the triple bond, different types of products are obtained. In all cases, however, the first step of the reaction is addition of the O-H group to the P=N center to yield the three-coordinate phosphine. In the case of α-acetylenic alcohols, this addition is followed by a rapid acetylene-allene rearrangement and a 1,3-silyl shift to afford the stable, distillable allene-phosphonates (14) in high yield. Both of these type of rearrangements are well documented in organophosphorus chemistry (4, 6), but, as applied here in concert, they lead very easily to some novel allene-phosphoranimines.

The β-acetylenic alcohol, 3-hexyne-1-ol, also reacted with the P=N species via addition of the O-H moiety. In this case, however, the next step is a simple proton migration from nitrogen to phosphorus to give the P-H bonded P^V isomer. Similar addition-migration processes are found in the reactions of the two-coordinate phosphinimine with simple aliphatic alcohols (3).

$$(Me_3Si)_2N \diagdown P{=}N{\diagup}^{SiMe_3}$$

$$HO-\underset{R}{\overset{R}{\underset{|}{\overset{|}{C}}}}-C{\equiv}C-R'$$

R, R' = Me, H

$$\longrightarrow$$

$$\left[\begin{array}{c} R' \diagdown C{\equiv}C \diagup^{R}_{\diagdown R} \\ (Me_3Si)_2N \diagup P-O \\ Me_3Si-N \\ | \\ H \end{array} \right]$$

(9)

$$\underset{Me_3Si-N}{\overset{Me_3SiN}{\diagdown}} \underset{|}{\overset{C{=}C{=}C}{\diagup}} \quad 14$$

$$\longleftarrow$$

$$\left[\begin{array}{c} R' \diagdown C{=}C{=}C \diagup^{R}_{\diagdown R} \\ (Me_3Si)_2N \diagup P{=}O \\ Me_3Si-N \\ | \\ H \end{array} \right]$$

Reactions with Diketones

Our results and related literature data (*1*) clearly show that the P=E double bond in two-coordinate phosphines undergoes facile addition of polar reagents to give three-coordinate P^{III} derivatives. It is also well known that the three-coordinate P^{III} center is relatively nucleophilic and that, in molecules containing an electrophilic site, further chemical transformations are usually observed. Thus, in certain reactions of the two-coordinate phosphines, it is possible to find both types of reactions: electrophilic addition to the P=E bond, followed by nucleophilic attack by the three-coordinate P atom on an internal electrophilic center. For example, in the enol form of a ketone, the O-H moiety is capable of undergoing addition to the P=E double bond of the two-coordinate phosphines. Acetylacetone and related reagents react rapidly with the phosphinimine at low temperature (ca. - 50 °C) and with the (methylene)phosphine at 0 °C to give the new 2,3-dihydro-1,2-oxaphospholes (**16**: eq 10). These reactions most likely proceed with initial addition of the O-H group to the P=E bond, followed by nucleophilic attack of the three-coordinate phosphorus intermediate (**15**) on the other carbonyl group. This intermediate could be observed by ^{31}P NMR spectroscopy (δ_P 35 - 37 ppm). Because the facile migration of a Me_3Si group from nitrogen to oxygen is a well established process (*6*), it is not surprising to find that the final product of these reactions is the siloxy-substituted phosphoranimine (**16**) which contains the P=N rather than the P=O bond to the four-coordinate P^V center. Overall, this type of reaction is a simple one-step process for converting the two-coordinate phosphorus species into a heterocyclic P^V derivative.

The same type of reaction was also found for a γ-diketone (eq 11) but, in this case, the product is a six-membered ring system (**17**) and the reaction proceeds at a much slower rate. It is interesting to note that this γ-ketone also appears to react via enol formation from the terminal CH_3 group rather than from the internal CH_2 moiety.

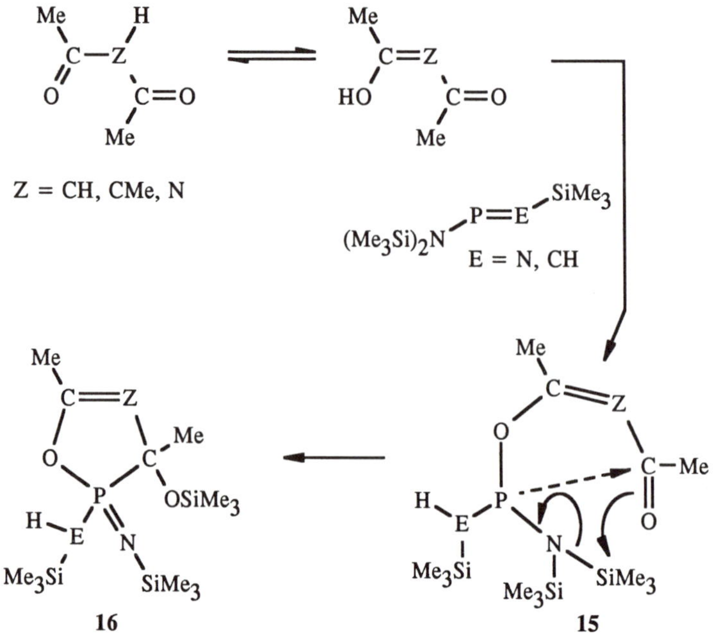

$$Z = CH, CMe, N$$

$$(10)$$

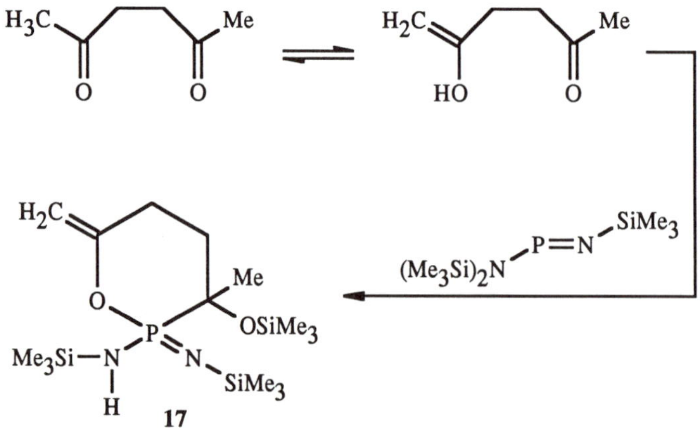

16

15

$$(11)$$

17

Conclusion

These investigations have produced a relatively large number of interesting, unexpected, and synthetically useful results by combining the high reactivity of the two-coordinate P=E moiety with unsaturated organic compounds including allenes, acetylenes, diketones, and acetylenic halides and alcohols. It seems likely that this work has barely scratched the surface in this relatively new area of low-coordinate phosphorus chemistry. The complete experimental details of these studies will be reported in forthcoming publications.

Acknowledgment. The authors thank the Robert A. Welch Foundation and the Texas Christian University Research Fund for financial support of this research.

Literature Cited

1. For some general reviews of the field of low-coordinate phosphorus compounds, see for example: (a) Niecke, E.; Gudat, D. *Angew. Chem., Intl. Ed. Engl.* **1991**, *30*, 217. (b) Appel, R.; Knoll, F. *Adv. Inorg. Chem.* **1989**, *33*, 259. (c) Regitz, M. *Chem. Rev.* **1990**, *90*, 191. (d) Cowley, A. H. *Polyhedron* **1984**, *3*, 389.
2. Neilson, R. H. *Inorg. Chem.* **1981**, *20*, 1969.
3. Niecke, E.; Flick, W. *Angew. Chem., Intl. Ed. Engl.* **1973**, *12*, 585.
4. Ignatiev, V. M.; Ionan, B. I.; Petrov, A. A. *J. Gen. Chem. USSR (Engl. Transl.)* **1967**, *37*, 1807.
5. Morton, D. W.; Neilson, R. H. *Organometallics* **1982**, *1*, 623.
6. Neilson, R. H.; Wisian-Neilson, P.; Wilburn, J. C. *Inorg. Chem.* **1980**, *19*, 413; and references therein.

RECEIVED November 27, 1991

Chapter 7

Factors Influencing the Chain Lengths of Inorganic Polyphosphates

Edward J. Griffith

Monsanto Company, 800 North Lindbergh Boulevard, St. Louis, MO 63167

The theory of formation of inorganic condensed phosphates is well established and has had many years of tests.(1) It has been demonstrated in both amorphous and crystalline systems that the M_2O-P_2O_5 ratio of a system is a dominant controlling factor, where M is one equivalent of cationic functions which may contain one or more metal functions. For any single molecule-ion the ratio precisely dictates the chain length. As the chain lengths of polyphosphates increase other factors play a greater and greater role but the M_2O-P_2O_5 ratio continues to dominate as a necessary but not sufficient condition to effectively control chain lengths. These other conditions include factors as the specific metals ions contained in the metal oxides, the phase state of the phosphate (crystalline or amorphous), the composition of the system containing the polyphosphate, seed crystals, and the thermal history of the system.

Alkali metal and alkaline earth polyphosphates crystallize as short chains, two to six phosphate groups per chain or very long chains with hundreds to thousands of PO_3^- per chain. All polyphosphates in the alkali metal and alkaline earth systems are amorphous in the intermediate chain lengths. Control of the short chain length polyphosphates, both crystalline or amorphous, is a function of R, the M_2O-P_2O_5 ratio. The control of the chain lengths of very long crystalline polyphosphates as Maddrell's salt, Kurrol's salt, and calcium phosphate fibers is not well understood.

0097–6156/92/0486–0086$06.00/0

This work was directed toward three primary goals:
1. To gain a better understanding of the variables controlling the chain lengths of very long crystalline polyphosphate molecule-ions.
2. Determine whether or not "cross-linked" potassium Kurrol's salt contains cross-links in the crystalline phase and is indeed a crystalline ultraphosphate or does the salt obey the phase diagram for the two component, $K_2O-P_2O_5$ system.
3. Gain more understanding of the function of seed crystals as templets in the growth of long chain polyphosphates and the type of phosphate segment delivering the PO_3^- from the melt phase to the crystal surface during crystallization.

Polyphosphate is defined in this work as any linear, condensed phosphate, in which the phosphate, but not necessarily the system containing the phosphate exhibits the conditions $1 \leq M_2O/P_2O_5 \leq 2$, where M is any single or mixed metals with a total equivalency of unity. The definition is required to differentiate the total composition of a system from the polyphosphates crystallizing in a melt. An example is a polyphosphate crystallizing from a melt of ultraphosphate composition where several metal oxides may be involved.

One of the long accepted concepts of polyphosphate chemistry is derived from the equation (2)

1.
$$R = \frac{\bar{n} + 2}{\bar{n}}$$
Eq.

where $R = M_2O/P_2O_5$, the molar ratio, and $\bar{n}$ = average number of phosphorus atoms per molecule; the number average chain length of the polyphosphate. The equation is exact for a single molecule or for an assemblage of molecules all of which have the same structure and composition. It predicts well for systems both amorphous and crystalline when the chain length or average chain length of the system of molecules is relatively small, fifty or less. When amorphous systems become complex with many small mole fractions of molecules with exactly the same chain length summing to an assemblage of many chain lengths the exactness of the equation can become compromised.(3, 4, 5) Other problems arise when the chain lengths of the polyphosphate become very long. The value of R is not very sensitive to a change in chain length from 1000 to 1001, for example, as the value of R approaches unity as a limit for the individual molecules. Very long chain polyphosphates (R = unity) can form and crystalize in a systems where the system's R is a fractional number, as in ultraphosphate melts. Factors influencing the formation of very long chain length polyphosphates are addressed.

Experimental

A large number of samples of $[Ca(PO_3)_2]_n$ were made in batches from a few grams to fifty pounds or more. In all cases the samples were heated to 1000°C and allowed to slowly crystallize from a sodium ultraphosphate melt. Under the best conditions fibers of three inches in length were prepared from the large melts where crystallization could be controlled. (6)

Potassium Kurrol's salts were prepared similar to those prepared by Pfansteil and Iler (7) who first reported the very interesting solution behavior of Kurrol's salts prepared from ultraphosphate melts. If cross-linking is responsible for the behavior of these phosphates when dissolved in water, for a $K_2O-P_2O_5$ molar ratio equal to 0.98, two percent of the phosphate groups should be involved in cross-linking. Two of every 100 phosphate groups should not have a matching K^+ ion and must be attached to another phosphate group to maintain neutrality. In a crystal if the phosphate chain lengths are 10,000 before cross-linking then three dimensional cross-linking would result in gigantic molecule-ions locked into a complex crystal lattice.

Clear melts with the following compositions were made by thoroughly mixing best grades of monopotassium orthophosphate, 85% phosphoric acid, or potassium carbonate and heating the mixtures to 850°C for thirty minutes. Lower temperatures can be used for the mixtures on either side of the metaphosphate composition.

$K_2O\backslash P_2O_5$ = 1.02, 1.00, 0.98, 0.94, 0.85, 0.75, 0.65, and 0.50

X-ray powder patterns were very carefully determined of the compositions 1.02, 1.00, 0.94, 0.75, and 0.65. The x-ray patterns were identical for ratios 1.02 through 0.94 except for minor variations in intensities that were probably a result of orientation of the very fibrous crystals. An amorphous phase was found mixed with normal Kurrol's salt pattern in the 0.75 and 0.65 samples while a sample with a $K_2O-P_2O_5$ ratio of 0.50 could not be crystallized.

Differential thermal analyses were determined on samples

with $K_2O-P_2O_5$ ratios ranging from 1.00 to 0.65. The KPO_3-$K_4P_2O_{10}$ phase diagram *(8)* was extended into the ultraphosphate region and is presented in Figure 1. The extension is not highly precise nor is the extension an equilibrium phase system. The eutectic behavior between potassium Kurrol's salt and a potassium phosphate was seen in all thermal analyses and had a melting temperature of 450°C. The phosphate mixed with the Kurrol's salt is an ultraphosphate glass that may be partially crystalline. The ultraphosphate absorbed water from a zeolite dried nitrogen gas purge during thermogravmatric analyses as shown by an increase in weight of the samples at temperatures between 100°C and 300°C.

The following experiment was performed to determine if it is possible to mimic the properties obtained when potassium Kurrol's salt with a 0.98 $K_2O-P_2O_5$ ratio is prepared from a melt in a furnace. The approach is to mix the potassium ultraphosphate with a ratio of 0.65 with potassium Kurrol's salt made with a ratio $K_2O-P_2O_5$ = 1.0.

Mix 0.5g of the 0.65 ratio ultraphosphate with 9.5g of 1.0 $K_2O-P_2O_5$ ratio potassium Kurrol's salt and dissolve the salts in 1 liter of water containing 5.0 grams of sodium pyrophosphate as a solubilizing agent. Compare this to a solution made with 10g of 1.0 $K_2O-P_2O_5$ ratio potassium Kurrol's without added ultraphosphate. The mixtures do not behave as the potassium polyphosphates grown in an ultraphosphate melt and are no more viscous than Kurrol's salt without the added ultraphosphate.

Visual observations of the crystallization products made from melts prepared in the $M_2O-P_2O_5$ range from 0.5 to 0.75 showed very definite indications of a two phase system. The melt with an $M_2O-P_2O_5$ ratio of 0.50 could not be crystallized at any temperature attempted as low as 400°C and was a clear transparent glass. The melt with a $M_2O-P_2O_5$ ratio of 0.65 could be crystallized to fine crystals contained in milky white solid. The melt with an $M_2O-P_2O_5$ ratio of 0.75 crystallized to a plate-like crystal that appeared to be crystalline throughout, behaved as an amorphous-crystal mix in DTAs. In all respects the potassium Kurrol's salt is similar to the calcium polyphosphate fibers, $[Ca(PO_3)_2]_n$, grown under similar conditions.

Short Chain Kurrol's Salts

Attempts were made to grow short chain length Kurrol's salt from melts with $M_2O-P_2O_5$ ratios between 1.04 ($\bar{n}$ = 50)

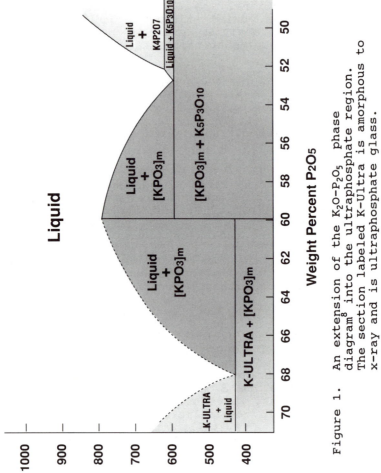

Figure 1.   An extension of the K_2O–P_2O_5 phase
diagram[8] into the ultraphosphate region.
The section labeled K-Ultra is amorphous to
x-ray and is ultraphosphate glass.

and 1.10 ($\bar{n}$ = 20). If a slight excess of Na_2O or K_2O behaved on the basic side of the metaphosphate composition as P_2O_5 is reported to behave on the acid side of the metaphosphate composition it should be possible to grow short chain Kurrol's salts as well as cross-linked Kurrol's salts. In all cases the required quantities of $Na_5P_3O_{10}$ and $K_5P_3O_{10}$ crystallized in the melts as required by the phase diagram while the Kurrol's salt was very similar to the salt obtained from melts with $M_2O-P_2O_5$ = 1.0.

Discussion and Results:

In the preparation of very long crystals (3 inches x 10 μ) of calcium polyphosphate fibers, the fibers are grown in sodium ultraphosphate melts.   Both the crystalline Kurrol's salt and ultraphosphate phases are detectable in the cooled fiber cake. The ultraphosphate phase is water soluble and can be quickly leached from the crystalline phase to produce free fibers. (9)   The polyphosphate chains grown in the calcium polyphosphate crystals are of ultra long length and showed no signs of cross-linking in the crystalline phase when single crystal x-ray structures were determined.

Cross-linked potassium Kurrol's is prepared in exactly the same manner as the calcium polyphosphate fibers. The only difference is potassium is substituted for calcium in the preparation of potassium Kurrol's salt. There is strong reason to question whether or not the strange solution behavior of cross-linked potassium Kurrol's salt is a result of cross-linking in the crystalline phase or is a result of ultra long polyphosphate chains mixed with an independent ultraphosphate phase that contains the required cross-linking. All of the observations which led to the belief that the molecule-ions in the crystalline phase were cross-linked can be explained based upon the two phase model where the ultraphosphate phase is rather quickly degraded in aqueous solutions to form acidic groups.

Attention is directed toward the polyphosphate anions as the structural element under consideration. The cationic functions are considered only as they influence the behavior of the anions.   Phase chemistry and thermodynamics become important considerations in devising schemes to control the grow of longer and longer polyphosphate chains. The cationic components of the phase systems cannot be ignored, but they will be relegated to a causal role.

Not only the composition and thermodynamics can influence the growth of polyphosphate chains, but the thermal history of the phosphate can often lead to metastable equilibria with polyphosphates that are not found in phase diagrams for the systems. Slow crystallization rates usually favor the formation of longer chain polyphosphates if the crystallizing salt is the thermodynamically stable crystal under the growth conditions. During crystallization impurities (chain growth terminators) are driven to the surface of a crystal if neither solid solutions nor double salts are formed between the impurities and the crystallizing polyphosphate. The slower growth rates provide longer times for impurities to migrate to the crystal surface.

In systems where the crystallizing salt is not the thermodynamically stable salt the system can be cooled too slowly to obtain maximum chain lengths of the crystallizing polyphosphate. Sodium Kurrol's salt is unstable with respect to sodium cyclic triphosphate under all conditions at atmospheric pressure. If a crystallizing melt is cooled too slowly it converts to sodium cyclic triphosphate. Proper seeding and rates of crystallization are important in obtaining high yields of sodium Kurrol's salt where the chain length may be 5,000 to 10,000, or more. (10)

Metal oxides (and water) are chain growth terminators of long chain polyphosphates as predicted by Equation 1. The long chain polyphosphate chains are very sensitive to chain breakers. The difference between a polyphosphate anion of chain length 1000 and one of chain length 2000 is only one M_2O group per two thousand PO_3 groups.

Long chain polyphosphates crystallize from melts to form straight molecule-ions aligned parallel to the longest axis of the crystal (Maddrell's salt) or as long helical chains (Kurrol's salt), aligned parallel to the longest axis of the crystal. In neither case are the molecules twisted nor folded in the crystal. This is important when considering how long chain molecule-ions grow from melts. (11)

Attempts to obtain amorphous polyphosphates with average chain lengths of 1000 directly by quenching melts have not been successful. Average chain lengths of amorphous polyphosphates of near 400 have been the maximum obtained directly from melts. Sodium Kurrol's salt of average chain lengths of several thousand have been melted to make quenched glass. The chain lengths are no longer than obtained by melting high purity sodium (cyclic

triphosphate) trimetaphosphate. Indirect methods of obtaining amorphous very long chain length glasses have been devised. (12)
If the calcium phosphate fibers grown from ultraphosphate melts are slurried with a solution of sodium ethylene diamine tartaric acid the calcium crystals can be converted to a sodium phosphate glass that dissolves very slowly yielding very long chains.

In the discussion to follow, the great differences between sodium and potassium condensed phosphate phase systems is highlighted. Where the sodium Kurrol's salt is metastable with respect to the cyclic sodium trimetaphosphate there is no cyclic metaphosphate in the potassium phosphate two component phase system and the Kurrol's salt is the only stable compound in the phase system with $K_2O-P_2O_5$ ratios near unity.

The sodium polyphosphate systems produce glasses very easily while it is difficult to make potassium glasses. The milky glassy masses of potassium metaphosphate compositions obtained by quenching the potassium melts are rich in potassium trimetaphosphate, though the phase diagram for the system contains none. Potassium cyclic triphosphate can be made thermally by dehydrating monopotassium orthophosphate with urea at temperatures less than 300°C,

$$3 \; KH_2PO_4 + 3 \; NH_2\overset{O}{\underset{}{C}}NH_2 \quad => \quad K_3(PO_3)_3 \; + \; 6 \; NH_3 + 3 \; CO_2$$

or in aqueous solution by an ion exchange of sodium for potassium in the cyclic triphosphate system.

Seed Crystals As Templets

If polyphosphates of chain lengths greater than about 350 are to be obtained a crystallization process is required in which a seed crystal is behaving as a templet for chain growth. The templet aids the crystallization to proceed without the chains of molten polyphosphate "sequestering" metal ions to terminate their growth or thermal activity rupturing the molecules.

To obtain high tensile strengths of calcium polyphosphate fibers crystals it is necessary to grow polyphosphate molecule-ions as long as possible. When a phosphate group from a high temperature melt system moves to a high temperature crystal lattice the phosphate group has transgressed a phase boundary. The melt system is a region where reorganization is chaotic. In the crystal lattice the environment is highly organized. The phosphate group or groups must not only fall to a lattice of lower chemical potential, but must also be accompanied by the correct number of cationic groups to exactly balance the negative charge on the phosphate. This must be done without the polyphosphate sequestering enough metal oxide to act as either chain terminators or breakers.

Longer chains of polyphosphates (n > 5,000) grow to even greater lengths in systems deficient in metal oxides or containing an excess of P_2O_5. Under these conditions the metal oxides will be attracted to the more acidic melt phase and the melt phase will exercise an element of control over the crystallizing polyphosphate. It is expected that longer chain polyphosphate molecules should grow in crystals where the melts have the ratio, M_2O/P_2O_5 < 1, but not so low as to move into a region of the phase diagram dominated by other compounds.

Condensed phosphate melts are composed of a huge number of phosphate species that are undergoing constant reorganization, as judged by analyses of the glasses formed by quenching these melts. When crystals form in high temperature systems the segments of phosphates that transfer from the melt phase to the crystalline phase at the boundry of crystallization are limited to four models.

1. The crystals may grow one PO_3 group at a time under which conditions a molecule-ion of 10,000 phosphate groups would require the formation of 9,999 P-O-P linkages on the surface of the crystal to create a single molecule-ion.

2. The PO_3 groups could be transported to the surface of the crystals as clusters of threes, as the stable triphosphate chains or rings.

3. The PO_3 groups could be transported as short to intermediate length of phosphate chains or rings as indicated by some physical data.

4. The crystals could grow from very large molecule-ions 10,000 PO_3 long in the melt phase and the crystals could grow from single segments.

To consider each model it is unlikely that single

monometaphosphate or orthophosphate groups are formed to be transported from the melt to the crystalline phase. Glasses formed from condensed phosphate melts contain no detectable orthophosphate when analyzed soon after they are prepared. (13)

There are several reasons why segments as short as three phosphate groups could be involved in the transfer of PO_3 groups from a melt phase to the crystalline phase of a long chain polyphosphate. As noted above when quenched potassium phosphate melts yield rings that are not found in the crystallized systems. The tripoly-phosphate that results from ring opening is too short to knot. It is well known that polyphosphates reorganize to cyclic trimetaphosphates in aqueous solutions, but this is yet another mechanism. (14)

Surface tension and density work suggest that the transfer unit may range from the dimer to as many as twelve phosphate groups. The most likely conditions are that the phosphate is transported as segments of six to eight phosphate groups. It was demonstrated by viscosity measurements of polyphosphate melts that the flow segment is between six and eight phosphate groups in length. (15) Segments of this length are short enough and rigid enough to be free of entanglement with other chains and should add to a chain growing on the surface of a crystal without encountering large stearic barriers.

Convincing experiments left little doubt that most amorphous polyphosphate glasses in the sodium system exhibit some branching. (16) It is unlikely that long chain branched structures could fit into a crystal lattice without first rupturing to destroy the branching point. It is well established that melts with $M_2O-P_2O_5 \geq 1$ resist branching. (13)

During crystallization the crystal surface grows to contact fresh melt while the molecule-ions of the melt are transported to the crystal surface by thermal motion. It is unlikely that chains of several thousand phosphate groups are formed in the liquid melt phase, then transported, in tact, to the surface of the growing crystal templet.

It has been claimed that long chain ammonium polyphosphates can be treated with potassium hydroxide to form long chain polyphosphates that crystallize from solution as the dihydrate. It was suggested that the phosphate was a Kurrol's salt, but only after the salt

had been heated to 600°C. (17)   There is doubt that long
chains can fit directly into a crystal lattice when
crystallized from an aqueous solution.   Admittedly the
chaotic   reorganization   that   takes   place   in   high
temperature melts definitely does not occur in aqueous
solutions at atmospheric pressure, but more evidence is
needed before it is accepted that long chain alkali metal
polyphosphates can crystallize from aqueous media.

The long chain polyphosphates grown as crystals are a
more   narrow   distribution   of   chain   lengths   than   the
amorphous systems when measure by intrinsic viscosity or
gel chromatography. (18)   The narrow distribution would
be required because the molecules must fit an orderly
lattice, but it has been pointed out that a single long
chain polyphosphate may reside in ten-thousand or more
unit cells. (20)   End groups occur so seldom that a
crystal can accommodate the small dislocations without
changing the x-ray pattern of the phosphate.
Cross-linked Potassium Kurrol's Salt

There is no reorganization in a crystal of potassium
Kurrol's   salt   even   at   temperatures   near   the   melt
temperature because there are no phase transitions in
potassium Kurrol's salt crystals.   In the sodium system
the   crystals   of   Kurrol's   salt   can   suffer   a   phase
transition.   At a phase transition the chaos of the melt
is recaptured over the interval of the transition.

The x-ray patterns of the crystalline long chain (1,000
or more phosphorus atoms per average chain) potassium
Kurrol's salt are highly reproducible despite the fact
that the phosphates can be crystallized from a variety of
melt   compositions.   An   outstanding   example   of   this
behavior is the growth of crystals of potassium Kurrol's
salt in the more basic polyphosphate phase regions and
the   potassium   Kurrol's   salt   grown   in   the   acidic
ultraphosphate phase regions are identical. (19)   The
solution properties of the long chain length phosphates
are very different depending on the composition of the
melt from which the crystals were grown and yet the x-
ray patterns are precisely the same.   It has been assumed
for   many   years   that   the   difference   between   the   two
preparations was cross-linking of the molecule-ions of
long chain polyphosphates grown in crystals from melts
with $M_2O-P_2O_5$ ratios less than unity. (20)

The evidence for cross-linking in crystalline inorganic
polyphosphates   to   convert   them   to   crystalline
ultraphosphates with small degrees of cross-linking is
circumstantial   rather   than   direct.   This   is   not   to
question   the   triply-linked   phosphate   groups   in   the

ultraphosphate region of the condensed phosphate systems. Crystalline phosphorus pentoxide, P_4O_{10}, is completely triply-linked. The question is whether or not the crystalline polyphosphates grown from ultraphosphate melts are very long chain polyphosphates or are ultraphosphates with small degrees of cross-linking.

The aqueous solution properties of potassium Kurrol's salt prepared from ultraphosphate melts are nothing less than spectacular. A 1% solution of potassium Kurrol's salt crystallized from a melt with a $M_2O-P_2O_5 = 1.0$ is obviously viscous, but a salt make with a $M_2O-P_2O_5 = 0.98$ is stringy breaking into threads when a rod is removed from the solution. The Kurrol's salts could derive their aqueous solution properties from cross-linking or from ultra-long single stranded linear molecules. When the calcium polyphosphate system is compared with the potassium Kurrol's salt systems the evidence is controversial.

Triply-linked Phosphate Groups

If the crystalline inorganic phosphates grown in ultraphosphate media contain triply-linked phosphate groups they are either isopolyphosphates or ultraphosphates. An example is a PO_3 groups bonded through an oxygen linkages to three other PO_3 groups. Three types of triply-linked phosphate groups are possible. They are:

Type I Cross-linking

              O O O O O O   (charges are ignored)
              -OPOPOPOPOPOP-
              O O O O O O
                  OPO
              O O O O O O
              -OPOPOPOPOP-
              O O O O O O   where there are two

triply-linked PO3 groups, but there are two other types of bonding for branching and cross-linking. In the structure shown above there is a single PO3 in the bridge. There is no reason that there could not be several PO_3 groups between the two longer chains. It is also possible that there could be no PO3 between the chains as in the following structure.

Type II   Cross-linking

```
              o o o o o o              (charges are
ignored)
              -OPOPOPOPOPOP-
              o o | o o o
                  o
              o o | o o o
              -OPOPOPOPOPOP-
              o o o o o o
```

where there are but two triply-linked PO3 groups.

There is then the branched isopolyphosphate that contains
a single triply-linked PO3 group as below.

Type III   Branching
```
              o o o o o o              (charges are
ignored)
              -OPOPOPOPOPOP-
              o o o o o o
                  OPO
                   o
                  OPO
                   o
                  OPO
                   |
```

In an attempt to obtain direct evidence for the presence
of cross-linking in the potassium polyphosphates
crystallized from ultraphosphates melts two tools were
chosen.   Cross-linking should in theory be observed by
either phase chemistry or x-ray.   Also the existence of
end and middle group phosphorus is well established in
nmr studies of inorganic phosphates while triply-linked
phosphate groups have been reported in organic phosphate
systems.   Cross-linking in polyphosphates should be
directly observable by tunneling microscopy, but time did
not permit the completion of this approach in this work.

The linear molecule-ions of Kurrol's salts pass through
a thousand or so unit cells before terminating.   This
suggests that in a Kurrol's salt with a $K_2O-P_2O_5$ ratio
equal to 0.98 on the average every fifty cells should be
a cross-linking cell of one of the two types listed
above.   The potassium Kurrol's salt cells contain
segments of two helical chains and four PO_3 groups per
unit cell.   Cross-linking within these cells should
create a very complex unit cell and an even more
complicated lattice.

It is unlikely that a crystal complex enough to contain
phosphate chains of several thousand phosphate groups

could tolerate molecules with simple branching groups as shown in Type III branching. There is no reason to expect that they are excluded from amorphous systems however. Assuming that either of the two types of cross-linking groups are equally probable, Type II should be accommodated by a cross-linking unit cell with less dimensional contortion. The dimensions of the cell would shrink by one K_2O unit per cross-link. If Type I were exhibited the cell would be required to increase by one PO_3 while decreasing by $\frac{1}{2}K_2O$. Either type should be seen in x-ray patterns.

Three possibilities exist when a salt with a polyphosphate x-ray pattern crystallizes from a melt containing an excess of phosphorus pentoxide. 1. The phosphorus pentoxide is incorporated into the polyphosphate chains converting the chains to crystalline ultraphosphates. 2. The excess phosphorus pentoxide does not enter the polyphosphate crystal structure, but forms an amorphous phase between the crystals of polyphosphate. The amorphous phase is not detected by x-ray. 3. The excess phosphorus pentoxide does not enter the crystal structure of the polyphosphate, but forms as an ultraphosphate between the crystalline polyphosphate crystals as a eutectic phase. (This latter case is precisely what happens in the calcium sodium ultraphosphate system from which calcium phosphate fibers are grown (*21*) and the phase diagram of Hill et. al. is obeyed as it should be.)

In the potassium Kurrol's salt phase system the crystalline analogues to $Ca_2P_6O_{17}$ or CaP_4O_{11} were not found by a literature search. (*22*) The amorphous potassium ultraphosphate systems have been studied. (*23*) Amorphous condensed phosphates are seldom , if ever, single compounds, but are a random mixtures of compounds and can be large and very complex. If the ultraphosphates were embedded between crystals of pure potassium phosphate fibers they would be very difficult to detect.

When dissolved in aqueous solutions of diverse ions potassium Kurrol's salt is a viscous solution at concentration of 1% polyphosphate. The "cross-linked" Kurrol's is even more viscous and exhibits a stringiness not seen in the Kurrol's salts prepared with K_2O-P_2O_5 ratios equal to unity. The aqueous solution behavior of cross-linked Kurrol's salt may be caused by the a second phase of the <u>system</u> superimposed upon the viscous behavior of the solutions of 1.0 K_2O-P_2O_5 ratio potassium Kurrol's salt solutions.
The evidence does not support cross-linking as the primary reason that potassium Kurrol's salt grown in ultraphosphate melts exhibits the very high viscosities and drag reducing properties at low concentrations in aqueous solutions. (*24*) It has never been adequately

demonstrated that the cross-linking is a part of the molecules in a crystalline phase. It was demonstrated, however, that even amorphous sodium polyphosphate glasses could be made to exhibit the same aqueous solution behavior as the potassium Kurrol's salt grown in melts with an $M_2O-P_2O_5$ ratio less than unity. (25) In the sodium ultraphosphate glasses cross-linking is required and is surely responsible for the behavior.

In the classic work of Hill, Foust and Reynolds they prepared the crystalline ultraphosphates, CaP_4O_{11} and $Ca_2P_6O_{17}$, in phase studies. (26) The preparation of the thermodynamically stable salts were elegantly confirmed. (27) Single crystal x-ray studies were used to confirm the structures of both salts. In the work of Glonic, Van Wazer, Kleps, Legeros, and Meyers, (28, 29, 30) also classics, they approached the chemistry of triply-linked phosphorus groups preparing the 1,5-μ-oxo-tetrametaphosphate anion in organic media. The 1,5-μ-oxo-tetrametaphosphate anion is the same anion discovered by Hill, Foust and Reynolds.

Conclusions:

It is concluded that the crystalline phase of potassium Kurrol's salt is not cross-linked. The mixture of very long chain polyphosphates and amorphous ultraphosphates dissolve to form highly viscous aqueous solutions. It is also concluded that the seed crystals aiding the growth of ultra long polyphosphate molecule-ions is built up from small segments of polyphosphate formed in a phosphate melt.

Tunneling microscopy is a tool currently available potentially capable of definitively resolving the question of cross-linking in crystalline potassium Kurrol's salt. The current evidence strongly suggests that the phase diagram for the system is obeyed. Short chain ($\bar{n} \leq 50$) Kurrol's cannot be prepared on the basic side of the metaphosphate compositions nor can cross-linking of the potassium Kurrol's salt chains occur in the crystals grown on the acidic side of the metaphosphate composition.

Literature Cited:

1. Van Wazer, J. R. and Griffith, E. J., J. Am. Chem. Soc. 77, 6140 (1955).
2. Van Wazer, J. R., J. Am. Chem. Sac. 72, 644; 647 (1950).
3. Westman, A. E. R. and Gartaganis, P. A., J. Am. Ceramics Soc. 40, 293 (1957).
4. Parks, J.R. and Van Wazer, J. R., J. Am. Chem. Sac. 79, 4890 (1957).

5. Griffith, E. J. and Buxton, R. L., <u>J. Am. Chem. Sac.</u> <u>89</u>, 2884 (1967).
6. Griffith, E. J., <u>Proc. Phosphorus Symposium, Duke University, 75</u>, 362, (1981).
7. Pfansteil, R. and Iler, R. K., <u>J. Am. Chem. Soc. 74</u>, 6059 (1952).
8. Morey, G. W., Boyd, F. R., England, J. L., and Chen, W. T., <u>J. Am. Chem. Soc. 77</u>, 5003 (1955).
9. Griffith, E. J., U. S. Patent 4,360,625, November 23, 1982.
10. Griffith, E. J. and Kodner, I. J., U. S. Patent 3,312,523, April 4, 1967.
11. Corbridge, D. E. C., <u>Structural Chemistry of Phosphorus</u>, pg. 143, Elsevier Scientific Publishing Company, Amsterdam 1974.
12. Crutchfield, M. M., Private communication, Monsanto Company; Griffith, E. J., <u>International Conference On Phosphate Chemistry</u>, Tokyo, Japan, July, 1991.
13. Van Wazer, J. R., <u>J. Am. Chem. Soc. 78</u>, 5709 (1956).
14. Osterheld, K., <u>Topics In Phosphorus Chemistry Vol.</u> 7, pg. 103 Ed. M. Grayson and E. J. Griffith, John Wiley and Sons, New York, N. Y. 1972
15. Callis, C. F., Van Wazer, J. R. and Metcalf, J. S., <u>J. Am. Chem. Sac. 77</u>, 1471 (1955).
16. Strauss, U. P., Smith, E. H. and Wineman, P.L., <u>J. am. Chem. Soc. 75</u>, 3935 (1953). Strauss, U. P. and Treitler, T. L., ibid. <u>77</u>, 1473 (1955).
17. Stahlheber N. E., U. S. Patent 3,723,602 March 27, 1973.
18. Filer, W. A., Unpublished Monsanto report.
19. Ngo, T. M., and May, F. L., Unpublished Monsanto Report done for this work.
20. Van Wazer, J. R., <u>Phosphorus and Its Compounds</u>, pg. 676, Interscience Publishers, Inc. New York, N. Y., 1958; ibid. pg. 762.
21. Griffith, E. J., <u>Proc. International Symposium On Phosphorus Chemistry</u>, pg. 361, Duke University, 1981.
22. Morey, G. W., <u>J. Am. Chem. Soc. 74</u>, 5783 (1952).
23. Kalmykov, S. I., Bekturov, A. B., Shevchenko, N. P., Malakohova, K. I., and Poletaev, E. V., <u>Izv. Akad. Nauk, Kaz. SSR, Ser. Khim, 23,</u> 1 (1973).
24. Hunston, D. L., Griffith, J. R., Little, R. C., <u>Nature 245</u>, 141 (1973).
25. McCullough, J. F., Unpublished data, private demonstration.
26. Hill, W. L., Faust, G. T. and Reynolds, D. S., <u>Am. J. Sci. 242</u>, 457, 542 (1944).
27. Bencher, M., <u>Mat. Res. Bull. 4</u>, 15 (1969).
28. Glonic, T., Meyers, T. C. and Van Wazer, J. R., <u>J. Am. Chem. Soc. 92</u>, 7214 (1940)._
29. Glonic, T. , Van Wazer, J. R., Kleps, R. A. and Meyers, T. C., <u>Inorg. Chem. 13</u>, 233 (1974).
30. Van Wazer, J. R., and Legeros, R. Z., <u>International Congress On Phosphorus Compounds</u>, Rabat, October, 1977. pg.95

RECEIVED December 30, 1991

Chapter 8

Mechanism of Reactions of Monosubstituted Phosphates in Water

Appearance and Reality

William P. Jencks

Graduate Department of Biochemistry, Brandeis University,
Waltham, MA 02254–9110

Studies of structure-reactivity parameters, ^{18}O isotope effects, solvent effects and other parameters have shown that the hydrolysis of monoanions and dianions of acyl phosphates and monoanions of phosphate esters proceeds through strongly dissociative transition states that closely resemble the monomeric metaphosphate monoanion. For example, acetyl phosphate monoanion undergoes hydrolysis in water with $\Delta S^{\neq}$ = +4 e.u., k_{H_2O}/k_{D_2O} = 1.0, small salt and solvent effects, and $\Delta V^{\neq}$ = -0.6 cc mol^{-1}. This might suggest that the reactions proceed through a metaphosphate intermediate. However, experiments are reviewed which show that monosubstituted phosphate derivatives react with water and other nucleophilic reagents through a fully coupled, concerted mechanism of displacement in aqueous solution. In particular, there is an interaction between the nucleophile and the leaving group in the transition state, which is not expected for a stepwise reaction through a metaphosphate intermediate.

In 1955 Westheimer, Bunton, Vernon and their coworkers suggested that the hydrolysis of the monoanions of phosphate monoesters proceeds through the formation of an intermediate monomeric metaphosphate monoanion, which reacts rapidly with water to give inorganic phosphate (equation 1)(*1,2*).

0097–6156/92/0486–0102$06.00/0

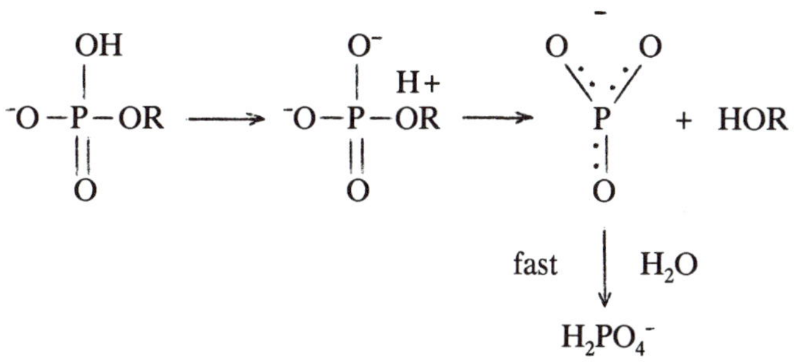

$$
\begin{array}{ccc}
\underset{\displaystyle O}{\overset{\displaystyle OH}{\underset{\|}{{}^-O-\overset{|}{P}-OR}}} & \longrightarrow & \underset{\displaystyle O}{\overset{\displaystyle O^-}{\underset{\|}{{}^-O-\overset{|}{P}-OR}}} \; H^+ \; \longrightarrow & \underset{\displaystyle O}{\overset{\displaystyle O\quad O}{P}} \; + \; HOR \quad (1)
\end{array}
$$

$$
\text{fast} \downarrow H_2O
$$

$$
H_2PO_4^-
$$

Such a mechanism might account for the rapid hydrolysis of phosphate monoester monoanions, which occurs some 10,000 times faster than that of the corresponding diester monoanions. It was suggested that the rapid reaction proceeds by transfer of a proton from the OH group to the leaving OR group of the monoester, followed by elimination to form metaphosphate, which then reacts rapidly with water as shown in equation 1. Such a mechanism is not possible for the slower hydrolysis of phosphate diesters. The dependence on pH of the rate of hydrolysis of the monoanion showed that most phosphate monoester monoanions undergo hydrolysis much faster than the dianion or the uncharged acid (*3-5*). Phosphate dianions with very good leaving groups, such as dinitrophenyl phosphate, acyl phosphates and phosphorylated pyridines, also undergo rapid hydrolysis; this might also proceed through monomeric metaphosphate (equation 2) (*6-8*).

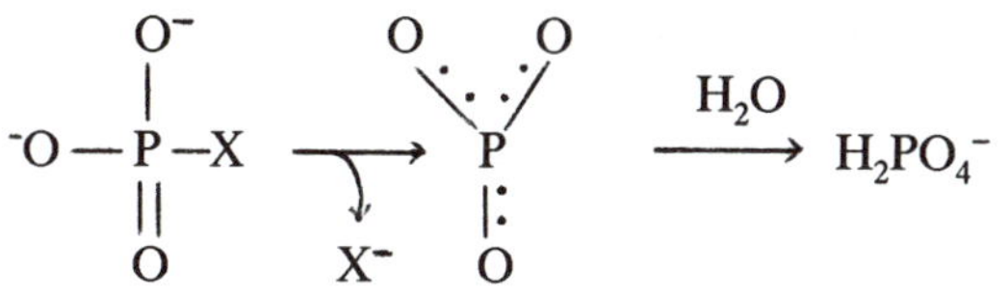

$$
\underset{\displaystyle O}{\overset{\displaystyle O^-}{\underset{\|}{{}^-O-\overset{|}{P}-X}}} \; \longrightarrow \; \underset{X^-}{\overset{\displaystyle O\quad O}{P}} \; \overset{H_2O}{\longrightarrow} \; H_2PO_4^- \qquad (2)
$$

The rates of hydrolysis of these compounds show a very large dependence on the pK_a of the leaving group, with values of β_{lg} in the range of -1.0 to -1.2 (*6-8*), and the hydrolysis of aryl-^{18}O-2,4-dinitrophenyl phosphate dianion with ^{18}O in the leaving oxygen atom shows a large oxygen isotope effect of 2.04% (*9*).

These results show that the reactions proceed with a large amount of bond-breaking in a transition state that resembles the monomeric metaphosphate monoanion.

The monoanions of substituted benzoyl phosphates undergo hydrolysis rapidly, with only a small dependence of the rate on the pK_a of the leaving group; the dianions also undergo rapid hydrolysis, with a value of $ß_{lg}$ = -1.2 (7). The rapid hydrolysis of the monoanions could be explained by intramolecular transfer of the proton to the leaving group (1), while the large dependence of the rate on the pK_a of the leaving group in the dianion series indicates a metaphosphate-like transition state (2). The entropies of activation are near

1   2

zero for the monoanion and dianion of acetyl phosphate, but $\Delta S^{\neq}$ is -28.9 e.u. for the monoanion of acetyl phenyl phosphate. There is no significant solvent deuterium isotope effect for the hydrolysis of the monoanion and dianion of acetyl phosphate, while the hydrolysis of the monoanion of acetyl phenyl phosphate shows an isotope effect of k_{HOH}/k_{DOD} = 2.5. The addition of acetonitrile has little effect on the rates of hydrolysis of acetyl phosphate monoanion and dianion, while it causes a large decrease in the rate for acetyl phenyl phosphate (7). The volumes of activation are -0.6 to -1.0 cm^3/mol for acetyl phosphate monoanion and dianion, while $\Delta V^{\neq}$ is -19 cm^3/mol for the monoanion of acetyl phenyl phosphate (10). It was concluded that "while no single one of these considerations should be taken alone as conclusive proof of the monomolecular or bimolecular nature of a reaction, taken together they constitute strong evidence that the neutral hydrolyses of acyl phosphates occur through a metaphosphate monoanion intermediate" (7).

This is a conclusion that we remember well, because it should never have been reached. These data, and data from many other laboratories (11), are certainly consistent with a metaphosphate mechanism, but they are also consistent with a concerted substitution reaction; they do not distinguish which mechanism is followed. Structure-reactivity coefficients, isotope effects, volumes and entropies of activation and similar parameters are all measures of the structure of the rate-limiting transition state. They provide information regarding the structure of the transition state, but they provide no information as to whether the reaction occurs in one or two steps. They tell us about the appearance of the rate-limiting transition state; they do not determine the reality of the mechanism. They provide no information as to whether or not there is a metaphosphate intermediate with a significant lifetime that reacts with water in a second step, with a significant barrier after the rate-limiting transition state. All of the other criteria described above that were applied to different reactions of phosphate compounds also tell us that monosubstituted phosphate derivatives tend to react through dissociative transition states. They establish that the rate-determining transition state of the reaction is strongly dissociative in nature, but they do not tell us whether this transition state goes on to form a metaphosphate intermediate with a significant lifetime (equation 2, top) or gives the product directly with no intermediate PO_3^- ion (equation 2, center).

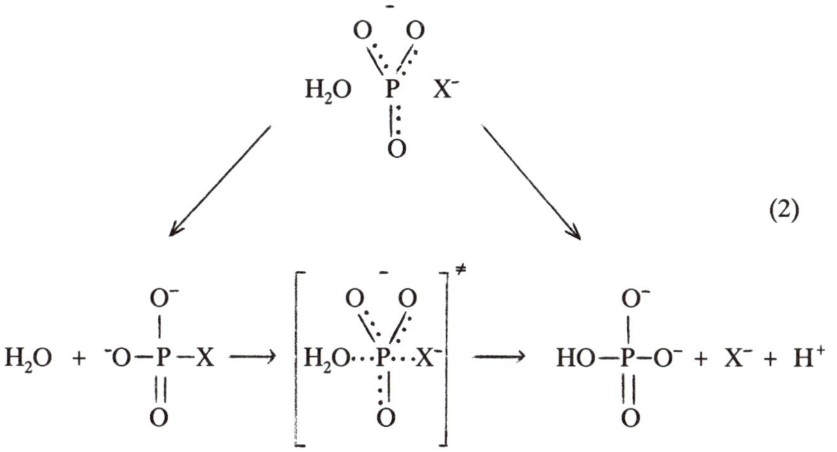

$$
\begin{array}{c}
\overset{-}{O} \quad O \\
H_2O \quad \overset{\cdots}{\underset{O}{P}} \quad X^- \\
\end{array}
\tag{2}
$$

$$
H_2O + {}^-O-\overset{\overset{O^-}{|}}{\underset{\underset{O}{\|}}{P}}-X \longrightarrow \left[H_2O\cdots\overset{\overset{\overset{-}{O}\; O}{\diagdown\diagup}}{\underset{\underset{O}{\vdots|}}{P}}\cdots X^- \right]^{\neq} \longrightarrow HO-\overset{\overset{O^-}{|}}{\underset{\underset{O}{\|}}{P}}-O^- + X^- + H^+
$$

Phosphoryl Transfer between Amines

The possibility that phosphoryl transfer between amines occurs through a metaphosphate intermediate was examined by determining the rate constants for reactions of phosphorylated pyridines with higher and lower pK compared with the leaving group (*12,13*). If the reaction proceeds by dissociation to pyridine and PO_3^-, followed by reaction with the acceptor pyridine (equation 3), there should be a change in the rate-limiting step for

$$\ce{\overset{\diagdown}{N} + {}^{=}O_3P-\overset{+}{\overset{\diagup}{N}}} \underset{k_{-1}}{\overset{k_1}{\rightleftharpoons}} \overset{\diagdown}{N} \overset{\overset{O^-}{\underset{|}{P}}}{\underset{\overset{\diagup\,\cdot\,\diagdown}{O\quad O}}{}} \overset{\diagup}{N} \overset{k_2}{\longrightarrow} \overset{+}{\overset{\diagdown}{N}}-PO_3^- + \overset{\diagup}{N}} \qquad (3)$$

pyridine nucleophiles with pK values that are higher and lower than the pK of the leaving group. With pyridines of higher pK than the leaving group the metaphosphate ion will almost always react with the attacking pyridine ($k_2 > k_{-1}$, equation 3), so that bond-breaking will be largely rate limiting and the rate will be almost independent of the pK of the attacking pyridine. However, if the attacking pyridine is less basic than the leaving group, the metaphosphate intermediate will generally be attacked by the leaving group to regenerate starting material ($k_{-1} > k_2$). Under these conditions, reaction with the attacking pyridine will be rate-limiting and the observed rate will depend on the basicity of the attacking pyridine.

Examination of this reaction series by Bourne and Williams (*12*) and by Skoog and Jencks (*13*) showed no evidence for curvature of the Brønsted-type plot of log k against the pK of the attacking pyridine with pyridines of higher and lower pK than the leaving group. Values of $ß_{nuc}$ describing the dependence of the rate on the pK of the attacking pyridine are small, close to 0.2, and the dependence on the pK of the leaving group is very large, with $ß_{lg}$ ~ -0.9. These values show that the transition state is strongly dissociative and resembles metaphosphate, as in the hydrolysis reaction.

Strong evidence in support of a concerted, coupled mechanism for this reaction, even though the transition state is strongly dissociative, is provided by the interaction coefficient p_{xy} (equation 4). An increase in the dependence

$$p_{xy} = \frac{\partial ß_{nuc}}{\partial pK_{lg}} = \frac{\partial ß_{lg}}{\partial pK_{nuc}} = 0.014 \qquad (4)$$

of the rate on the basicity of the nucleophile, $ß_{nuc}$, is observed with increasing pK_a of the leaving group. This corresponds to a coefficient of $p_{xy} = 0.014$. The same coefficient describes the change in the dependence of the rate on the leaving group, $ß_{lg}$, with changing pK_a of the nucleophile. This represents a "Hammond effect", or an expression of the "Bema Hapothle" (*14,15*), in which the sensitivity to the basicity of the nucleophile increases with poorer leaving groups. The observed increase in $ß_{nuc}$ from 0.17 to 0.22 as the pK of the leaving group increases from pK 5.1 to 9.0 corresponds to a p_{xy} coefficient of 0.014.

Similar behavior was observed previously by Jameson and Lawler for the reactions of a series of amines with phosphorylated pyridines (*8*) and by Kirby and Varvoglis (*16*) for the reactions of substituted pyridines with *p*-nitrophenyl phosphate and 2,4-dinitrophenyl phosphate. The values of β_{nuc} = 0.13 for the *p*-nitrophenyl and zero for the dinitrophenyl leaving group correspond to a p_{xy} coefficient of 0.043. These interaction coefficients show that both the nucleophile and the leaving group are involved in the rate-determining transition state, so that the reactions are concerted. Although metaphosphate is not an intermediate in the reaction, the fact that the transition state closely resembles metaphosphate indicates that the interaction of phosphorus with the attacking and leaving groups is weak in the transition state. It may be appropriate to compare metaphosphate to the proton - neither species exists as a free intermediate in water, but both are identifiable chemical species that are transferred from one nucleophilic reagent to another.

The Role of Nucleophile Solvation. The value of β_{nuc} = 0 for the reaction of substituted pyridines with 2,4-dinitrophenyl phosphate (*16*) is puzzling. If the value of β_{nuc} is a measure of the amount of the bond formation to the nucleophile in the transition state, this value might be taken to mean that there is no bond formation to the nucleophile in the transition state. This is obviously not the case, because there is a large increase in the rate of disappearance of the phosphate ester with increasing concentration of the nucleophile; the reactions follow simple second-order kinetics.

The answer to this question only became apparent when it was observed that the reactions of substituted quinuclidines with 2,4-dinitrophenyl phosphate, *p*-nitrophenyl phosphate and phosphorylated pyridine show a decrease in rate with increasing basicity of the attacking quinuclidine (*17*). These second-order reactions clearly cannot have a negative amount of bond formation with the nucleophile in the transition state, so that this result forced us to think harder about what could cause zero or negative slopes in Brønsted-type plots against the pK of the nucleophile.

The answer is that the amine nucleophile must lose solvating water that is hydrogen bonded to the lone-pair electrons of the nitrogen atom before it can react with the substrate. The strength of the hydrogen bond to the solvating water is expected to be stronger as the basicity of the amine increases, so that the concentration of the free amine will be smaller for more basic amines. The value of β for desolvation was estimated to be -0.2 for substituted quinuclidines (*17*). In a subsequent study the rate constants for dissociation of hydrogen-bonded water from the nitrogen atom of substituted quinuclidines were determined by examining inhibition of the exchange of protons on protonated quinuclidines by acid, according to the Swain-Grunwald mechanism. This work showed that the rate of dissociation of water decreases with increasing basicity of the amine. The decrease corresponds to a value of ß = 0.25 for the formation of a hydrogen bond between the substituted quinuclidine and water (18). Therefore, the observed values of ß for reactions of substituted quinuclidines should be corrected by 0.25 for reactions in which the ß value is small; the correction becomes progressively smaller as the observed ß value becomes larger (17).

Hydrolysis of Phosphates. The conclusion that phosphoryl transfer between amines is concerted does not exclude the formation of a metaphosphate intermediate in reactions with weaker nucleophiles, including water. Pyridine is a much stronger nucleophile than water and it is possible that phosphoryl transfer between pyridines is concerted because the metaphosphate ion does not have a significant lifetime in the presence of pyridine; the metaphosphate ion might exist for a short time in water. However, it is much more difficult to determine whether or not the mechanism is concerted for a reaction with water than for a reaction with pyridine.

Buchwald, Friedman and Knowles succeeded in preparing 2,4-dinitrophenyl phosphate in which the three free oxygen atoms on phosphate were labeled stereospecifically with different isotopes of oxygen. Solvolysis of this compound in methanol and analysis of the methyl phosphate product showed that the reaction had proceeded with inversion of configuration at phosphorus (19). This remarkable experiment supports a concerted bimolecular displacement mechanism, with no metaphosphate intermediate, for the solvolysis of 2,4-dinitrophenyl phosphate in methanol. However, it does not rigorously exclude a stepwise mechanism in which a metaphosphate intermediate with a very short lifetime is formed and reacts with methanol faster than it rotates, and it does not provide direct evidence for a bimolecular, concerted reaction with solvent water.

Racemization was observed for the same reaction in t-butanol. This may well represent the transient formation of a metaphosphate intermediate in this solvent. However, it is conceivable that racemization arises from slow proton removal from the phosphorylated tertiary butanol, so that the phosphoryl group is transferred to the hydroxyl group of another molecule of tertiary butanol, with resulting racemization (20).

If a series of compounds reacts by one reaction mechanism and another compound, B, reacts by a different mechanism, the rate constant for reaction of compound B is not expected to fall on the structure-reactivity correlation that fits the first series of compounds. If the reaction of compound B is fast enough to be observed, it will usually be faster than expected from the correlation of rate constants for the other mechanism.

Phosphorylated γ-picoline was found to react with a series of anionic oxygen nucleophiles in a bimolecular concerted reaction, with a dependence on the pK of the nucleophile that is described by $\partial \log k / \partial pK = \beta_{nuc} = 0.25$. Figure 1 shows that the bimolecular rate constant for the reaction of this compound with water falls on the correlation line that is defined by the rate constants for reaction of the anionic nucleophiles; there is no positive deviation that might suggest a different, monomolecular mechanism for the reaction with water (21). This result is particularly striking because it might be expected that an anionic substrate would react relatively slowly with anionic nucleophiles, compared with water, because of electrostatic repulsion. Evidently, electrostatic repulsion is not large at the anionic strength of 1.5 (KCl) in which these experiments were carried out. This result is consistent with reaction of phosphorylated γ-picoline with water through the same bimolecular, concerted mechanism that is followed for the anionic nucleophiles.

Figure 2 shows that there is also no positive deviation of the rate constant for reaction with water in a correlation of log k for the reactions of oxygen nucleophiles with the monoanions of phosphorylated γ-picoline and with methyl dinitrophenyl phosphate (21). Methyl dinitrophenyl phosphate reacts with RO⁻ by a concerted bimolecular substitution. The fit of the rate constant for water to this correlation is consistent with a concerted bimolecular reaction mechanism of water with phosphorylated γ-picoline.

The reactions of anionic nucleophiles with a series of phosphorylated pyridines show an increase in selectivity with decreasing reactivity of the phosphorylated pyridine, as the pK_a of the leaving pyridine is increased. This is shown in Figure 3, in which the ratios of the rate constants for reaction of oxygen nucleophiles and with water are plotted against the pK_a of the leaving pyridine (21). The sensitivity of the rate to the base strength of the nucleophile increases as the substrate becomes less reactive, with increasing pK_a of the leaving group, and there is a corresponding increase in the sensitivity of the rate to the leaving group, $-\beta_{lg}$, as the nucleophile becomes less reactive. This behavior corresponds to a positive coefficient of p_{xy} = 0.013 (equation 5). This

$$\frac{\partial \beta_{nuc}}{\partial pK_{lg}} = \frac{\partial \beta_{lg}}{\partial pK_{nuc}} = 0.013 \tag{5}$$

interaction between the nucleophile and the leaving group in the transition state shows that both the nucleophile and the leaving group are involved in the rate-limiting step of the reaction. This interaction is not expected for a dissociative reaction mechanism with a metaphosphate intermediate, because the nucleophile and the leaving group are not both reacting in the rate-limiting transition state of the stepwise mechanism. Therefore, this interaction provides evidence for a concerted reaction mechanism. Figure 4 shows the dependence of $-\beta_{lg}$ on the pK_a of the nucleophile for reactions of phosphorylated pyridine monoanions with oxygen nucleophiles. The value of $-\beta_{lg}$ decreases with increasing pK_a of the nucleophile with a slope of p_{xy} = 0.013, as described by eq 5. The fact that the value of β_{lg} for the reaction with water fits the correlation supports a concerted mechanism for the reaction with water. In fact, the interaction coefficient of p_{xy} = 0.013 for oxygen nucleophiles, including water, does not differ significantly from the value of p_{xy} = 0.014 for the reaction of pyridines with a phosphorylated pyridine (equation 4).

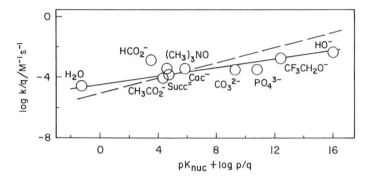

Figure 1.   Brønsted-type plot of log k against the pK$_a$ of oxygen nucleophiles for reactions with phosphorylated γ-picoline monoanion. The solid line has a slope of ß$_{nuc}$ = 0.13 and the dashed line has a slope of 0.25 (*19*).

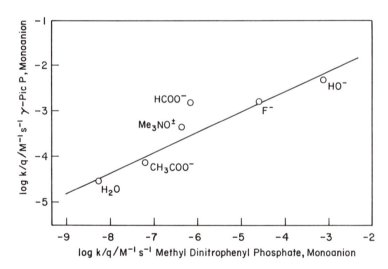

Figure 2.   Correlation of the rate constants for the reaction of nucleophilic reagents with phosphorylated γ-picoline monoanion and methyl 2,4-dinitrophenyl phosphate monoanion. The line has a slope of 0.51 (*19*).

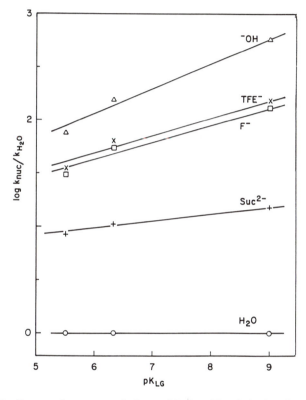

Figure 3. Brønsted-type correlations with the pK_a of the leaving group of the ratio of the second-order rate constants for reactions of phosphorylated pyridines with anionic nucleophiles and water (Suc = succinate, TFE = trifluoroethanol) *(19)*.

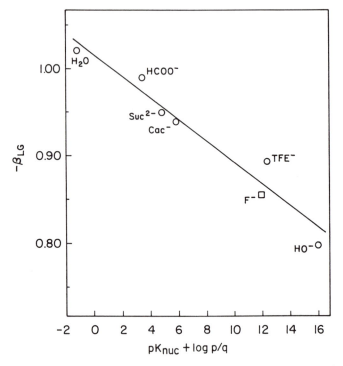

Figure 4. Plot of -ß_{lg} against the pK_a of the nucleophile for reactions of phosphorylated pyridine monoanions with oxygen nucleophiles (Suc = succinate, Cac = cacodylate, TFE = trifluoroethanol) (19).

If the reaction with water were stepwise, the value of β_{lg} would be expected to be -1.25, which corresponds to the value of β_{lg} that is expected for complete breaking of the P-N bond to form the unsolvated pyridine (equation 6) (*21*). The reaction of the putative metaphosphate intermediate with pyridine

$$\text{(6)}$$

$$+ \text{H}_2\text{O}$$

is expected to be faster than with water; i.e. $k_{-1} > k_2$, so that the rate-limiting step of the reaction would be the reaction of PO_3^- with water. The equilibrium constants for the hydrolysis of phosphorylated pyridines follow a value of $\beta_{eq} = -1.25$ and the same value is expected for the observed rate constant for the reaction of equation 6, in which k_2 is rate-limiting after the reversible cleavage of the phosphorylated pyridine (*21*). The observed value of $\beta_{lg} = -1.02$ is inconsistent with $\beta_{lg} = -1.25$. It is consistent with the observed values for the bimolecular reactions of pyridines and it supports a concerted bimolecular mechanism for the reaction with water.

It was suggested in 1961 (*7*) that the formation of pyrophosphate from the anions of acetyl phosphate and inorganic phosphate in concentrated aqueous sodium perchlorate solutions occurs through the generation of a metaphosphate intermediate that reacts with inorganic phosphate to give pyrophosphate. No formation of pyrophosphate is observed in the absence of added salt. However, reexamination of this result has shown that pyrophosphate is formed in a bimolecular reaction that is first-order with respect to acetyl phosphate and inorganic phosphate and fourth-order with respect to Na^+ (*22*). This result suggests that the reaction occurs through bimolecular displacement, with binding of the reacting anions by sodium ions. No pyrophosphate formation is observed at low salt concentrations because of electrostatic repulsion between the reacting anions.

We can conclude that phosphate esters and other phosphate compounds react with water through bimolecular substitution in a concerted S_N2, or A_ND_N, mechanism with no metaphosphate intermediate. The appearance of the transition state is that it resembles metaphosphate monoanion, but the reality of the mechanism is that the reaction is a one-step bimolecular substitution. The metaphosphate ion can be formed in the gas phase (*23*) and there is evidence that metaphosphate can exist briefly in non-nucleophilic solvents (*11,24*). The reason that it is not an intermediate in water is presumably that there is no significant barrier for its reaction with water.

When the bond to the leaving group breaks, a water molecule attacks the incipient metaphosphate ion before bond breaking is complete, so that no intermediate is formed. The concerted mechanism is enforced, because a stepwise mechanism is not possible.

Literature Cited

1. Butcher, W.; Westheimer, F. H. *J. Am. Chem. Soc.* **1955**, *77*, 2420.
2. Barnard, P. W. C; Bunton, C. A.; Llewellyn, D. R.; Oldham, K. G.; Silver, B. L.; Vernon, C. A. *Chem. Ind. (London)* **1955**, 760.
3. Desjobert, A. *C. R. Hebd. Seances Acad. Sci.* **1947**, *224*, 575.
4. Desjobert, A. *Bull. Soc. Chim. Fr.* **1947**, *14*, 809.
5. Bailly, M. C. *Bull. soc. Chim. Fr.* **1942**, *9*, 314, 340, 421.
6. Kirby, A. J.; Varvoglis, A. G. *J. Am. Chem. Soc.* **1967**, *89*, 415.
7. DiSabato, G.; Jencks, W. P. *J. Am. Chem. Soc.* **1961**, *83*, 4400.
8. Jameson, G. W.; Lawler, J. M. *J. Chem. Soc. B* **1970**, 53.
9. Gorenstein, D. G.; Lee, Y.-G.; Kar, D. *J. Am. Chem. Soc.* **1977**, *99*, 2264.
10. DiSabato, G.; Jencks, W. P.; Whalley, E. *Can. J. Chem.* **1962**, *40*, 1220.
11. Westheimer, F. H. *Chem. Rev.* **1981**, *81*, 313.
12. Bourne, N.; Williams, A. *J. Am. Chem. Soc.* **1984**, *106*, 7591.
13. Skoog, M. T.; Jencks, W. P. *J. Am. Chem. Soc.* **1984**, *106*, 7597.
14. Hammond, G. S. *J. Am. Chem. Soc.* **1955**, *77*, 334.
15. Jencks, W. P. *Chem. Rev.* **1985**, *85*, 511.
16. Kirby, A. J.; Varvoglis, A. G. *J. Chem. Soc. B* **1968**, 135.
17. Jencks, W. P.; Haber, M. T.; Herschlag, D.; Nazaretian, K. L. *J. Am. Chem. Soc.* **1986**, *108*, 479.
18. Berg, U.; Jencks, W. P. *J. Am. Chem. Soc.* **1991**, *113*, 6997.
19. Buchwald, S. L.; Friedman, J. M.; Knowles, J. R. *J. Am. Chem. Soc.* **1984**, *106*, 4911-4916.
20. Friedman, J. M.; Freeman, S.; Knowles, J. R. *J. Am. Chem. Soc.* **1988**, *110*, 1268.
21. Herschlag, D.; Jencks, W. P. *J. Am. Chem. Soc.* **1989**, *111*, 7679.
22. Herschlag, D.; Jencks, W. P. *J. Am. Chem. Soc.* **1986**, *108*, 7938.
23. Harvan, D. J.; Hass, J. R.; Busch, K. L.; Bursey, M. M.; Ramirez, F.; Meyerson, S. *J. Am. Chem. Soc.* **1979**, *101*, 7409.
24. Satterthwaite, A. C.; Westheimer, F. H. *J. Am. Chem. Soc.* **1981**, *103*, 1177.

RECEIVED December 17, 1991

Chapter 9

Ring Fragmentation Approaches to Low-Coordinate Phosphorus Species

L. D. Quin, S. Jankowski, G. S. Quin, A. Sommese, J.-S. Tang, and X.-P. Wu

Department of Chemistry, University of Massachusetts, Amherst, MA 01003

The photochemical fragmentation of phosphine oxides with the 2-phosphabicyclo[2.2.2]octene framework constitutes a useful method for the generation of methylene(oxo)phosphoranes, $R-P(O)=CH_2$. The previously uncharacterized and highly reactive P-phenyl and P-methyl derivatives were prepared and trapped with alcohols as phosphinates. The fragmentation also occurred when alkoxy substituents were present on phosphorus, but significant amounts of dialkyl phosphites were formed as by-products. Some new observations on the photochemical fragmentation of the 7-phosphanorbornene system have also been made. Contrary to literature reports, this system is resistant to fragmentation and does not extrude the 2-coordinate phosphoryl species. Fragmentation with loss of the phosphorus bridge does occur readily when an alcohol is also present in the medium. The data support a mechanism in which an initial photo-promoted addition reaction of the 7-phosphanorbornene and the alcohol occurs to form a 5-coordinate species, which undergoes the fragmentation. Thermal fragmentation of the 7-phosphanorbornenes is slow, but does appear to release the 2-coordinate phosphoryl species.

We have demonstrated in several recent studies (*1-4*) that certain unsaturated bicyclic frameworks containing phosphorus functions can be fragmented in solution to release a simple phosphorus-containing moiety. The nature of the bond breaking process leaves phosphorus in these fragments in a condition of low coordination (number of attached atoms) for the particular oxidation state concerned, and the method has proved of practical value for the synthesis of several low coordination species. As a general rule, low-coordination phosphorus species are of very high reactivity and some types cannot be directly observed or isolated even when stabilization through kinetic or thermodynamic effects is attempted. In such cases, their intermediacy in chemical processes is usually demonstrated by trapping reactions. We continue to explore applications of the ring fragmentation approach in the low-coordination chemistry of phosphorus, and in this paper we will describe the first extension of our methods to the generation of two types of the 3-coordinate methylene(oxo)-

0097–6156/92/0486–0115$06.00/0

phosphorane species (Y-P(O)=CH$_2$), as well as to the 2-coordinate phosphenite species of structure RO-P=O. In the latter study, we have been led into a re-examination of some published work on the photochemical generation of related types of 2-coordinate species from heterocycles and will provide a different interpretation of the data obtained.

Present Scope of the Heterocycle Fragmentation Method

Our previous work has established techniques for the generation of four different types of low-coordination phosphorus species by ring fragmentation [some contributions made by other laboratories are summarized in a recent book (5)].

The Generation of Metaphosphoric Acid Derivatives. These are among the most reactive of low-coordinate phosphorus species, and are of great importance as reactive intermediates in some phosphorylation processes. The metaphosphate ion receives study as a possible intermediate in biological processes, a concept introduced in 1955 by the laboratories of F. H. Westheimer (6) and of C. A. Bunton (7). The 2,3-oxaphosphabicyclo[2.2.2]octene ring system, as embodied in derivatives of the basic structure **1**, is a valuable source, on fragmentation, of several types of metaphosphoric acid derivatives (1-4). A typical synthetic approach to such compounds is shown in Scheme 1.

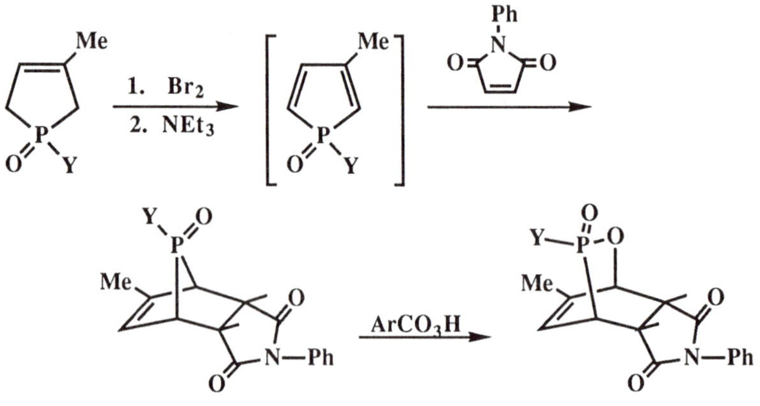

Scheme 1

These compounds undergo retrocycloaddition (8) on heating in inert solvents, and are subject to fragmentation on irradiation at 254 nm as well. Both are high-yield processes, easily monitored by ^{31}P NMR spectroscopy. The released metaphosphates are routinely trapped by the addition reaction with alcohols or amines, or by electrophilic substitution on the pyrrole ring. In some cases, trapping has been accomplished by reaction with the OH groups on the surface of suspended silica gel (9); the resulting phosphorylated silica gel is finding use in special HPLC applications (10).

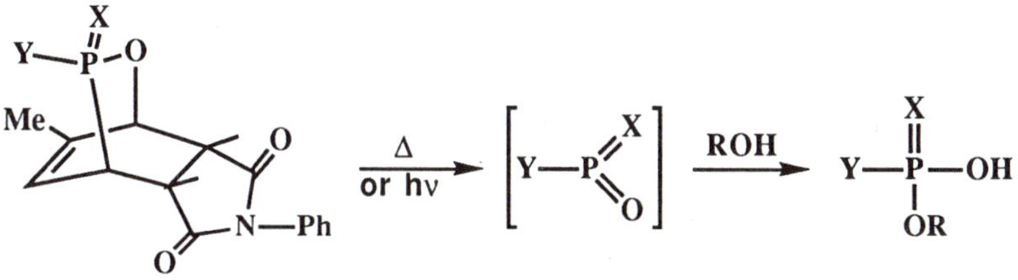

2, Y = OR, X = O
3, Y = OR, X = S
4, Y = NR$_2$, X = O

The value of the new approach to metaphosphoric acid chemistry is seen in the range of the species that can be generated, which includes so far alkyl metaphosphates (2), alkyl metathiophosphates (3), and metaphosphoramides (4). The latter two are of a structural type never before synthesized.

Metaphosphonic Acid Anhydrides. In a related process, the fragmentation of compounds **5** (*11*) and **7** (*1*) by thermal or photochemical means releases fragments (**6** or **8**) that are the metaanhydrides of phosphonic acids. The properties of these species are very similar to those of the metaphosphates.

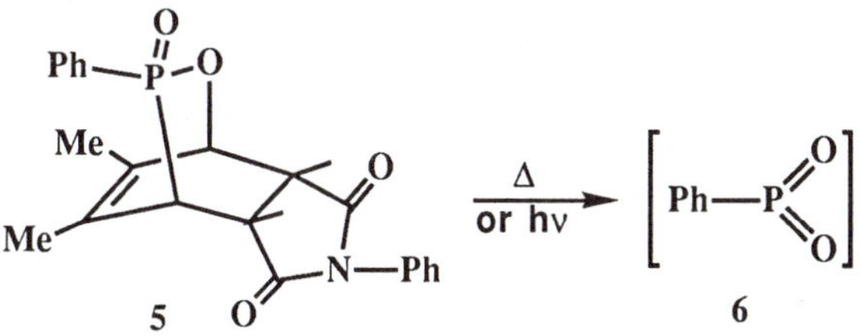

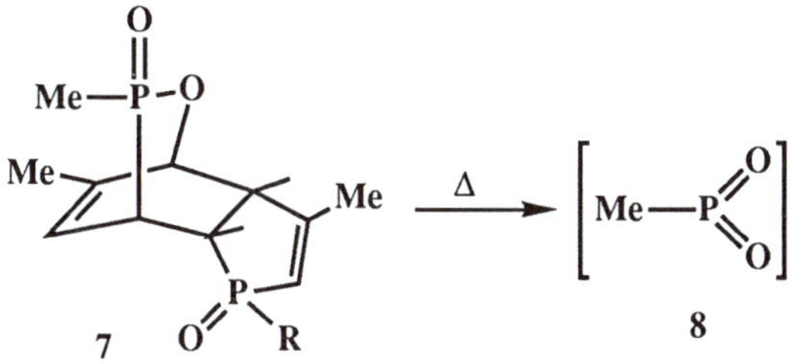

Phosphaalkenes. There is a much larger literature on 2-coordinate species containing the P-C double bond, and many examples are known of stabilized derivatives (*5*). The thermal fragmentation process that we have devised for the generation of phosphaalkenes employs the 2-phosphabicyclo[2.2.2]octadiene ring system. Here the synthetic advantage is in the ability to generate highly reactive, *unstabilized* species in a simple, unimolecular reaction in an inert solvent. The simple species **11** was generated for the first time by the ring fragmentation approach, using compounds **9** (*2*) or **10** (*12*). It was characterized by the addition reactions that occurred with alcohols or dienes. The phenyl analogue Ph-P=CH$_2$ was prepared similarly (2).

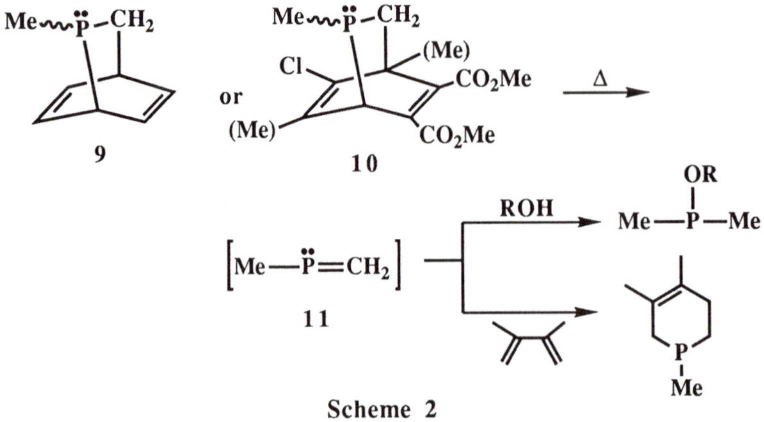

Scheme 2

Thionophosphenous Acid Esters and Amides. Another ring system that we have found (13) to serve as a precursor of low-coordination species is the 7-phosphabicyclo[2.2.1]heptene system (also known as 7-phosphanorbornene). We have used this sytem for the thermal generation of derivatives of two previously unknown 2-coordinate species, the thionophosphenous amides **12** and the thionophosphenites **13** (Scheme 3). These species are trapped by a cycloaddition reaction with a diene to form the 5-membered 3-phospholene ring, as shown for **12**.

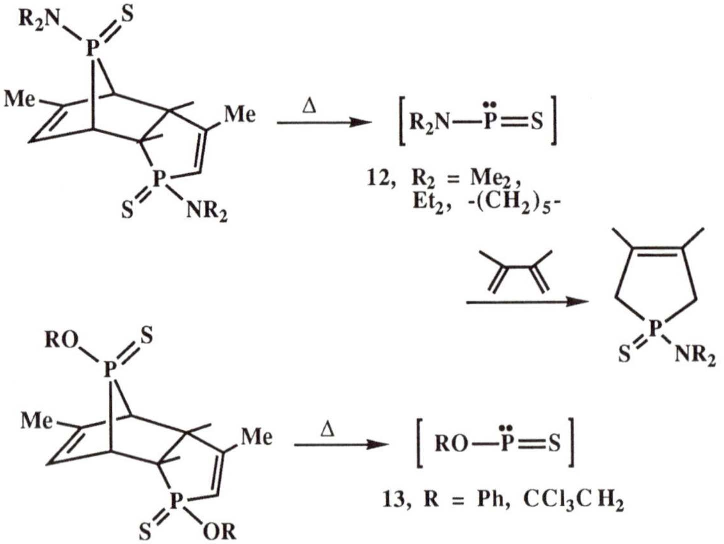

Scheme 3

The Generation of Methylene(oxo)phosphoranes by the Ring Fragmentation Approach

Methylene(oxo)phosphoranes are formally anhydrides of phosphinic acids; a recent review (*14*) has summarized the existing knowledge on methods of generation of the species and their reactivity. Two cases (**14** and **15**) are known of compounds with sufficient stability to allow isolation, but as in the case of other metaanhydride substances, they are usually far too reactive to allow observation or isolation.

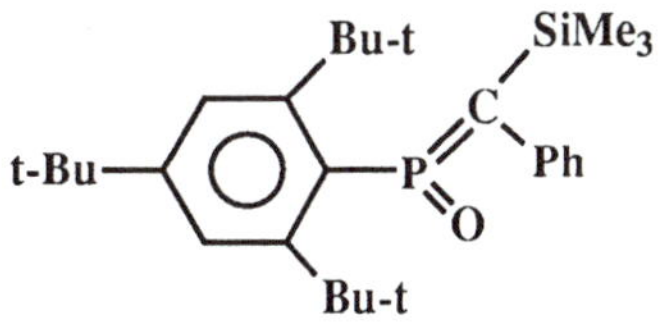

14

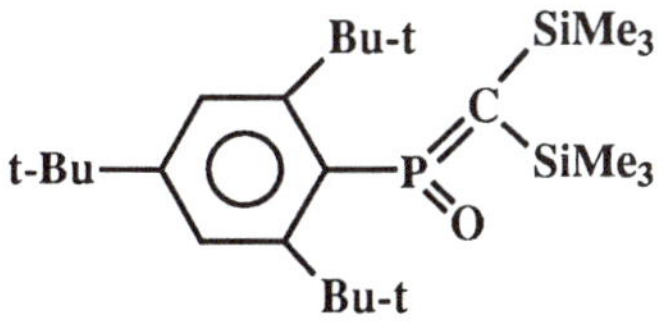

15

We have now been able to show that the ring fragmentation approach is quite successful for the generation of methylene(oxo)phosphoranes. The precursors for the P-phenyl (**18**) and P-methyl (**19**) derivatives are **16** and **17**, respectively; both are obtained as 4-component isomer mixtures from the synthetic route used (Scheme 4), but all isomers fragment to the methylene(oxo)phosphorane. The P-methyl compounds (**17**) have been used (*12*) as precursors, after silane reduction, of the 2-coordinate phosphaalkene **11** (Scheme 2).

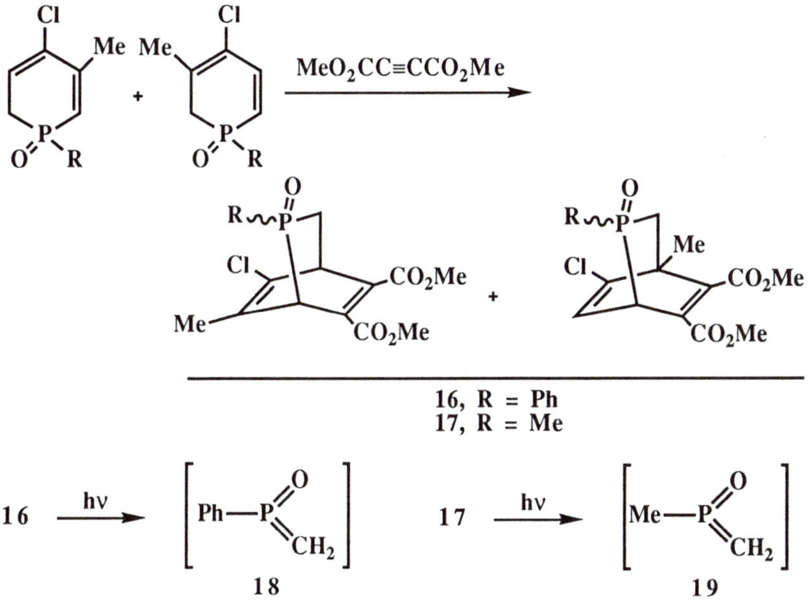

16, R = Ph
17, R = Me

Scheme 4

Neither of these simple methylene(oxo)phosphines has been generated previously, and it is doubtful if the few existing synthetic methods (14) could be applicable to their formation. We first found that the thermal degradation of the ring system is not easy; for example, 16 requires about 10 hours at 150°C (xylene, sealed tube) to achieve about 90% fragmentation. However, we found that the photochemical fragmentation occurred very readily on irradiation of acetonitrile solutions either with light of the specific length of 254 nm as supplied in an actinometric apparatus, or with broader-spectrum 254 nm light as supplied by a medium-pressure Hanovia lamp. Small-scale photolyses (20-50 mg) are conveniently conducted in quartz NMR tubes held closely to the irradiation unit, and are generally complete in 1-2 hours. That species 18 and 19 were indeed released was proved by their trapping as phosphinates (20 and 21) when an alcohol was included in the photolysis medium. The reactions were very clean, and the mixtures gave single ^{31}P NMR signals, with the expected shifts for the phosphinates (20, δ (CDCl$_3$) 42.3; 21, δ (CDCl$_3$) 52.6).

The identity of the phosphinate 21 was confirmed by NMR comparisons with the known compound; GC-MS was used to confirm the identity of the P-phenyl compound 20. These results clearly demonstrate the value of the ring fragmentation approach for the generation of simple methylene(oxo)phosphoranes.

We have proceeded to apply the ring fragmentation approach to the generation of yet another type of previously unknown 3-coordinate phosphoryl derivative, where the methylene(oxo)phosphorane has an alkoxy substituent on phosphorus. Such structures (26-29) can be considered as anhydrides of monoalkyl alkylphosphonates. The precursors for these new species have the structures 22-25 (isomer mixtures); they are known compounds (12) from our previous work and are synthesized by the cycloaddition reaction of dihydrophosphinines with dimethyl acetylenedicarboxylate, as used in Scheme 4.

When these bicyclic phosphinates were irradiated at 254 nm in acetonitrile containing ethanol as a trapping agent, fragmentation was complete in 1-2 hours. Examination of the reaction solutions by ^{31}P NMR (CDCl$_3$) showed that the major products were the expected ethyl alkyl methylphosphonates; in one case (30, R = Et, δ 30.7), the identity was confirmed by ^{31}P NMR comparison with a known sample. However, in

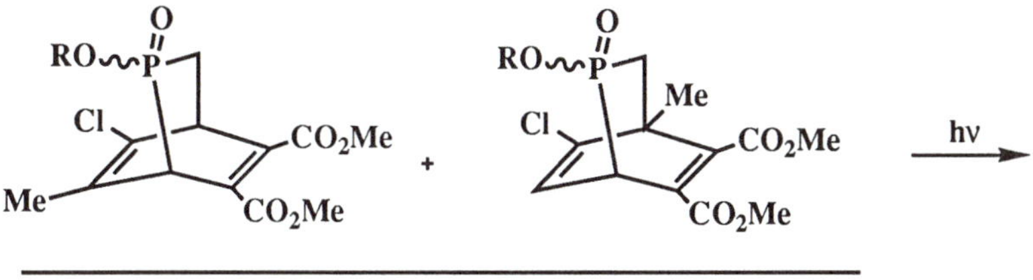

22, R = Me
23, R = Et
24, R = i-Pr
25, R = n-Pr

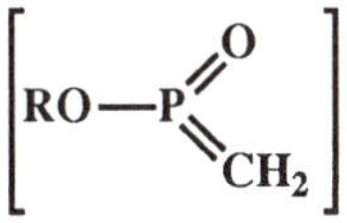

26, R = Me
27, R = Et
28, R = i-Pr
29, R = n-Pr

all of these photolyses, there was a significant amount of a by-product with a shift around δ 7-10 that accompanied the major product. These by-products contained a hydrogen atom attached to phosphorus, as was readily evident when the ^{31}P NMR spectra were obtained with proton coupling present. The coupling constants and the chemical shifts were consistent with an assignment of a dialkyl phosphite structure (**31**) to the products. For example, the by-product in the photolysis of the ethyl ester **23** in the presence of ethanol had δ ^{31}P NMR 7.5 with $^{1}J_{PH}$ = 690 Hz; diethyl phosphite gave the same spectrum. We established that the by-product does not originate from P-C bond cleavage of the major methylphosphonate reaction product, by showing that the same photolysis conditions applied to a sample of the phosphonate **30** caused no change. It must be concluded that the by-product originates from a second reaction that occurs with the bicyclic starting material. This would represent the only instance in all of the many types of ring fragmentation photolyses we have performed where a reaction other than the desired extrusion of the P-containing bridge as a low-coordinate species has taken place to a significant extent. There is no obvious reason at this time why the P-C bond in the bicyclic esters **22-25** has some lability, while the same bond in the phosphine oxides with the same framework (**16, 17**) is stable.

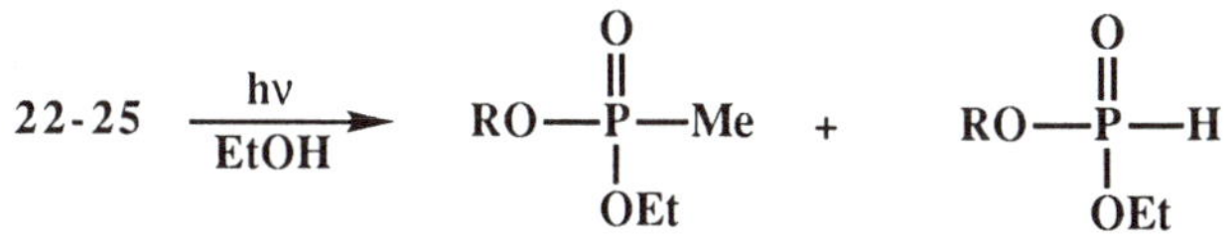

30   31

It may be possible to employ a different wavelength in the photolysis that will give greater selectivity for the desired extrusion process, but we have not yet performed such experiments.

New Observations on the Fragmentation of the 7-Phosphanorbornene System

The literature on the generation of 2-coordinate phosphoryl and thiophosphoryl species, which has recently been reviewed (*15*), suggests that they may be generated in useful fashion by the photochemical fragmentation of the 7-phosphanorbornene ring system. The photochemical method is potentially valuable for the attempted generation of such species at low temperature, where an anticipated longer lifetime might permit the previously unaccomplished feat of their direct spectral observation in solution. An additional factor that could impart kinetic stabilization would be the presence on phosphorus of a sterically large substituent, a device used throughout studies on low-coordinate species. This line of thought led us to synthesize phosphinates in the 7-phosphanorbornene system with the highly space-demanding 4-methyl-2,6-di-*tert*-butylphenyl (**32**) and adamantyl (**33**) substituents, for use in photolyses at temperatures around -75°C as we had employed earlier in the stabilization of metaphosphoramide derivatives (*16*).

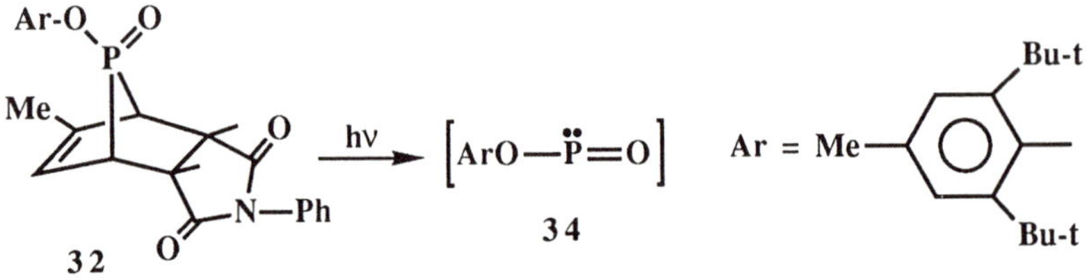

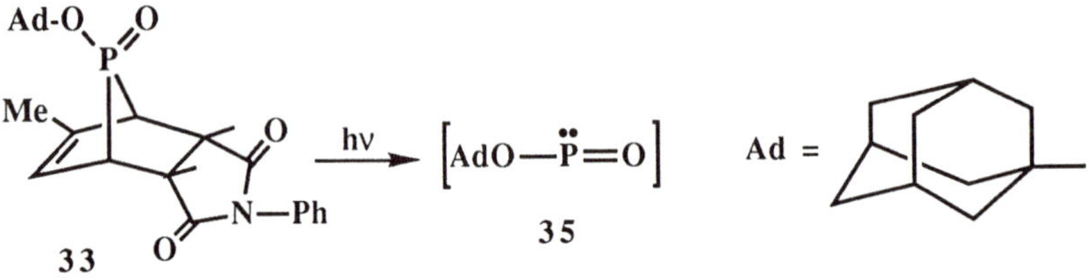

We were surprised to find, however, that photolysis (254 nm) of these compounds was quite slow and was incomplete even after several hours of irradiation; more importantly, no condensation products to suggest that a monomeric phosphenite had been released were observed. Thus, it is known (*17*) for 4-alkyl-2,6-di-*tert*-butylphenyl phosphenites, when generated by a quite different method involving partial hydrolysis of the phosphorodichloridite **36**, that a cyclic trimer (**38**) of the monomeric phosphenite **37** is formed. The trimer on being heated is partly converted to a cyclic dimer (**39**) (*18*). These compounds give highly characteristic ^{31}P NMR spectra (**38**, R = Me, δ 127.9 (t, J$_{PP}$ = 10 Hz) and δ 120.0 (d, J$_{PP}$ = 10 Hz); **39**, R = Me, δ = 176.5).

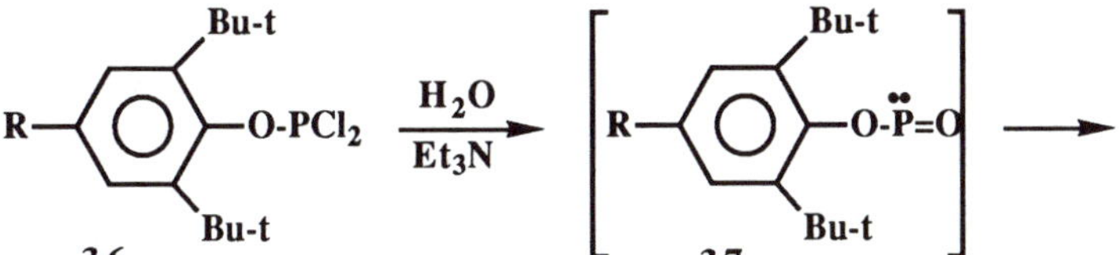

No signals for the dimer or trimer (or a monomer) were seen in the solution following the photolysis of **32**. Always, however, was present the signal for some 4-methyl-2,6-di-*tert*-butyl phosphonate [**40**, δ 5.6, $^1J_{PH}$ = 720 Hz, as found for a known sample (*17*)], which could only arise from the presence of a trace of water in the medium. Attesting to the difficulty in obtaining complete removal of water, we observed a small amount of product **40**, along with a product that could be an anhydride of this substance (**41**, δ 4.0, $^1J_{PH}$ = 690 Hz), even when vacuum line techniques were used for drying and transfers. The same result was obtained for the adamantyl derivative; never was there any indication of the formation of the dimer or trimer of the expected phosphenite, and only the phosphonate **42** was observed (δ 4.0, $^1J_{PH}$ = 698 Hz). We then discovered that, when ethanol was present in the photolysis media, the decomposition of the 7-phosphanorbornenes **32** and **33** proceeded quite rapidly and gave complete conversion in 1-2 hours of the phosphorus into the form of the phosphonates **43** and **44**.

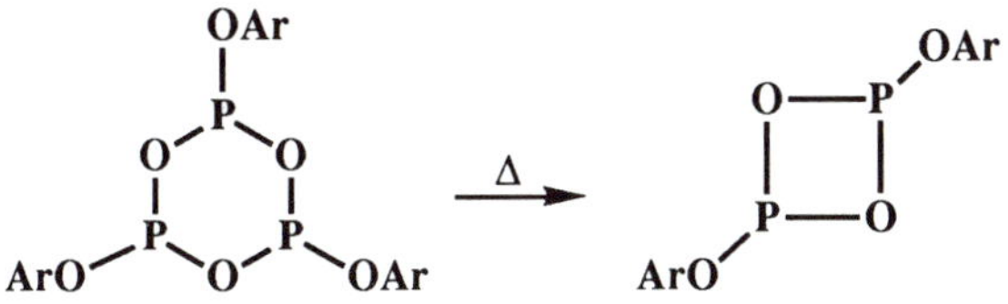

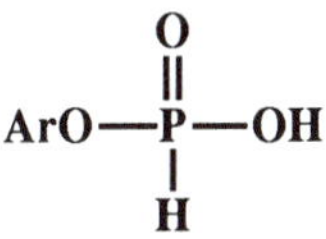

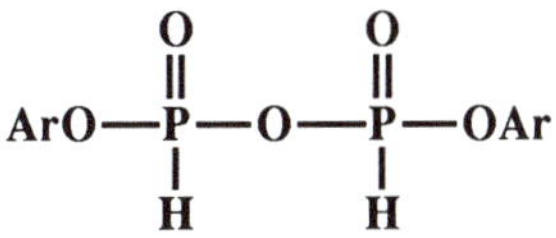

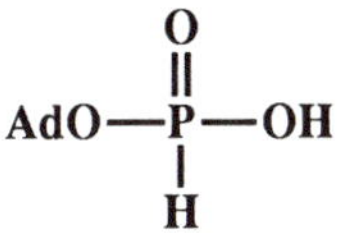

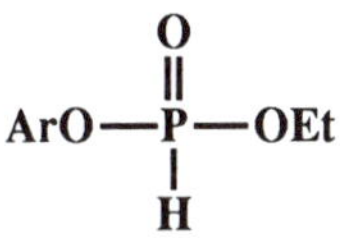

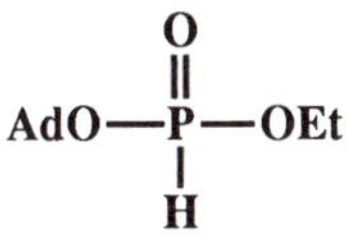

Addition of water to the solution before photolysis also caused a more rapid destruction of the starting material. Puzzled by these results, we re-examined the original literature (summarized in ref. 15) on the photolysis of 7-phosphanorbornene derivatives and noticed that in no case had the photolytic fragmentation been performed in the absence of a hydroxylic species as a trapping agent. We then attempted experiments resembling those in the literature (*19*) on the photolysis of P-phenyl derivatives of this ring system, so as to generate the presumably known species Ph-P=O and Ph-P=S. Compounds **45** and **46** were photolyzed in 1,2-dichloroethane in the presence of methanol, and each gave a rapid fragmentation with complete conversion of phosphorus into the phosphonate (**47**, δ 26.8, $^1J = 568$ Hz) or thionophosphonate (**48**, δ 67.7, $^1J_{PH} = 528$ Hz) form, as expected.

However, when no alcoholic trapping agent was present, no fragmentation of the heterocyclic system occurred. Furthermore, the fragmentation of **45** was found to be accelerated by increasing amounts of added alcohol. The conclusion seems inescapable that the alcoholic trapping agent is intimately involved in the initiation of the fragmentation process, and that if this is so, there can be no release of a free 2-coordinate phosphenite species under these conditions. Literature reports (*15*) therefore seem to require reconsideration with regard to the proposal that the formation of alcohol reaction products proves that a 2-coordinate intermediate was involved. The literature similarly records the fragmentation of some monocyclic 3-phospholene oxides in the presence of alcohols, and suggests that these compounds too decompose with the formation of 2-coordinate intermediates (*20,21*). We have not yet checked these experiments, but it seems likely that here too no fragmentation reaction will occur in the absence of an alcohol.

We propose, to explain our observations, that a species with 5-coordinate phosphorus is formed as an intermediate in the irradiation of a 7-phosphanorbornene oxide in the presence of a hydroxylic reactant. The tendency of the 7-phosphanorbornene system to achieve the 5-coordinate state is strong, and many examples are known (*22*). (We later found that this had been noted in a recent review (*23*) article, and was suggested to be implicated in the photolytic fragmentation, without experimentation.) The highly-contracted C-P-C internal bond angle, which is around 83° rather than the customary approximately tetrahedral value, is responsible for this effect. The 5-coordinate state is usually unstable and collapses with elimination of the phosphorus bridge; it is proposed that the 5-coordinate species achieved on UV irradiation collapses in this manner to give the phosphorus product observed. The product is the same as would be formed from a 2-coordinate intermediate on reaction with the hydroxylic reactant, but since the 4-coordinate 7-phosphanorbornene oxide alone shows no sensitivity to UV irradiation, this pathway can be ruled out. While our proposal for the formation of a 5-coordinate intermediate on irradiation lacks photochemical detail at present, it is consistent with the facts that are now available. We do note that the experimental quantum yield for the reaction of **45** with ethanol at various concentrations is about 0.4, indicating a reasonably efficient passage of the excited state to final product.

We noted that a small amount of anhydride **41** was formed in the photolysis of **32** when no added hydroxylic reactant was present. Presumably a trace of water is responsible for the small amount of fragmentation that occurred. The anhydride may arise from the photo-addition of the initially formed phosphonic acid **40** to the phosphoryl group, followed by fragmentation.

In contrast to the photochemical process, the thermal fragmentation of the 4-methyl-2,6-di-*tert*-butylphenyl phosphinate **32** does appear to occur with release of the 2-coordinate phosphenite. When the decomposition was conducted on a neat sample of **32** at 250°C at a pressure of 0.03 mm in a Kugelrohr apparatus, substantial amounts of the trimer **38** as well as some of dimer **39** were collected in the cold receivers. Authentic samples prepared by the reported procedures (*17,18*) gave identical ³¹P NMR spectra. There was also noted a small (about 5% by intensity measurements) signal at δ 240, which disappeared when 2-propanol was added to the product. The small signal was also present in the pyrolysate of the trimer performed according to the literature (*18*). It is tempting to attribute this signal to the highly desired monomeric species **34**, but additional proof of this structure is required before it can be accepted. The fragmentation of **32** was also conducted in solution at 150°C in toluene in a sealed tube. The reaction was quite slow, and was only about 60% complete after 100 hours. The solution contained signals in the expected region for the trimer (δ 120-130).

These new results reiterate the need for great caution in the use of trapping agents to prove a transient intermediate in a mechanism. It must be established that the agent does not *cause* the observed reaction. This has been a point of concern in the processes we have described for the generation of 3-coordinate metaphosphate species, but we have resolved it by performing kinetics experiments (*8*) that show the thermal reactions to be independent of the amount or identity of the alcohol trapping agent, and by measuring quantum yields to prove that the photochemical excited state is achieved without involvement of the alcohol.

Literature Cited

1. Quin, L. D.; Marsi, B. G. *J. Am. Chem. Soc.* **1985**, *107*, 3389.
2. Quin, L. D.; Hughes, A. N.; Pete, B. *Tetrahedron Lett.* **1987**, *28*, 5783.
3. Quin, L. D.; Pete, B.; Szewczyk, J.; Hughes, A. N. *Tetrahedron Lett.* **1988**, *29*, 2627.
4. Quin, L. D.; Sadanani, N. D.; Wu, X.-P. *J. Am. Chem. Soc.* **1989**, *111*, 6852.
5. *Multiple Bonds and Low Coordination in Phosphorus Chemistry*; Regitz, M.; Scherer, O. J., Eds.; George Thieme Verlag: Stuttgart, 1990.
6. Butcher, W. W.; Westheimer, F. H. *J. Am. Chem. Soc.* **1955**, *77*, 2420.
7. Barnard, D. W. C.; Bunton, C. A.; Llewellyn, D. R.; Oldham, K. G.; Silver, B. L.; Vernon, C. A. *Chem. Ind. (London)* **1955**, 760.
8. Jankowski, S.; Quin, L. D. *J. Am. Chem. Soc.*, in press.
9. Quin, L. D.; Wu, X.-P.; Breuer, E.; Mahajna, M. *Tetrahedron Lett.* **1990**, *31*, 6281.
10. Quin, L. D.; Wu, X.-P.; Quin, G. S.; Dickinson, L. C.; Uden, P. C.; Stewart, C. Northeast Regional Meeting, American Chemical Society, Amherst, MA, June 24, 1991; Abst. 134.
11. Quin, L. D.; Wu, X.-P. *Heteroatom Chem.* **1991**, *2*, 359.
12. Quin, L. D.; Tang, J.-S.; Keglevich, G. *Heteroatom Chem.* **1991**, *2*, 283.
13. Quin, L. D.; Szewczyk, J. *J. Chem. Soc., Chem. Commun.* **1986**, 844.
14. Heydt, H. In *Multiple Bonds and Low Coordination in Phosphorus Chemistry*; Regitz, M.; Scherer, O. J., Eds.; Georg Thieme Verlag: Stuttgart, 1990, pp 381-386.
15. Quin, L. D. *ibid.*, pp 352-366.
16. Quin, L. D.; Bourdieu, C.; Quin, G. S. *Tetrahedron Lett.* **1990**, *45*, 6473.
17. Chasar, D. W.; Fackler, J. P.; Mazany, A. M.; Komoroski, R. A.; Kroenke, W. J. *J. Am. Chem. Soc.* **1986**, *108*, 5956.

18. Chasar, D. W.; Fackler, J. P., Jr.; Komoroski, R. A.; Kroenke, W. J.; Mazany, A. M. *J. Am. Chem. Soc.* **1987**, *109*, 5690.
19. Holand, S.; Mathey, F. *J. Org. Chem.* **1981**, *46*, 4386.
20. Tomioka, H.; Hirano, Y.; Izawa, Y. *Tetrahedron Lett.* **1974**, 1865.
21. Tomioka, H.; Izawa, Y. *J. Org. Chem.* **1977**, *42*, 582.
22. Quin, L. D. *Rev. Heteroatom Chem.* **1990**, *3*, 39.
23. Mathey, F. *Chem. Rev.* **1988**, *88*, 429.

RECEIVED October 21, 1991

Chapter 10

Phosphoranide Anions (10-P-4) with Electronegative Apical and Electropositive Equatorial Ligands Coordinated to Metals To Form Symmetrical 10-P-5 Species

Chester D. Moon, Suman K. Chopra[1], and J. C. Martin

Department of Chemistry, Vanderbilt University, Nashville, TN 37235

Hypervalent species, such as the 10-P-4 phosphoranide potassium anion **1**, are most stable when the apical ligands are very electronegative with two CF_3 groups adjacent to the apical oxygens. They also require electropositive ligands in the equatorial positions, such as the two carbons. These stabilize the anions on the apical positions and of **1**, the cation on the hypervalent main-group element at the center. Compound **1**, with these proper ligands, makes the phosphoranide anion coordinate just to the phosphorus, while many other 10-P-4 species coordinate by metal to both phosphorus and the apical ligand. The gold metal species **5** has an X-ray structure here that makes it understandable. The five-membered rings containing the apical and the equatorial groups are also helpful to make the phosphoranide **1** stable.

Phosphoranide anion **1** is a stable 10-P-4 (*1*) hypervalent anionic species that could be considered as a transition state for a nucleophilic substitution reaction at a three coordinate phosphine center (*2*). The stability of **1** (*3*) can be attributed to several factors.

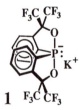

1

To begin with, the three-center four-electron (3c-4e) O-P-O bond is stabilized by placing the electronegative hexafluorocumyl groups adjacent to the oxygens at the apical positions of the pseudo-trigonalbipyramidal geometry (Ψ-TBP) of **1**.

[1]Current address: Colgate Palmolive Company, 909 River Road, Piscataway, NJ 08855

0097–6156/92/0486–0128$06.00/0

These electronegative groups serve to stabilize the negative charge build up at the apical positions. This negative charge build up is due to two electrons of the hypervalent bond being placed into the non bonding orbital of the (3c-4e) bond as shown in the orbital diagram below. Placing two electrons in this orbital clearly would put considerable electron density in the terminal positions of the hypervalent bond.

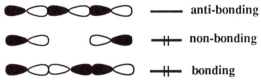

Orbital Diagram for (3c-4e) Hypervalent Bonding

Although there could be some phosphorus d-orbitals contributing to 10-P-4 or 10-P-5 species, these are in a high enough position to have a very little amount of an electron in the d-orbitals. This hypervalent bonding was suggested by Musher (4) and it was found by others to have a very small amount of an electron in the d-orbitals, as suggested by Musher and by others (5) with calculations.

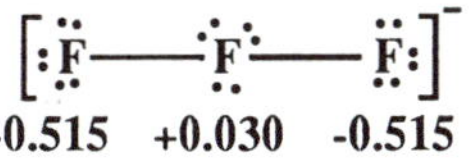

$$\left[:\ddot{F}\!\!-\!\!\cdot\ddot{F}\cdot\!\!-\!\!\ddot{F}: \right]^{-}$$
-0.515   +0.030   -0.515

Calculated Charges for the 10-F-2 Trifluoride Anion

The reactions of F^- with F_2 formed $(F_3)^-$ with reasonable evidence for the formation of the 10-F-2 species for the first-row element (6). Calculations performed on the 10-F-2 trifluoride anion (7) confirm that there is considerable negative charge build up at the apical positions of the hypervalent bond. It is also evident from these calculations that there is considerable positive charge build up on the central position of the hypervalent bond. Placing electropositive groups on the central position therefore serves to stabilize the hypervalent bond. In phosphoranide **1** there are two carbons and an electron pair that are equatorial, stabilizing the positive charge build up on phosphorus. Another factor that stabilizes the hypervalent bond is the connection of the apical and equatorial positions of the (Ψ-TBP) geometry by a five-membered ring. This stability is plainly seen when comparing the stability of sulfuranes **2** and **3**. Acyclic sulfurane **2** is very hygroscopic and is rapidly hydrolyzed by atmospheric moisture. In contrast, the bicyclic ligands of sulfurane **3** are not hydrolyzed even in acidic water (8).

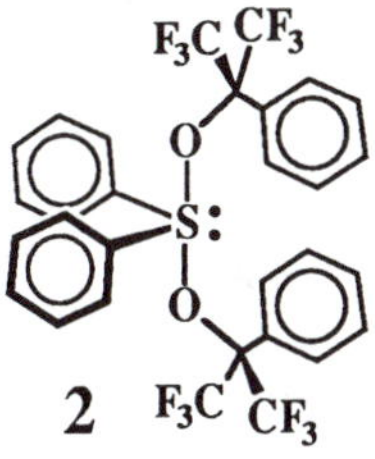

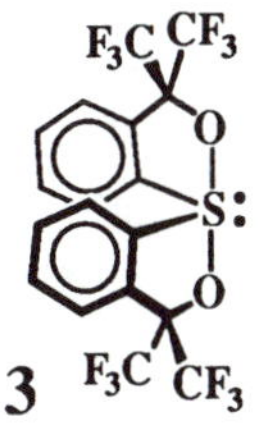

Results

Phosphoranide **1** is prepared by reacting the neutral hydridophosphorane **4** with KH (*3,9*). The KH serves to deprotonate **4** leaving the potassium salt of **1** with the evolution of hydrogen (Scheme 1). The ^{31}P NMR of **1** exhibits a septet at (δ 32.95). This shift is downfield from the neutral hydridophosphorane **4** (δ -48.2). One would expect an upfield shift for the anionic species when compared to the neutral species. The downfield shift can be explained by the elongation of the P-O bonds of **1** to put more positive charge on phosphorus. It could be possible that **1** is unsymmetrical having one of the P-O bonds broken. However, the ^{19}F NMR does show only a doublet of quartets, even at very low temperatures. If **1** would be unsymmetrical, it could have a rapid equilibrium in which the two oxygens are therefore alternating coordination to the phosphorus. It is probably symmetrical as shown by **1**.

Scheme 1

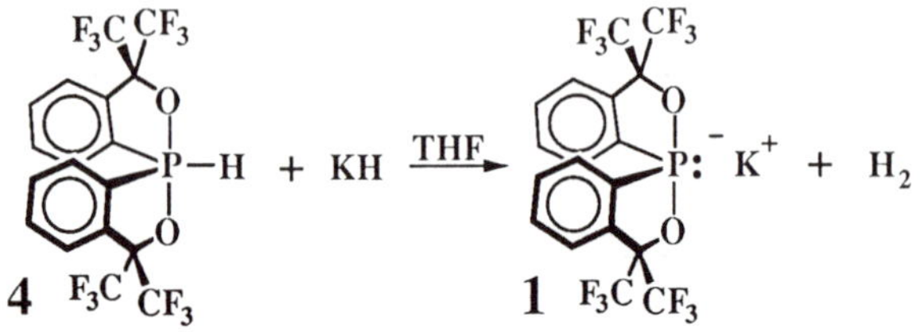

Phosphoranide **1** forms phosphoranes **5** and **6** when treated with the transition metal complexes AuBrPEt$_3$ and Fe(CO)$_2$CpI respectively (Scheme 2). This has been found to work with a large number of metals, providing the 10-P-5 species like **5** and **6**.

Scheme 2

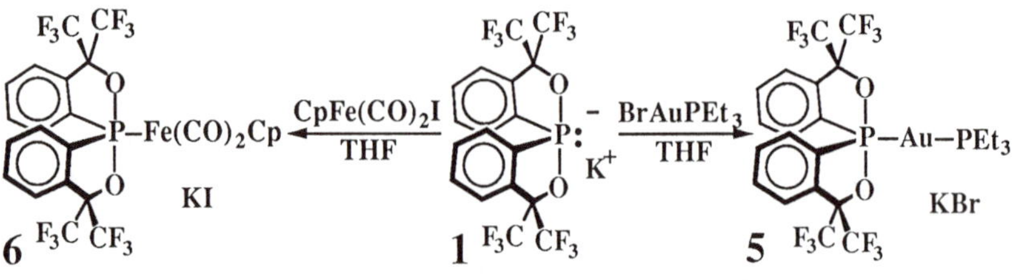

The X-ray crystal structure of **6** was obtained earlier (*9*). The structure of **6** was found to be a TBP geometry with the two oxygens, adjacent to the hexafluorocumyl groups, in the apical positions. The equatorial positions are occupied by the phenyl ring carbons of the bidentate ligands and the iron metal. As mentioned earlier, the positioning of the electronegative groups in the apical position and the electropositive groups in the equatorial positions help to stabilize the complex. Compound **6** is stable both to air and moisture.

We now report the X-ray crystal structure of complex **5**. Similar to **6**, compound **5** exhibits a TBP geometry. The apical O-P-O bond angle of **5** was found to be 173.7°. The two equatorial C-P-Au angles are 118.8° and 120.5° and the

Table I. Structure Determination Summary for 5[*]

Crystal Data	
empirical formula	$C_{24}H_{23}O_2F_{12}P_2Au$
color; habit	colorless prism
crystal size (mm)	0.40 x 0.40 x 0.30
crystal system	orthorhombic
space group	Pbca (#61)
unit cell dimensions	a = 16.494 (6) Å
	b = 30.063 (3) Å
	c = 11.632 (3) Å
volume	5768 (2) cubic Å
Z	8
formula weight	830.34
density (calc)	1.912 g/cm^3
F(000)	3200

Data Collection	
diffractometer used	Rigaku AFC6S
radiation	CuKα (λ = 1.54178 Å)
temperature (K)	296.2 °C
2θ max	119.6°
scan type	$\omega = 2\theta$
scan speed	4°/min (in omega) (3 scans)
scan width	$(1.78 + 0.30 \tan \theta)°$
reflections collected	5314

Solution and Refinement	
structure solution	Patterson method
refinement	full-matrix least-squares
function minimized	$\Sigma w (F_o - F_c)^2$
least-squares weights	$4 F_o^2/\sigma^2(F_o^2)$
p-factor	0.03
anomalous dispersion	all non-hydrogen atoms
no. observations (I>3.00σ (I))	3329
no. variables	370
reflection/parameter ratio	9.00
residuals: R; R$_w$	0.068; 0.095
goodness to fit indicator	3.89
max shift/error in final cycle	0.27
maximum peak in final diff. map	1.92 eÅ^{-3}
minimum peak in final diff. map	-2.86 eÅ^{-3}

[*]The total amount of the Crystal Data is being deposited with the Cambridge Crystallographic Data Centre, University Chemical Laboratory, Lensfield Road, Cambridge CB2 1EW England.

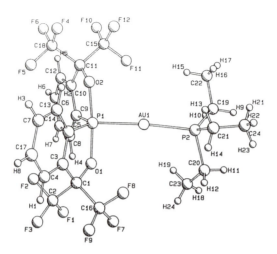

Table II. Selected Bond Lengths and Bond Angles of 5

Bond	Length (Å)	Atoms Defining Angle	Angle (°)
P1-O1	1.818 (9)	O1-P1-O2	173.7 (5)
P1-O2	1.80 (1)	O1-P1-C5	86.6 (5)
P1-Au1	2.336 (4)	O1-P1-C9	89.7 (5)
P1-C9	1.79 (1)	O2-P1-C5	90.5 (6)
P1-C5	1.85 (2)	O2-P1-C9	87.0 (6)
Au1-P2	2.287 (5)	O1-P1-Au1	91.2 (3)
P2-C19	1.82 (3)	O2-P1-Au1	95.1 (4)
P2-C20	1.86 (3)	P1-Au1-P2	175.4 (2)
P2-C21	1.89 (6)	Au1-P1-C5	118.8 (4)
		Au1-P1-C9	120.5 (5)
		C5-P1-C9	120.6 (6)

equatorial C-P-C angle is 120.6°. The O-P-C apical-equatorial angles of 5 within the five-membered rings are 86.6° and 87.0°, while the other O-P-C angles are 89.7° and 90.5°. There is a smaller angle within either of the five-membered rings. The O-P-Au angles on the other side are 91.2° and 95.1°, with the total of 6.3° more than 180°. The O-P-O angle is 186.3° which is 6.3° more than 180° as the P-O bonds are moved directly away from the Au. The equatorial hydrogen of 4 is smaller, and the P-O bonds are moved toward the hydrogen with the O-P-O angle being 178.47°, toward the hydrogen. The equatorial O⁻ of 7 is larger than H and negative. It does move the O-P-O angle away from the O⁻ to 188.7°, more than the O-P-O 186.3° of 5. The O-P-O angle of 6 is moved away from the equatorial Fe to 189.7°. The Fe coordinated phosphoranide, **6**, has 2 CO ligands and C_5H_5 on the Fe making it much larger than the Au of **5** and the O^- of **7** (*2*). The structural comparisons of the compounds with equatorial hydrogen (**4**), gold (**5**), iron (**6**), and O^- (**7**) (*10*) have different apical P-O bond lengths. These are 1.746 Å for **4** (H), 1.809 Å for **5** (Au), 1.824 Å for **6** (Fe), and 1.80 Å for **7** (O^-).

The equatorial O^- of **7** provides anionic electrons, as shown in **7b**, into the antibonding orbital of the O-P-O bond, making the apical P-O bonds longer than that for the equatorial H of **4**. The Au compound, **5**, and the Fe compound, **6**, do have even longer apical P-O bonds than those for **7**. The electrons on the iron section of **6** are more removed into the antibonding orbital, making its P-O bonds longer than those of **5**. The P-O bonds of **5** are longer than those of **4** or **7**, made from the electrons of the $P(Et)_3$ P-C bonds moving across the Au to the O-P-O antibonding orbital. The Fe of **6** has electrons in its d-orbitals and p-orbitals because of coordination to its three ligands. The ^{13}C NMR for the C_5H_5 ligand of **6** is at δ 87.52 for the five carbons. The compound with the hypervalent phosphorus of **6** replaced by a single iodine, which was used to prepare **6**, has the ^{13}C NMR for the C_5H_5 on the Fe as δ 85.58. The C_5H_5 of **6** is thought to donate electrons to the antibonding O-P-O bond (3c-4e), making the ^{13}C NMR of the C_5H_5 of **6** to have a position lower by δ 1.94, relative to the single iodine species.

The apical P-O bond lengths are longer for **6**, with 6>5>7>4. The O-P-O angles mentioned earlier made **6** have the largest angle, relative to the equatorial Fe, with 6(189.7°)>7(188.7°)>5(186.3°)>4(178.47°). It is probable that the longer P-O bonds of **6** make the oxygens, in each of the five-membered rings, further away from the equatorial Fe. Compound **4** has the shortest P-O bonds and the oxygens are closer to the equatorial H. The other two compounds provide 5>7 for the P-O bond lengths, while the O-P-O bond angles away from the Au or O^- are 7>5. The P-O bond lengths of **5** and **7** are not the only method to change the O-P-O angle. It is probable that both the P-O bond lengths and the size of the equatorial ligand make the O-P-O angle change its position relative to the equatorial ligand.

Discussion

The phosphoranide anion **1** is stabilized by the bidentate ligands. These ligands contain the stabilizing factors mentioned earlier for stabilizing hypervalent bonding. This ligand system renders the phosphoranide symmetrical. When compared to compound **8a**, it is clear that the ligand system of **1** is more suitable for stabilizing phosphoranides. Compound **8a** is in equilibrium with **8b**. The bidentate ligand system of **8a** places two electronegative oxygens in the equatorial positions of the

(Ψ-TBP) geometry (*11*). Therefore, the bidentate ligand system of **8a** provides less stabilization than the ligand system of **1**.

The X-ray crystal structure of **5** shows the elongation of the P-O bonds when compared to **4**. This suggests that the gold atom does donate electron density into the antibonding orbital of the (3c-4e) bond. These results are consistent to those seen earlier for **6**.

The bidentate ligand system does allow the phosphoranide to react with metals to give monodentate complexes. Other phosphoranides such as **9** (*12*) act as a bidentate ligand to the metal as seen in **10** (*13*) and **11** (*14*). However protonation of the apical nitrogens of **11** make them more electronegative and the phosphoranide

becomes monodentate complex **12**. This protonation makes the bidentate ligand of **11** similar to that of **1**. This indicates that the bidentate ligand of **1** can be used to prepare phosphoranide-metal complexes in which the phosphoranide acts as a monodentate ligand to the metal.

The trilithio derivative (**13**) reacts with D_2O to give a large amount of deuterium (**14**) replacing the C-Li. We have made a number of hypervalent main-group element species from **13**, by reaction with $SOCl_2$, $SeCl_4$, $TeCl_4$, etc. A few attempted other reactions at **13** did not work, however, for us in this way.

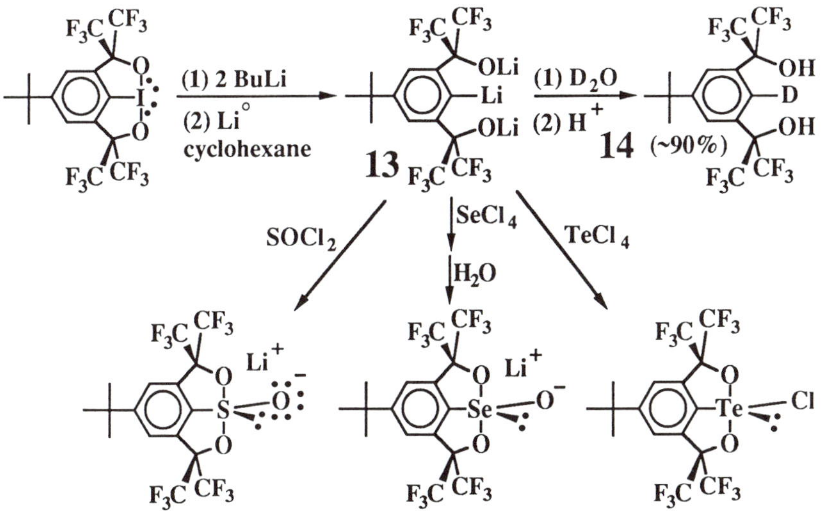

Recent attempts by Milne (Milne, S. Graduate Chemical Research Assistant now, Vanderbilt University.) to have **13** react with PCl₅, PCl₃, or a mixture of PCl₅ and PCl₃, failed to have the trilithio species **13** replace three chlorines. The trilithio species is not as reactive as most lithium species, since it has the central lithium coordinated to the carbon and to the two apical oxygens probably forming a 3c-4e O-Li-O bond (**15**).

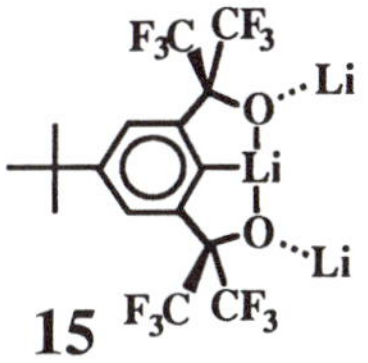

Needed hypervalent phosphorus species have not yet been formed with trilithio species **13**. Recently considered methods for making needed phosphoranide species are now being tried, and will be reported later.

Acknowledgments. Parts of this work were supported by the National Science Foundation (CHE-8910896) and by the Department of Health and Human Services (GM 36844-06).

Literature Cited

(1) The *N-X-L* classification scheme characterizes species in terms of the number (*N*) of formal valence shell electrons about an atom *X* and the number of ligands (*L*) bonded to *X*. Perkins, C. W.; Martin, J. C.; Arduengo, A. J., III; Lau, W.; Alegria, A.; Kochi, J. K. *J. Am. Chem. Soc.* **1980**, *102*, 7753.
(2) Wittig, G.; Maercker, A. *J. Organomet. Chem.* **1967**, 491.
(3) Granoth, I.; Martin, J. C. *J. Am. Chem. Soc.* **1979**, *101,* 4623.
(4) Musher, J. I. *Angew. Chem., Int. Ed. Engl.* **1969**, *8*, 54.

(5) Francl, M. M.; Pietro, W. J.; Hehre, W. J.; Binkley, J. S.; DeFrees, D. J.; Pople, J. A. *J. Chem. Phys.* **1982**, *77*, 3654. Pietro, W. J.; Francl, M. M.; Hehre, W. J.; DeFrees, D. J.; Pople, J. A.; Binkley, J. S. *J. Am. Chem. Soc.* **1982**, *104*, 5039. Baybutt, P. *Mol. Phys.* **1975**, *29*, 389. Kutzelniggin, W. *Angew. Chem., Int. Ed. Engl.* **1984**, *23*, 272.
(6) Ault, B. S.; Andrews, L. *J. Am. Chem. Soc.* **1976**, *98*, 1591. Ault, B. S.; Andrews, L. *J. Org. Chem.* **1977**, *16*, 2024.
(7) Cahill, P. A.; Dykstra, C. E.; Martin, J. C. *J. Am. Chem. Soc.* **1985**, *107*, 6359.
(8) Hays, R. A.; Martin, J. C. *Organic Sulphur Chemistry: Theoretical and Experimental Advances*, I. G. Csizmadia, A. Mangini, and F. Bernardi, Eds.; Elsevier Scientific Publishing Company: Amsterdam, 1985; chap. 8, pp 408-485.
(9) Chopra, S. K.; Martin, J. C. *Heteroatom Chem.* **1991**, *2*, 71.
(10) Perozzi, E. F.; Michalak, R. S.; Figuly, G. D.; Stevenson, W. H., III; Dess, D. B.; Ross, M. R.; Martin, J. C. *J. Org. Chem.* **1981**, *46*, 1049.
(11) Schomburg, D.; Storzer, W.; Bohlen, R.; Kuhn, W.; Roschenthaler, G. V. *Chem. Ber.* **1983**, *116*, 3301.
(12) Lattman, M.; Olmstead, M. M.; Power, P. P.; Rankin, D. W. H.; Robertson, H. E. *Inorg. Chem.* **1988**, *22*, 3012.
(13) Lattman, M.; Chopra, S. K.; Cowley, A. H.; Arif, A. M. *Organometallics* **1986**, *5*, 677.
(14) Khasnis, D. V.; Lattman, M.; Siriwardane, U. *Inorg. Chem.* **1989**, *28*, 681.

RECEIVED November 12, 1991

Chapter 11

Photorearrangements of Allyl Phosphites

Wesley G. Bentrude

Department of Chemistry, University of Utah, Salt Lake City, UT 84112

Photoarrangements of allyl phosphites to the corresponding allylphosphonates are being studied under both direct irradiation and triplet-sensitization conditions. Triplet sensitized processes proceed with a regiochemistry consistent with the intervention of triplet, cyclic, 1,3-biradical phosphoranyl species. Direct irradiation gives phosphonate non-regiospecifically by an undefined process. The incorporation of phosphorus in a five-, six-, or seven-membered ring results in large effects on the apparent relative efficiencies of these processes.

Phosphoranyl radical intermediates ($Z_4P\bullet$ e.g., Z = alkyl, aryl, dialkylamino, aryloxy and combinations thereof) have been well-studied over the past two decades. The subject has been reviewed several times (1-8). The formation of these radical intermediates by oxidative addition processes and their subsequent scission processes to yield the products of oxidation and substitution are shown in Scheme I (R = alkyl, benzyl). Somewhat less well characterized are bimolecular processes in which intact phosphoranyl radicals ($Z_4P\bullet$ = a bicyclic or spiro species) are trapped, as illustrated in Scheme I by the reaction with a disulfide (9).

It occurred to us some time ago that since electronically excited states of functional groups such as ketone carbonyls and alkenes undergo reactions very similar to those of alkoxy and alkyl radicals, it might be possible to observe reactions of these excited states with three-coordinate phosphorus. Of particular interest would be attack at phosphorus in an *intramolecular* context as shown conceptually in the equations of Scheme II. In fact we observed the first example of a photorearrangement that may well proceed by such a mechanism over twenty years ago (10).

Perhaps one of the most striking examples of carbon radical-like reactivity involving an electronically excited alkene is found in the photoreaction of Scheme

0097–6156/92/0486–0137$06.00/0

Scheme I

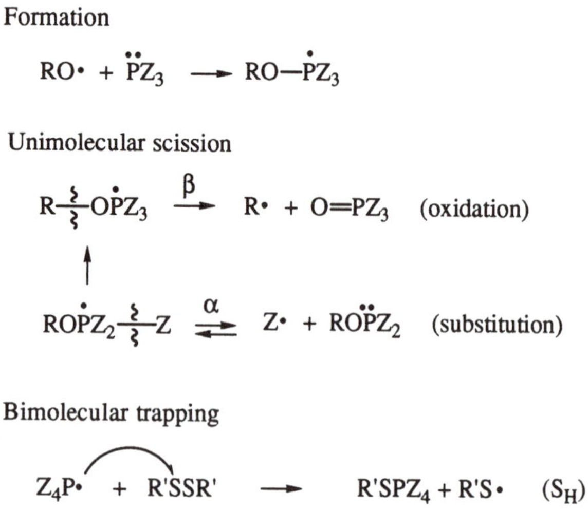

Formation

$$RO\cdot + \overset{\cdot\cdot}{P}Z_3 \longrightarrow RO\overset{\cdot}{-}PZ_3$$

Unimolecular scission

$$R\text{---}O\overset{\cdot}{P}Z_3 \xrightarrow{\ \beta\ } R\cdot + O{=}PZ_3 \quad \text{(oxidation)}$$

$$\big\uparrow$$

$$RO\overset{\cdot}{P}Z_2\text{---}Z \rightleftharpoons Z\cdot + RO\overset{\cdot\cdot}{P}Z_2 \quad \text{(substitution)}$$

Bimolecular trapping

$$Z_4P\cdot\;\overset{\frown}{+}\;R'SSR' \longrightarrow R'SPZ_4 + R'S\cdot \quad (S_H)$$

III (*11*). The triplet excited state abstracts hydrogen to initiate a sort of Norrish Type

Scheme II

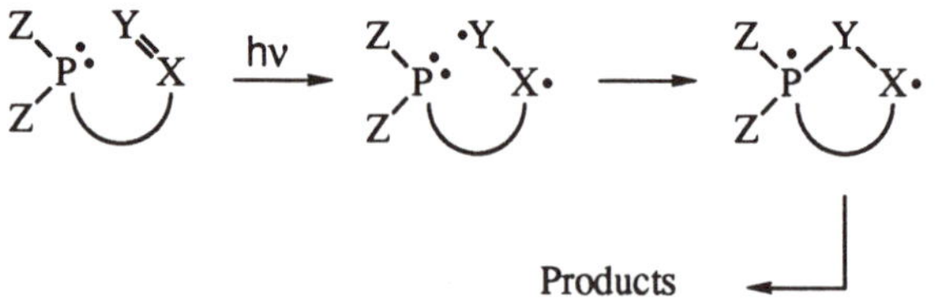

Products

II alkene photochemical fragmentation. The singlet, by contrast, undergoes chemistry similar to that of a benzyl cation. Examples of hydrogen abstraction by singlet alkene excited states are also well documented (*12,13*).

Allyl Phosphite Rearrangements

One of the photorearrangements studied in this laboratory is the conversion of allyl phosphites to the the corresponding phosphonates (*14*). Although this rearrangement occurs thermally at 200 °C (*12*), the photochemical process proceeds readily at <u>room temperature</u> in a matter of hours with light at 254 nm from a Rayonet photoreactor, Scheme IV.

Regiochemistry. The process in benzene indeed proceeds with the regiochemistry shown in Scheme V. This result is at least consistent with the presence of the 1,3-biradical intermediate of Scheme V. This biradical is formed on attack of the electronically excited alkene functionality according to the concepts of Scheme II.

Scheme III

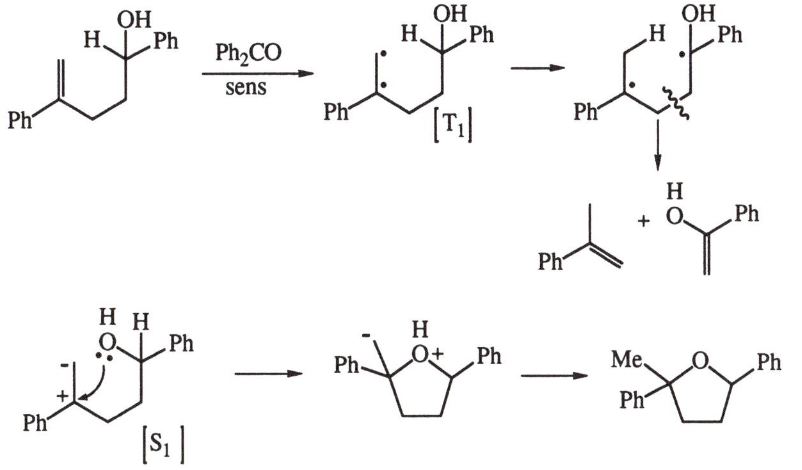

Rapid β scission of the 1,3-biradical results in the formation of the C=O and P=O π bonds and yields the allylphosphonate, a product of the order 40 kcal/mol more

Scheme IV

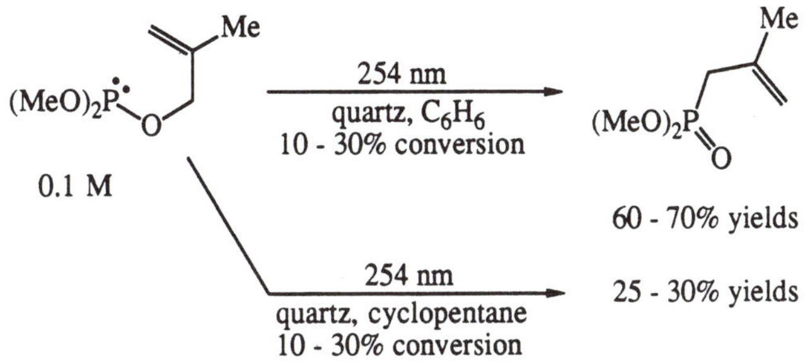

stable enthalpically than the starting allyl phosphite. The above experiment also excludes rearrangement via caged radical pair intermediates $(MeO)_2P(O)\bullet$ and $CH_2=CH-CH_2\bullet$ with normal lifetimes, which would give product with deuterium scrambled between positions 1 and 3 of the allyl group.

Possible Radical Chain Mechanism. Potentially, the regiochemistry observed in Scheme V could result from a free-radical chain mechanism, Scheme VI. However, a crossover experiment utilizing phosphites **1** and **2** failed to give so much as one percent of cross products (*14*).

$(MeO)_2P\text{-}O\text{-}CH_2\text{-}CH=CH_2$ **1** $(MeO)_2P\text{-}O\text{-}CH_2\text{-}C(CH_3)=CH_2$ **2**

Further evidence against kinetically free radical pairs is also provided by this finding.

Scheme V

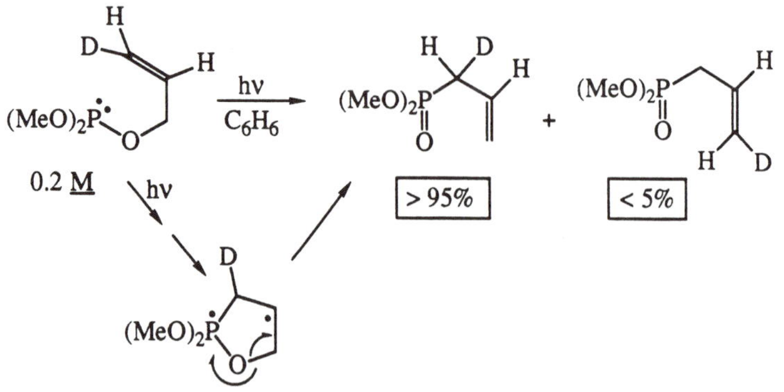

Significantly, 0.1 M dimethyl 2-methylallyl phosphite, irradiated at 254 nm in cyclopentane, reached 40% conversion about four times faster in the presence of

Scheme VI

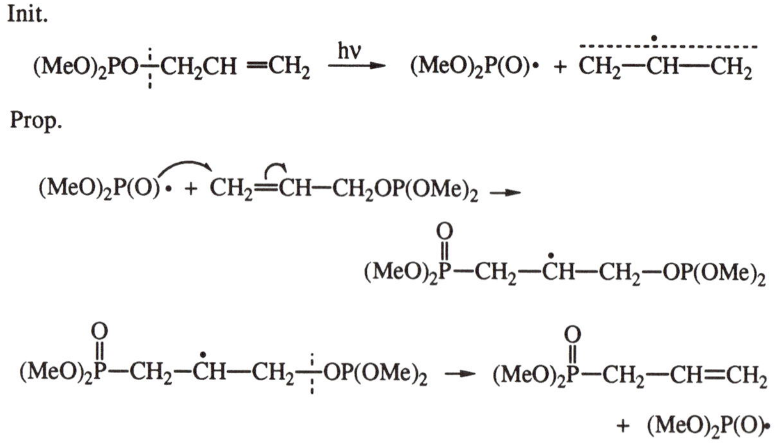

Init.

$$(MeO)_2PO\dot{-}CH_2CH=CH_2 \xrightarrow{h\nu} (MeO)_2P(O)\cdot + \overset{\cdot}{CH_2-CH-CH_2}$$

Prop.

$$(MeO)_2P(O)\cdot + CH_2\!\!=\!\!CH-CH_2OP(OMe)_2 \longrightarrow$$

$$(MeO)_2\overset{O}{\overset{\|}{P}}-CH_2-\overset{\cdot}{CH}-CH_2-OP(OMe)_2$$

$$(MeO)_2\overset{O}{\overset{\|}{P}}-CH_2-\overset{\cdot}{CH}-CH_2\dot{-}OP(OMe)_2 \longrightarrow (MeO)_2\overset{O}{\overset{\|}{P}}-CH_2-CH\!\!=\!\!CH_2$$

$$+ (MeO)_2P(O)\cdot$$

0.6 M p-xylene, and also gave a 70% yield of phosphonate in the presence of p-xylene versus a 25% yield in its absence.   This suggested (14) that two mechanistically different processes are operative under the two conditions. Indeed p-xylene is a favorite triplet sensitizer of photochemists who study alkenes possessing high triplet energies (15).

When the deuterium-labeled allyl phosphite of Scheme V was irradiated directly in cyclohexane with 254 nm light, the deuterium in the phosphite was nearly totally scrambled indicating that the singlet state rearrangement involves a combination of what are formally 1,2- and 2,3-sigmatropic shifts. The phosphite did not undergo deuterium scramble under conditions of the photorearrangement nor did the product phosphonate (S.G. Lee, unpublished results, this laboratory).

Effects of Sensitizer and Suggested Mechanism. The 2-phenylallyl functionality of the phosphite of Scheme VII typically has a triplet energy of about 62 kcal/mol

Scheme VII

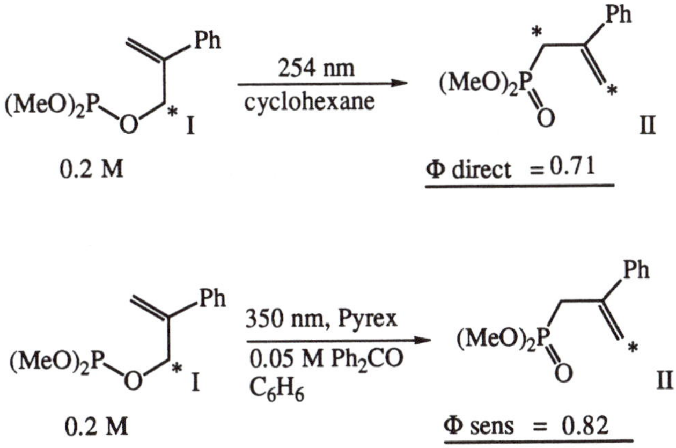

(16, 17) and is efficiently populated by benzophenone sensitization. Phosphite deuterium labeled at the CH$_2$O position of the allyl moiety gave the labeling results of Scheme VII and the quantum yields recorded. These results, previously published (14), appear to confirm the suggestions of the above work with benzene and *p*-xylene as photosensitizers. A possible mechanism for the triplet-sensitized photorearrangements is given in Scheme VIII. In the benzophenone-sensitized

Scheme VIII

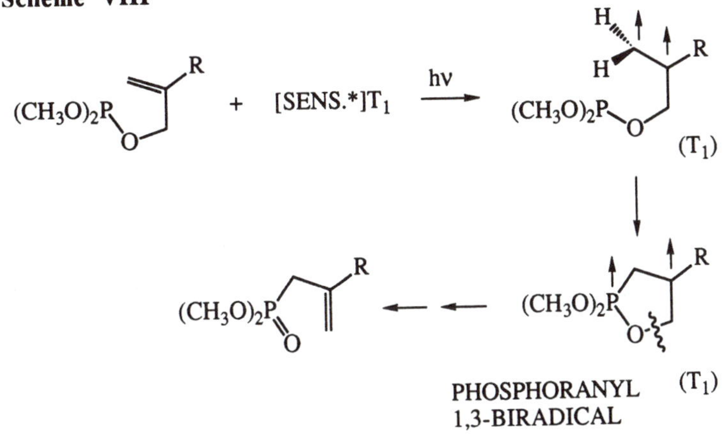

reaction, the 1,2-scheme VII biradical-like triplet attacks phosphorus to form a phosphoranyl 1,3-biradical that after spin inversion gives product phosphonate by scission of the C-O bond. A sensitizer with a very high triplet energy, such as benzene or *p*-xylene, is required for the unsubstituted or 2-methylallyl phosphites (Scheme V), whereas benzophenone is ideal for the dimethyl 2-phenylallyl case of Scheme VIII. Phosphites undergo reaction with ketones (sometimes sluggishly) when an efficient energy transfer is not available (*18, 19*). Direct irradiation of dimethyl 2-phenylallyl phosphite at 254 nm led to product with scrambled label (Scheme VII), a result analogous to that observed with dimethyl allyl phosphite (S.G. Lee, unpublished results, this laboratory).

Related Photoreactions

The above examples apppear to be the first cases of intramolecular reactions of electronically excited alkenes with three-coordinate phosphorus. The intermolecular reaction of triplet 1,1-diphenylethene with a diphosphine is represented in Scheme IX (*20*). In this process the primary carbon of the radical-like triplet is responsible

Scheme IX

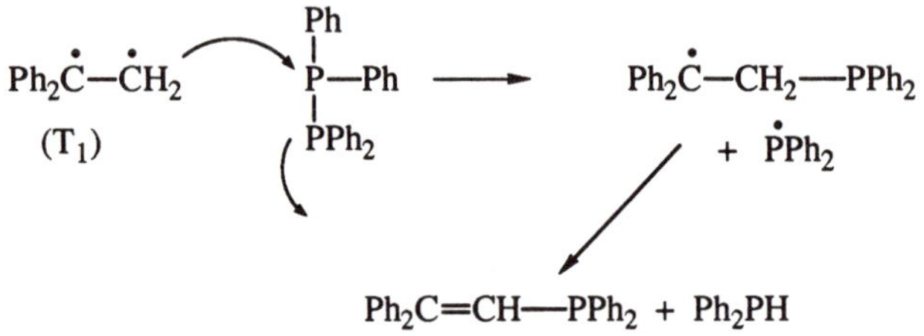

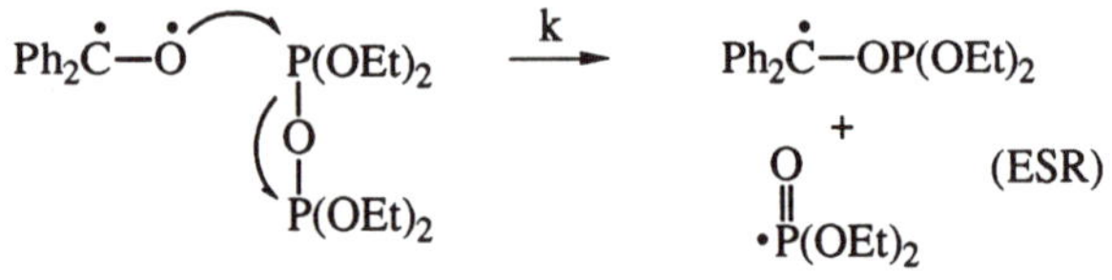

$$Ph_2\overset{\cdot}{C}-\overset{\cdot}{C}H_2 \quad \overset{Ph}{\underset{PPh_2}{\overset{|}{P}-Ph}} \longrightarrow Ph_2\overset{\cdot}{C}-CH_2-PPh_2$$

(T$_1$)   + $\overset{\cdot}{P}Ph_2$

$$Ph_2C{=}CH-PPh_2 + Ph_2PH$$

$$Ph_2\overset{\cdot}{C}-\overset{\cdot}{O} \quad \overset{P(OEt)_2}{\underset{P(OEt)_2}{\overset{|}{\underset{|}{O}}}} \overset{k}{\longrightarrow} Ph_2\overset{\cdot}{C}-OP(OEt)_2$$

$$+$$

$$\overset{O}{\underset{\cdot P(OEt)_2}{\overset{\|}{}}} \quad (ESR)$$

k = 8.0 x $10^8M^{-1}s^{-1}$ at 300 °K (*21*)

for a substitution reaction at phosphorus which is followed by a disproportionation. A phosphoranyl monoradical may be a transient intermediate. Several reactions of excited states of ketones (see Scheme IX (*21*)), quinones and thioketones with three-coordinate phosphorus are now known (*22-26*) in addition to the above-mentioned photorearrangements of β-ketophosphites (*10*).

Attempted Extensions to Other Unsaturated Phosphites

Attempts to extend the photorearrangements depicted in Schemes IV, V and VII to the phosphite featuring methyl substitution, 3, led to a very sluggish reaction yielding both regioisomers in overall yields less than 10% (W.Z. Ye, unpublished results, this laboratory). The result is reminiscent of the failure of isopropyl radicals to give

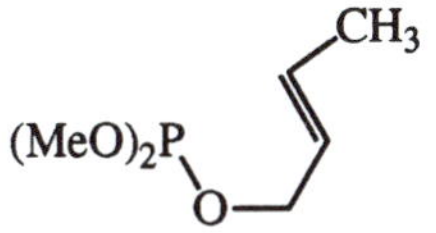

3

net reaction with even a benzyl phosphite via the potential process shown below (E.R. Hansen, P.E. Rogers, unpublished results, this laboratory). By contrast primary ethyl radicals react to give $EtP(O)(OEt)_2$, because a very rapid β scission step traps the intermediate phosphoranyl radical (E.R. Hansen, P.E. Rogers, unpublished results, this laboratory, cited in (27)).

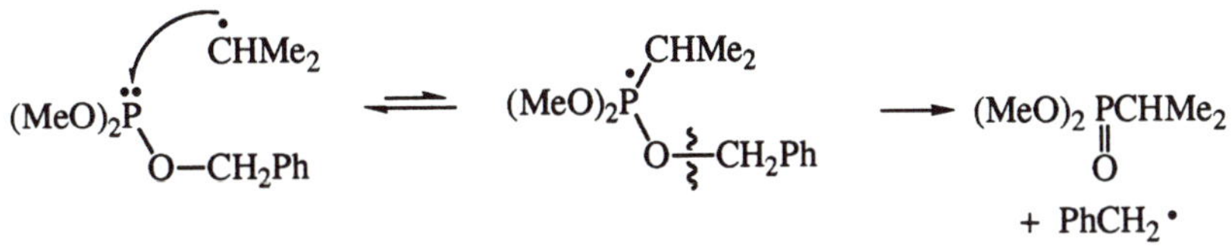

(no net reaction)

Scheme X shows the result of phenyl rather than methyl substitution. Sensitization led to no net reaction. Strikingly, direct irradiation yielded both

Scheme X

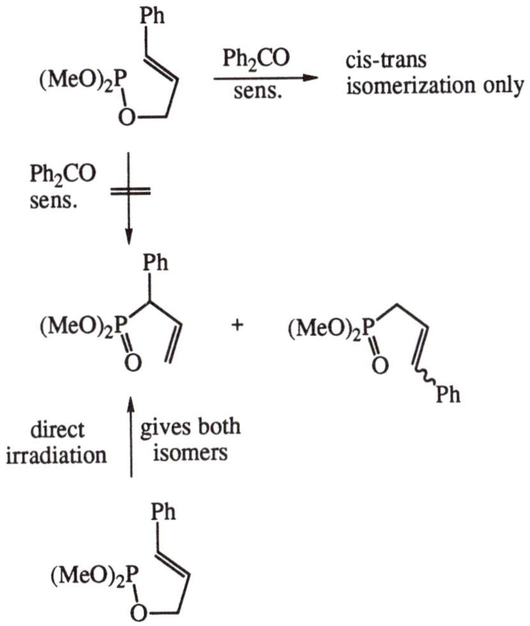

regioisomers in a ratio of approximately 70/30 in favor of the phosphonate with primary carbon bonded to phosphorus (S.G. Lee unpublished results, this laboratory). The mechanism of this reaction, i.e. whether it involves a radical pair or competitive concerted processes, is unknown. Disappointingly, attempts at photorearrangement of phosphites containing homologated alkene chains did not reveal any new process analogous to that depicted above (S.G. Lee, unpublished results, this laboratory). This was true even when the chain was phenyl-substituted in attempt to favor the β scission reaction to give a 1,3-carbon biradical which is benzylic at one position (Scheme XI). Instead of the cyclopropyl-containing phosphonate, the product of a 1,2-shift resulted cleanly. This is an example of the photo-Arbuzov rearrangement of benzyl phosphites discovered in this laboratory (28, 29). The unsaturation in the chain is not needed for the reaction.

Scheme XI

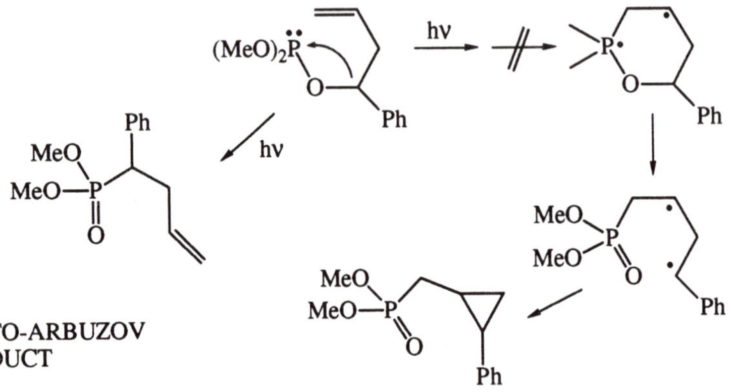

PHOTO-ARBUZOV
PRODUCT

Cyclic Allyl Phosphites

Reaction Scope. Extension of the allyl phosphite photorearrangment to five- six- and seven-membered ring compounds of the sort shown below with R = H or Me gave some surprising results. Five- and six-membered ring compounds yielded no allylphosphonates at all and were consumed only extremely slowly in either benzene or cyclohexane. Substitution of a phenyl group at carbon 2 of the allyl chain rendered the six-membered ring phosphite susceptible to a sluggish rearrangement at 254 nm in benzene. The compound, however, unlike the acyclic compound, $(MeO)_2POCH_2C(Ph)=CH_2$, underwent extremely inefficient photosensitization by benzophenone. Increase of the ring size to seven resulted in a 2-phenylallyl, 2-methylallyl and allyl phosphites that were photorearranged by direct photoirradiation at 254 nm, though still somewhat more slowly than the acyclic case.

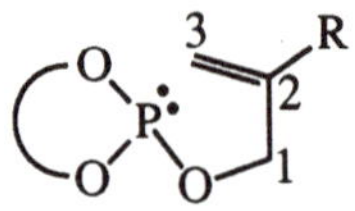

Moreover, benzophenone photosensitization with the seven-membered, 2-phenyl material succeeded, but relatively inefficiently. These are obviously only qualitative results and must be verified by careful quantum yield determinations. Nonetheless, the apparent order of reactivity in both direct and benzophenone-sensitized reactions is: acyclic>seven ring>six ring>five ring. (M. Tabet, W. Baik, Y.W. Wu, unpublished results, this laboratory).

The regiochemistry of rearrangement of the six-membered ring 2-phenyl molecule under direct irradiation at 254 nm was not selective, giving essentially complete scrambling of deuterium. This mirrors the findings with the acyclic 2-phenylallyl molecule. However, the seven-membered ring phosphite slowly gave product phosphonate regiospecifically under benzophenone-sensitized conditions. The latter result is consistent with the operation of an electrocyclic process similar to that depicted in Scheme VII.

The photorearrangement of the seven-membered ring 2-phenylallyl phosphite displayed a regiochemistry on direct irradiation in benzene or CH_3CN that was strongly wavelength dependent. Thus, by use of lamps with intensity peaking at 350 nm in a Rayonet reactor, phosphonate arising from regiospecific rearrangement resulted. Observed was what is formally a 2,3-sigmatropic rearrangement via a five-membered ring intermediate or transition state. At 254 nm in benzene the label in the phosphonate was nearly completely scrambled. Controls evidenced the lack of scramble of label in starting phosphite or product phosphonate. (M. Tabet, unpublished results, this laboratory).

Mechanistic Implications. If the transfer of triplet energy from benzophenone to the various 2-phenylallyl phosphites is highly efficient in all cases, then the order of reactivity of the triplet allyl functionality decreases when two oxygens of the phosphite are part of a ring and as the ring becomes smaller. A possible understanding of the above may be gained from knowledge concerning the permutations of phosphoranyl monoradicals. Exchange of substituents between apical and equatorial positions in near-trigonal-bipyramidal phosphoranyl radicals, such as those shown below, involving the ligands that are not part of the ring (I ⇌ II), occurs

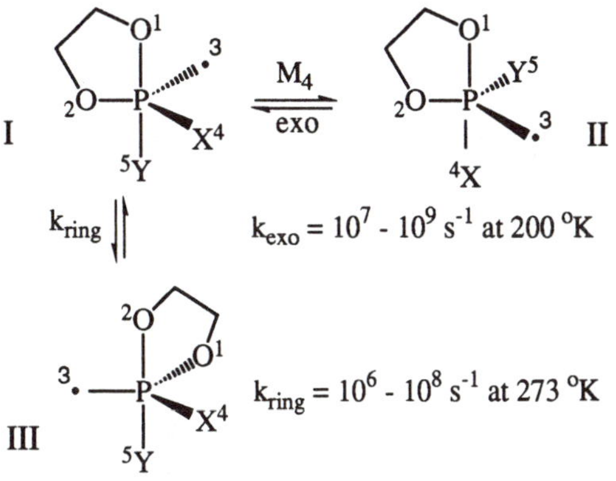

$k_{exo} = 10^7 - 10^9 \, s^{-1}$ at 200 °K

$k_{ring} = 10^6 - 10^8 \, s^{-1}$ at 273 °K

rapidly by a mode 4 process (M4-exo), as measured by ESR (X, Y = halo, alkoxy, amino, etc.). Rate constants are in the range 10^7-10^9 s^{-1} at about 200 °K (*1-4*). The same sort or exchange for oxygens attached to phosphorus as part of a five-membered ring (M4-ring, I $\rightleftharpoons$ II) is slower with k_{ex} 10^6-10^8 s^{-1} at 273 °K (*1-4*).

By contrast for the spiro radical, shown below (IV), ligand exchange is relatively slow. No exchange is observed even at 393 °K from which it has been estimated that k_{ex} is less than 10^6-10^7 even at 393 °K (*1-4*).

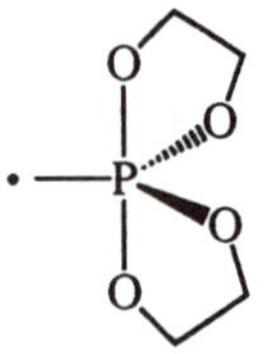

$$k_{ring} < 10^6 - 10^7 \text{ at } 393 \text{ }^oK$$

IV

In Scheme XII triplet alkene functionality adds oxidatively to tricoordinate phosphorus to close the ring at the apical position of II. Based on knowledge of monophosphoranyl radicals (*1-4*), one expects that permutamer III is thermodynamically more stable than II, because the ring carbon is equatorial, and the ring oxygen is apical. If indeed adduct II is insufficiently stable to trap I* irreversibly, the conversion of II to III may be too slow, because of the spiro structure, to compete with the reversion of II to ground state I. These ideas are speculative and subject to further studies now in progress.

Scheme XII

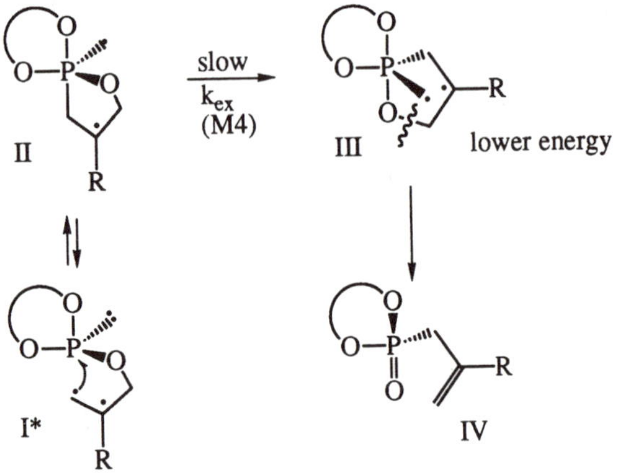

Wavelength Effects

Little has been said to now about the change in regiochemistry of the photorearrangements of the acyclic dimethyl 2-phenylallyl phosphite and its seven-membered ring counterpart with variation in wavelength on direct irradiation. In the former case the wavelength was varied by using a medium pressure Hg lamp and cutoff filters. The cyclic phosphite was irradiated by lamps with maximum intensity at 350 nm and those with only the Hg resonance line at 254 nm. It is well known that the are several singlets accessible for alkenes (*15*), and two at accessible wavelengths for styrenes (*30*). The higher levels of course are populated by absorption of shorter wave lengths. Indeed, variations in the photochemistry of olefins as a function of wavelength are well established (*31,32*). Quite possibly a lower energy singlet, populated at longer wavelengths, undergoes exclusively what is formally a 2,3-sigmatropic shift. (No conclusion as to whether there is a discrete intermediate in the process is intended.) Higher energy singlets perhaps then give rise to the formal equivalent of a 1,3-sigmatropic shift or a combination of 1,3- and 2,3-processes.

Summary

With the provision that a number of the above reaction systems are subject to more careful study (for example the determination of quantum yields and the efficiency of triplet sensitization), the following conclusions arise, some of which are tentative. For what appear to be rearrangements via triplet excited states: 1) the regiospecificity is essentially complete resulting in what is formally a 2,3-sigmatropic shift; 2) since such a process must be stepwise, the postulation of the intermediacy of a phosphoranyl 1,3-biradical is reasonable; 3) substitution of the terminal alkene position with Ph or Me leads to very sluggish (Me) phosphonate formation or none at all (Ph); 4) reaction efficiency is reduced with phosphorus in a ring in the order: 7-ring>6-ring>5-ring. The presumably singlet state, direct irradiation processes are regiospecific at longer wavelengths (2-Ph cases) but give increased amounts of 1,2-sigmatropic shift product at shorter wavelengths (254 nm, unsubstituted and 2-Ph cases). The mechanistic details of the singlet processes are yet to be defined.

Acknowledgments. I would like to express my thanks to the following members of my research group who have carried out the studies described: Drs. Y. Charbonnel, K. Akutagawa, S.G. Lee, W. Baik, and Y.W. Wu; Professor W.Z. Ye; and Mr. M. Tabet. Generous support has been provided by the National Science Foundation and the National Cancer Institute of the Public Health Service, Grant CA11045.

Literature Cited

1. Bentrude, W.G. In *The Chemistry of Organophosphorus Compounds*; Hartley, F.R., Ed.; John Wiley & Sons: New York, 1990; pp. 531-566.
2. Bentrude, W.G. In *Reactive Intermediates*; Abramovitch, R.A., Ed.; Plenum Press: New York and London; 1983; Vol. 3; pp. 199-298.
3. Bentrude, W.G. *Acc. Chem. Res.*, **1982**, *15*, 117.
4. Roberts, B.P. In *Advances in Free Radical Chemistry*; Williams, G.H., Ed.; Heyden: London; 1979; Vol. 6; pp. 225-284.
5. Soldovnikov, S.P.; Bubnov; N.N.; Prokof'ef; A.I. *Russ. Chem. Rev.* **1980**, *49*, 1.
6. Schipper, P.; Janzen, E.H.J.M.; Buck, H.M. *Top. Phos. Chem.* **1977**, *9*, 407.
7. Bentrude, W.G. In *Free Radicals*; Kochi, J.K, Ed.; Wiley-Interscience: New York; 1973; Vol. 2; Ch. 22.

8.   Walling, C.; Pearson, M.S. *Top. Phosphorus Chem.* **1966**, *3*, 1.
9.   Bentrude, W.G.; Kawashima, T.; Keys, B.A.; Garroussian, M.; Heide, W. Wedegaertner, D.A. *J. Am. Chem. Soc.* **1987**, *109*, 1227.
10. Griffin, C.E.; Bentrude, W.G.; Johnson, G.M. *Tetrahedron Lett.* **1969**, 969.
11. Hornback, J.M.; Proehl, G.S. *J. Am. Chem. Soc.* **1979**, *101*, 7367.
12. Scully, F.; Morrison, H. *J. Chem. Soc., Chem. Commun.* **1973**, 529.
13. Unpublished results of P.J. Kropp, recorded in ref. 15.
14. Bentrude, W.G.; Lee, S.-G.; Akutagawa, K.; Ye, W.; Charbonnel, Y. *J. Am. Chem. Soc.* **1987**, *109*, 1577.
15. See examples recorded in the review of Kropp, P.J. *Org. Photochem.* **1979**, *4*, 1.
16. Crosby, P.M.;, Dyke, J.M.; Metcalfe, J.; Rest, A.J.; Salisbury, K.; Sodeau, J.R. *J. Chem. Soc. Perkin Trans 2* **1977**, 182.
17. Lamola, A.A.; Hammond, G.S. *J. Chem. Phys.* **1965**, *43*, 2129.
18. Fox, M.A. *J. Am. Chem. Soc.* **1979**, *101*, 5339.
19. Chow, Y.L.; Marciniak *J. Org. Chem.* **1983**, *48*, 2910.
20. Okazaki, R.; Hirabayashi, Y.; Tamura, K.; Inamoto, N. *J. Chem. Soc., Perkin Trans 1* **1976**, 1034.
21. Alberti, A.; Griller, D.; Nazran, A.S.; Pedulli *J. Org. Chem.* **1986**, *51*, 3959.
22. For a review see Creber, K.A.M.; Chen, K.S.; Wan, J.K.S. *Rev. Chem. Intermed.* **1984**, *5*, 37.
23. For a review see Pedulli, G.F. *Rev. Chem. Intermed.* **1986**, *7*, 155.
24. Okazaki, R.; Tamura, Y.; Hirabayashi, Y. Inamoto, N.J. *J. Chem. Soc., Perkin Trans. 1* **1976**, 1924.
25. Alberti, A.; Hudson, A. Pedulli, G.F.; McGimpsey, W.G.; Wan, J.K.S. *Can. J. Chem.* **1985**, *63*, 917.
26. McGimpsey, W.G.; Depew, M.C.; Wan, J.K.S. *Phosphorus Sulfur* **1984**, *21*, 135.
27. Bentrude, W.G.; Fu, J.J.L.; Rogers, P.E. *J. Am. Chem. Soc.* **1973**, *95*, 3625.
28. Omelanzuk, J.; Sopchik, A.E.; Lee, S.G.; Akutagawa, K.; Cairns, S.M.; Bentrude, W.G. *J. Am. Chem. Soc.* **1988**, *110*, 6908.
29. Cairns, S.M.; Bentrude, W.G. *Tetrahedron Lett.* **1989**, *30*, 1025.
30. Crosby, P.M.; Dyke, J.M.; Metcalfe, J.; Res, A.J.; Salisbury, K.; Sodeau, J.R. *J. Chem. Soc. Perkin Trans 2* **1977**, 182.
31. Inoue, Y.; Dainal, Y.; Hagiwara, S.; Nakamura, H.; Hakushi, T. *J. Chem. Soc., Chem. Commun.* **1985**, 804.
32. Inoue, Y.; Mukai, T.; Hakushi, T. *Chemistry Letters* **1983**, 1665.

RECEIVED November 12, 1991

Chapter 12

Mechanisms of the Wittig Reaction

W. E. McEwen[1], F. Mari[2], P. M. Lahti[1], L. L. Baughman[1], and W. J. Ward, Jr.[1]

[1]Department of Chemistry, University of Massachusetts, Amherst, MA 01003
[2]Program in Molecular Medicine, University of Massachusetts Medical Center, Worcester, MA 01605

The metal ion-catalyzed reactions of benzaldehyde with benzylidene-triphenylphosphorane and benzylidenemethyldiphenylphosphorane, respectively, have been examined by molecular mechanics calculations (MMX89) and the computed results compared with experimental results for reactions carried out in tetrahydrofuran at temperatures of - 15°C or lower. Mechanisms for these reactions are suggested and compared with results obtained previously for a variety of salt-free Wittig reactions modeled by use of the MNDO-PM3 molecular orbital method.

In order to explain the origins of stereoselectivity in the Wittig reaction, various workers have considered two fundamentally different mechanisms throughout the years: (i) a stepwise ionic type of mechanism (2) and (ii) a direct cycloaddition mechanism (3). Wittig reactions of unstabilized ylides (alkylidenetriphenylphos-phoranes) with aldehydes are mainly Z-stereoselective with respect to alkene formation, whereas the corresponding reactions of stabilized ylides are E-stereoselective (4). These stereochemical results are usually not greatly influenced by solvent effects or even by the presence of lithium salts in the reaction media (4, 5, 6). By contrast, the stereoselectivity of Wittig reactions of semistabilized ylides (e.g., benzylidenetriphenylphosphorane) with aldehydes are extremely sensitive to solvent effects (2e, 2k, 4a) and to the presence of metal ions (7). These solvent and metal ion effects may be indicative of an ionic reaction pathway, whereas the lack of them, especially in the case of the reaction of stabilized ylides, may indicate the operation of a different mechanistic path.

The latest proposal by Vedejs (3d) is that the Wittig reaction proceeds via a concerted but asynchronous puckered 4-center cycloaddition pathway in which the stereoselectivity is determined by multiple steric effects and varying degrees of rehybridization at the phosphorus atom in the transition state. At present, there is not definitive evidence to prove that the reaction must proceed in this manner. Recent MNDO-PM3 computations by us (1, 8) and somewhat related MNDO computations by Yamataka et al. (9) do not support the puckered 4-center cycloaddition hypothesis for the reactions of unstabilized ylides with aldehydes (3d). Instead, the MNDO-PM3 computations indicate that such Wittig reactions proceed through an essentially planar transition state (TS) with respect to the four central atoms, P-C-C-O. This process is

0097–6156/92/0486–0149$06.00/0

best described as a very asynchronous cycloaddition, a borderline two-step mechanism in which the C-C bond is about 30 to 50% formed and with little evidence of P-O bond formation. This hypothetical "gas phase" reaction is a 2-center reaction with a planar, U-shaped TS because of the strong attractive ionic interaction of the P and O atoms (1, 8). Furthermore, it has been computed that steric effects exerted by the substituents at the phosphorus, the ylidic carbon and aldehyde have little effect on the geometries of Wittig reaction transition states, which are found to be planar in all cases (1, 9). In solution, however, where solvation of the dipole centers at the phosphorus and oxygen atoms is expected, or in which the presence of metal ions may lead to complexation at the oxygen (or at least to a strong electrostatic interaction), a nonplanar 2-center TS is conceivable.

We have chosen the Wittig reaction of two semistabilized ylides, benzylidenetriphenylphosphorane (1) and benzylidenediphenylmethylphosphorane (2) with benzaldehyde and pivalaldehyde, as representative cases where a wide range of stereoselectivities, depending on the reaction conditions, might be observed (7b). We and others have previously shown that stereochemical drift (10) does *not occur* in the conversion of *erythro*-(2-hydroxy-1,2-diphenylethyl)methyldiphenylphosphonium iodide to Z-stilbene (>99.9%, 88-97% yield) in THF at -78°C by the action of *n*-BuLi, NaHMDS or KHMDS (7b, 11). Since stereochemical drift is always faster for the erythro isomer than for the threo isomer (10), we can safely assume that the corresponding Wittig reactions of 2 with benzaldehyde under the same conditions are kinetically controlled and not subject to stereochemical drift by way of retro-Wittig half-reactions, through reversal of oxaphosphetane formation to form the aldehyde and ylide. Similar observations have been made by other workers for the deprotonation of erythro- and threo-(2-hydroxy-1,2-diphenylethyl)triphenylphosphonium salts in aprotic solvents at low temperatures. Thus, the use of benzaldehyde in reactions with 1 and 2, respectively, does not lead to atypical results. As further evidence of this, reactions of 1 and 2, respectively, with pivalaldehyde gave results that are similar to those observed in reactions with benzaldehyde (12).

Results and Discussion

The results of the reactions of the ylides with benzaldehyde in THF solution under a variety of conditions are given in Table I.

It can be seen that at temperatures ranging from -78°C to -15°C in the presence of soluble lithium or sodium salts, high yields of stilbenes are obtained in a ratio of about 67% of Z-stilbene to 33% of E-stilbene. In the absence of salts (or when insoluble salts have been formed in the generation of the ylides), the Z/E ratio becomes less than one.

Whereas the effects of Li salts on Wittig reaction stereoselectivity are well known, effects of sodium salts have received less attention. As in the case of the Li salts, the presence of Na salts tends to increase the Z/E ratio in alkene formation. However, this effect is observed only in those cases where the sodium salts are soluble and available for complexation with the reactants (NaI and $NaB(C_6H_5)_4$). In cases where the metal ion is sequestered by an additive such as 15-crown-5 or DMSO or when the metal salt is not soluble in the reaction media (NaCl, NaBr), lower Z/E ratios in alkene formation are observed.

These marked metal ion effects (7), in conjunction with well known solvent and substituent effects, suggest that an ionic mechanism is operative in the metal ion catalyzed Wittig reaction (7b). An ionic mechanism for the Wittig half-reaction of benzylidenetriphenylphosphorane with aldehydes can be envisioned as shown in Figure 1.

Table I. Metal Ion Effects on the Z/E Ratios of the Wittig Reactions of Benzylidenetriphenylphosphorane (1) and Benzylidenemethyldiphenylphosphorane (2) with Benzaldehyde in THF

| Salt Present | Z/E Ratios: | |
	Ylide 1	Ylide2
None[a]	36/64[b]	18/82[c]
LiCl	57/43[b]	66/34[c]
LiBr	63/37[b]	---
LiI	60/40[b]	63/37[c]
NaCl	64/36[b]	69/31[c]
NaBr	51/49[b] [54/46][c]	---
NaI	77/23[b] [76/24][c]	69/31[c]
NaI + 15-crown-5	48/52[b]	---
NaB(C_6H_5)$_4$	83/17[b]	---

[a]Salt-free conditions. See reference 7b.
[b]Reaction carried out at -15°C.
[c]Reaction carried out at -78°C.

A nucleophilic attack of the ylide on the solvated metal-aldehyde complex could generate four possible stereoisomeric oxaphosphetanes, i.e., two pairs of enantiomers. This process is likely to occur *via* a 2-center reaction, where the transition state structure involves the formation of a partial C-C bond and with no P-O bond formation. Since no restriction is placed on the direction of approach of the reactants, all four stereoisomeric transition states can be formed. Ring closure would yield the respective oxaphosphetanes as two racemates.

There is the possibility that the 2-center transition state described earlier would lead to the formation of metallated betaines, which are not unreasonable intermediates for Wittig reactions carried out in the presence of metal ions or polar solvents capable of stabilizing such highly polar intermediates. Recent results of reactions carried out under Li salt free conditions suggest that, in the reactions of P-isopropylidenephenyldibenzophosphorane with hydrocinnamaldehyde, intermediate betaines are not involved in the process (3f). However, this evidence does not pertain to metal-ion catalyzed Wittig reactions, and no stereocontrol issues can be addressed directly with the use of this particular system. Furthermore, ylides bearing the dibenzophosphole moiety exhibit different behavior from standard triaryl or trialkyl phosphoranes regarding P rehybridization and pseudorotation, which are key issues to be considered in the mechanism of the Wittig reaction. Thus, caution must be exercised in extrapolating the results of Wittig reactions of ylides that contain the dibenzophosphole moiety to all Wittig reactions.

In order to examine the stereochemical results of the reaction of benzylidenetriphenylphosphorane with benzaldehyde in the presence of Li+, we have adopted a systematic approach to evaluate the relative stabilities of the different possible configurations and conformations that groups on the transition states might assume, depending on the direction of the approach of the reactants and the intramolecular interactions within the system. We have performed Molecular Mechanics calculations (MMX89) (13) for the two diastereomeric racemates that could be generated in the reaction of benzylidenetriphenylphosphorane with benzaldehyde by different approaches of the two reacting prochiral centers involved in the carbon-carbon bond formation. Since the use of molecular mechanics computations does not permit the direct evaluation of transition states, we have assumed the Hammond postulate (*14*) to estimate their relative energies from comparisons of the computed relative energies of the related metallated betaines.

Molecular mechanics force fields provide excellent geometrical agreement between calculated and experimentally determined chemical structures for phosphorus containing compounds (*15*), including those involved in the Wittig reaction (*16*). The use of molecular mechanics computations as steric probes in molecular modeling has been well established (*17*). However, the possibility of multiple conformations in the system to be modeled has to be standardized in order to obtain results that are comparable. With the use of a Montecarlo-Metropolis approach to simulated annealing (*18*), we performed a multitorsional global optimization of dihedral angles involved in the system. Similar calculations were performed on the corresponding oxaphosphetanes, for which the configuration of each carbon that becomes part of the oxaphosphetane ring is preserved from the betaine stage to the formation of the respective oxaphosphetanes. The minimizations were carried out until self consistency in the total MMX89 energy was achieved among the different starting initial structures. As results of these calculations, the geometry of the structure of each global minimum among all possible conformations was found. However, several local minima gave total MMX89 energies very close to their correspondent global minima. The structures of the global minima of the Z-oxaphosphetane generating metallated betaines and their E-counterparts could be considered as representative of the most likely orientations of the substituents on the phosphorus atom, ylidic carbon and carbonyl group of the system. However, the direction of approach of the ylide to the aldehyde, and the effects on the interactions of these substituents, is better evaluated by a conformational analysis around the forming carbon-carbon bond. Such an analysis was accomplished with the use of the dihedral angle driver option (*17a*) implemented in the MMX89 program and applied to the P-C-C-O angle. Figure 2 shows the changes of the total MMX89 energy during the rotation about each of the forming C-C bonds of the Z and E-olefin generating betaines.

The conformational profile shown in Figure 2 presents three clearly defined minima for the E-oxaphosphetane generating metallated betaine, whereas the Z-counterpart shows only two well defined minima in the whole conformational profile.

Table II depicts the compilation of the total MMX89 energies of the minima found in the conformational profiles of the Z and E-oxaphosphetane generating structures and the Z and E oxaphosphetanes themselves.

Table II. Total MMX89 Energies of the Minima Found in the Conformational Profile Shown in Figure 3 [a]

Oxaphosphetane generated	syn	Metallated betaines gauche-syn1	anti	gauche-syn2
Z	65.9(0)	58.1(75)	56.5(165)	---
E	62.2(0)	57.3(55)	62.4(160)	58.2(305)

[a]Energies are given in kcal/mole. The 'syn' conformation of the metallated betaine (<PCCO = 0) has been given for reference. In parentheses are the corresponding PCCO angles.

The reported total MMX89 energy in each case represents the sum of the stretching, bending, torsional and van der Waals energies, which within the molecular mechanics framework (*17*) depicts the steric strain of the system according to the parameterization of the MMX89 force field. With the use of these calculations, we are

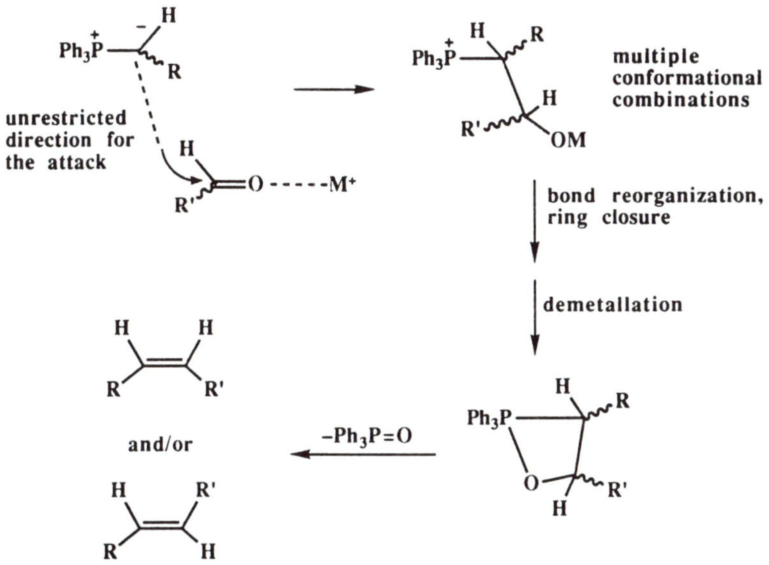

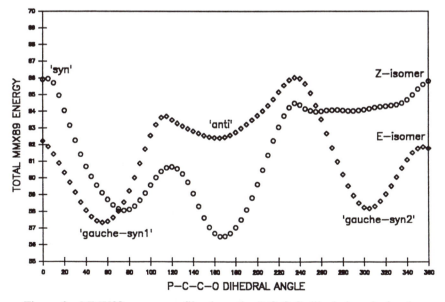

Figure 1.   Proposed mechanistic model for the Wittig reaction of nonstabilized and semistabilized ylides with aldehydes in the presence of soluble metal ions. Formation of the metallated betaines represents the step that determines the stereocontrol of the reaction.

Figure 2.   MMX89 energy profile along the P-C-C-O dihedral angle for the global minima of the Z and E-oxaphosphetane generating metallated betaines of the reaction of benzylidenetriphenylphosphorane (1) with benzaldehyde.

trying to evaluate effects caused by steric interactions only (*17*). However, since some of the structures that we are comparing are diastereoisomeric racemates, comparisons between them are valid only for those isomers where solvation, intermolecular aggregation, entropic factors and external polarization effects are equally important. In the present case solvation effects might be highly dependent upon the particular geometry under consideration, since dipole alignments within a molecule are going to be different subject to whether a "syn" or an "anti" conformation is considered for these metallated betaines. Furthermore, the calculated global minima are model representations of a continuum of thermally accessible conformations. For all these reasons, only qualitative inferences can be drawn from the present study. In Figure 2 and Table II, the energy content of the metallated betaine-like structure that generates the E-oxaphosphetane (the precursor of the E-olefin) is significantly higher than that of the Z-oxaphosphetane generating counterpart, if an anti approach with <PCCO ~180° is considered. The corresponding oxaphosphetanes themselves show the opposite trend; the Z-oxaphosphetane is higher in energy content than the diastereomeric E-oxaphosphetane, with an energy difference of up to 2 kcal/mole. These observed energy differences in the oxaphosphetanes are in agreement with our previous findings by MO calculations and MMX calculations where E-olefin generating oxaphosphetanes have been calculated to be more stable than their Z-counterparts (*16*). The same conclusions can be drawn from the analysis of the conformational profile shown in Figure 2. Eclipsed and gauche ('syn') conformations closely resemble the geometries of the oxaphosphetanes, and, as in the case of the oxaphosphetanes, these conformations show the Z-oxaphosphetane structures to have higher energy contents than their E counterparts.

   These results show an energy preference for the Z-oxaphosphetane generating metallated betaines that go on to form the Z-olefin only in the cases where the geometries of these structures indicate that they were generated by an antiperiplanar or "anti" approach of the P^+-C^- nucleophile to the $C^+-O^-M^+$ group of the metallated aldehyde. We have shown previously, with the use of MNDO-PM3 SCF MO calculations, that a "syn" approach cannot explain the unusual Z-stereoselectivity of the Wittig reaction under consideration (*1*). The "syn" structures located in the conformational analysis shown above confirm our MO results where the E-stereoisomer always shows an energy preference over its Z-counterpart in this type of geometry, even in the present case where these structures are shown to have highly puckered P-C-C-O conformations (*1*).

   In Figure 2 and Table II, the Z-oxaphosphetane generating "anti" conformation is shown to have a slightly lower energy content than the E-oxaphosphetane generating "syn" conformation. Although we have already discussed the assumptions involved in the use of the present approach, the magnitude in the energy difference between the eventual Z-olefin generating metallated betaine and its E-olefin generating counterpart is similar to the energy difference between their transition states (Hammond postulate) and is roughly consistent with the Z/E ratio of stilbenes actually observed. Both of the metallated enantiomeric pairs of betaines are formed, but the lower energy Z-olefin generating pair is formed in the greater amount.

   In an attempt to investigate the effect that a substituent other than phenyl on the phosphorus atom may exert on the stereochemical outcome of the Wittig reaction (*3b*, *f*, *g*), we have made calculations with benzylidenemethyldiphenylphosphorane as the reactant ylide. The examination of this reactant highlights the complication that, at the product stage of the Wittig half-reaction (oxaphosphetane formation), additional isomeric oxaphosphetanes can arise from the possibility that the methyl group on the phosphorus atom might occupy either of two equatorial or axial positions in the nascent trigonal bipyramidal geometry about the phosphorus atom. However, previous studies by Trippett (*11*) indicate that a phenyl group is more apicophilic than an alkyl group. Therefore, we restricted the methyl group only to occupancy of the two possible equatorial positions in the oxaphosphetanes.

Applying the same methodology as before, the reaction of benzylidenemethyldiphenylphosphorane with benzaldehyde shows the conformational profiles depicted in Figure 3.

Unlike the conformational profile shown before, both isomers show clearly the presence of three conformational minima, two "syn" or gauche conformations and one "anti" conformation. Table III shows the total MMX89 energy of these conformations and the total energies of their correspondent oxaphosphetanes. It is apparent that, as in the previous case, the Z-oxaphosphetanes are higher in energy content than their E-counterparts (Table III). However the "anti" Z-oxaphosphetane generating metallated betaines have lower energy contents than the "anti" E-oxaphosphetane generating metallated betaines. As in the previous study involving benzylidenetriphenylphosphorane, the "syn" E-oxaphosphetane generating structures have a lower energy content than their Z counterparts, but still, under the assumptions described above, the "anti" Z-oxaphosphetane generating metallated betaine shows the lowest energy content of all of the conformational possibilities. Again, the magnitudes of the differences in energies of the intermediate metallated betaines and the corresponding transition states by application of the Hammond postulate are roughly consistent with the ultimate Z/E ratio of stilbenes formed.

Table III.   Total MMX89 Energies of the Minima Found in the Conformational Profile Shown in Figure 4 [a]

Oxaphosphetane generated		Metallated betaines		
	syn	*gauche-syn1*	*anti*	*gauche-syn2*
Z	57.6(0)	50.3(45)	46.6(175)	51.8(280)
E	52.7(0)	49.0(60)	51.6(185)	48.9(305)

[a]Energies are given in kcal/mole. The 'syn' conformation of the metallated betaine ($<$PCCO = 0) has been given for reference. In parentheses are the corresponding PCCO angles.

In order to rationalize the greater stability of the ultimate Z-olefin generating structures, steric analyses of the calculated geometries can be assessed. An examination of space filling models of the two different types of geometries (Figure 4), the typical antiperiplanar or "anti" approach and the typical "syn" approach, reveals that the orientations of the phenyl groups are playing the predominant role in the steric interactions of the structures. The typical "anti" structure shows a parallel orientation of the phenyl groups which adopt a skew conformation around the C-C bond to minimize the steric interactions by a tight "packing" of the phenyl groups. On the other hand, the typical "syn" structure shows that the phenyl groups have orthogonal orientations that induce stronger interactions of the *ortho* hydrogens of these groups with other groups in the molecule.

Besides the steric considerations, the antiperiplanar approach of the reactants could be influenced by other factors. The "anti" geometry would leave the metal ion and oxygen atoms exposed at one end of the molecule (Figure 4), making these polar centers more open to solvation. A water surface analysis using the Lee and Richards algorithm (*19*), showed that the "anti" structures have a larger polar surface than their "syn" counterparts, thus confirming the above observation. Also, the positive pole of the phosphonium group and the positive pole of the metal ion adjacent to the negative oxygen are more separated in the "anti" structure, thus suggesting an energy content

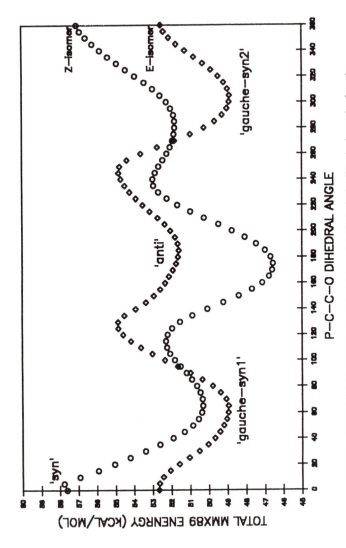

Figure 3. MMX89 energy profile along the P-C-C-O dihedral angle for the global minima of the Z and E-oxaphosphetane generating metallated betaines of the reaction of benzylidenemethyldiphenylphosphorane (2) with benzaldehyde.

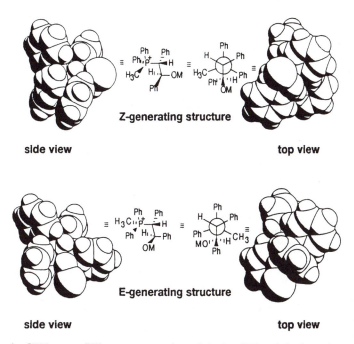

Figure 4. CPK space filling representation of the 'anti' Z and the 'gauche-syn1' E generating metallated betaines of the reaction of benzylidenemethyl-diphenylphosphorane with benzaldehyde. The side view representation indicates that the P-C ylidic bond is parallel to the plane of the figure, whereas the top view representation indicates that such a bond is perpendicular to the plane of the figure.

less than that of the "syn" structure, owing to both considerations of electrostatic interactions and the most advantageous alignment of dipoles. Another possibility to justify further the greater stability of the "anti" structure is that the tightly packed phenyl groups present a barrier that diminishes the repulsive interaction of the positive phosphorus atom and the positive metal ion.

The model herein presented has a resemblance to the ionic mechanism proposed by Bergelson et al. (*2b, c, d, e*) to explain the unusual Z-stereoselectivity of certain Wittig reactions. This mechanism calls for an antiperiplanar approach of the reactants to form an "anti" betaine that, in turn, by ring closure would generate the appropriate oxaphosphetane. The Bergelson et al. hypothesis was justified by the effects of Lewis bases (Br⁻, I⁻, amines or another ylide molecule by aggregation), solvent effects and the electropositivity of the phosphorus atom. All these factors were considered by these workers to favor dipole alignments in such a manner that an "anti" betaine was favored over its "syn" counterpart (*2c*). The main objection to this mechanism remains the unproven involvement of betaines as intermediates in salt free Wittig reactions (*3f*).

Whether or not betaines are actually involved in Wittig reactions as obligatory intermediates when metal ions are present, our previous work using the mythical Wittig half-reaction and the Wittig half-reaction of unstabilized ylides with aldehydes indicates that the step leading to oxaphosphetane formation has a 2-center transition state with very important ionic character, where the C-C bond is 30-50% formed and there is no evidence of P-O bond formation (*1, 8*). This process is best described as a very asynchronous cycloaddition. The process may be further classified as a symmetry-forbidden (F) nonradical (N) process (*19*), a borderline two-step mechanism. These transition states are "syn" pseudo betaine structures in which the "syn" geometry is partly a consequence of a strong interaction of the positively charged phosphorus atom and the negatively charged oxygen atom. For reactions carried out in solution, external polarizing effects by metal ions in conjunction with their solvation or complexation of the positively charged phosphorus atom by Lewis bases present in the medium may well upset the attractive electrostatic interaction of the phosphorus and oxygen atoms.

With the use of the present molecular modeling approach, we have presented a justification for the greater stability of Z-oxaphosphetane generating metallated structures, which do indeed show an antiperiplanar geometry in the global selection of conformational minima as originally suggested by Bergelson et al. Whether the structure under consideration is analogous to a betaine intermediate or an "anti" pseudo betaine transition state, external polarizing effects will definitely stabilize these structures.

We, like others, have found that Molecular Orbital calculations predict that betaines are high energy intermediates in the Wittig half-reaction (*20*). However, since these calculations neglect external polarizing effects by solvation, aggregation or metal salts effects, these results are not unexpected. For real Wittig reactions carried out in homogeneous solutions of polar solvents, under salt free conditions, unstabilized ylides are known to give Z-stereoselective products; stabilized ylides, E-stereoselectivity products; and semistabilized ylides, mixtures of Z and E products. The only apparently reasonable rationalization for Z stereoselectivity would be the result of an anti "approach" of the reactants. The formation of an "anti" transition state with the same pseudo betaine characteristics described by the use of MNDO-PM3 calculations is conceivable for the reactions of unstabilized ylides in the presence of metal ions or sufficiently polar solvents through polarization by these external elements (*21*), as depicted in Figure 5.

We have recently carried out MNDO-PM3 calculations, as previously described, to model the Wittig reaction of stabilized ylides with aldehydes. Unlike the case of the Wittig half-reaction of unstabilized ylides, an evaluation of the transition

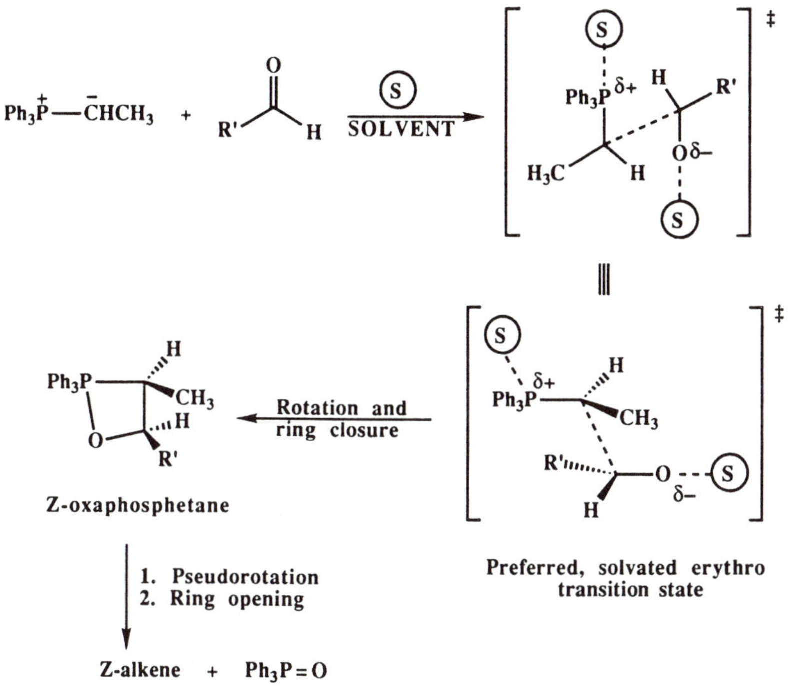

Figure 5.   Possible mechanism for the preferred formation of the Z-oxaphosphetane in solution.

states of these reactions provides evidence of both partial P-O bonding (5-10%) and partial C-C bonding (30-40%). This process is rather like a synchronous cycloaddition, where the planar "syn" geometry of the transition state is forced by the occurrence of concomitant P-O covalent bonding. As previously discussed, this "syn" geometry invariably leads to preferential formation of E-oxaphosphetanes and thus to E-stereoselectivity of the Wittig reaction.

Two mechanistic variations can be envisioned to be operative in some Wittig reactions. In some cases (Z-stereoselective reactions) a very asynchronous reaction with a 2-center "anti" pseudo betaine transition state is involved. The other mechanism involves a more synchronous concerted reaction with a 4-center "syn" transition state. These two mechanisms may compete, e.g., in the reactions of semistabilized ylides with aldehydes where mixtures of Z- and E-alkenes are obtained. Further details of these studies will be provided in future publications, as well as additional data about solvent effects.

Acknowledgment

The work was enabled by a grant of resources by the University of Massachusetts Computer Services.

Literature Cited

1. (a) Molecular Modeling of the Wittig Reaction, 4. (b) Previous paper in this series: Marí, F.; Lahti, P. M.; McEwen, W. E. *J. Am. Chem. Soc.*, in press.
2. (a) Wittig, G.; Schollkopf, U. *Chem. Ber.* **1954**, *87*, 1318. (b) Bergelson, L. D.; Shemyakin, M. M. *Tetrahedron* **1963**, *19*, 149. (c) Bergelson, L. D.; Shemyakin, M. M. *Angew. Chem. Inter. Ed.* **1964**, *3*, 250. (d) Bergelson, L. D.; Vaver, V. A.; Barsukov, L. I.; Shemyakin, M. M. *Tetrahedron Lett.* **1964**, *38*, 2669. (e) Bergelson, L. D.; Barsukov, L. I.; Shemyakin, M. M. *Tetrahedron* **1967**, *23*, 2709. (f) House, H. O.; Jones, V. K.; Frank, G. A. *J. Org. Chem.* **1964**, *28*, 3327. (g) Schlosser, M. Christmann, K. F. *Justus Liebigs Ann. Chem.* **1967**, *1*, 708. (h) Schneider, W. P. *Chem. Comm.* **1969**, 785. (i) Thacker, J. D.; Whangbo, M. H.; Bordner, J. *Chem. Comm.* **1979**, 1072. (j) Schlosser, M.; Tuong, H. B. *Ang. Chem. Int. Ed. Engl.* **1979**, *18*, 633. (k) Allen, D. W. *J. Chem. Res.* **1980**, 384. (l) Bestmann, H. J. *Pure & Appl. Chem.* **1980**, *52*, 771. (m) Bestmann, H. J.; Vostrowsky, O. *Top. Curr. Chem.* **1983**, *109*, 85.
3. (a) Vedejs, E.; Snoble, K. A. *J. Am. Chem. Soc.* **1973**, *95*, 5778. (b) Allen, D. W.; Ward, H. *Tetrahedron Lett.* **1979**, 2707. (c) Vedejs, E.; Meier, G. P.; Snoble, K. A. *J. Am. Chem. Soc.* **1981**, *103*, 2823. (d) Vedejs, E.; Marth, C. F. *J. Am. Chem. Soc.* **1988**, *110*, 3948. (e) Vedejs, E.; Fleck, T. J. *J. Am. Chem. Soc.* **1989**, *111*, 5861. (f) Vedejs, E.; Marth, C. F. *J. Am. Chem. Soc.* **1990**, *112*, 3905. (g) Schlosser, M.; Schaub, B. *J. Am. Chem. Soc.* **1982**, *104*, 3989. (h) Schlosser, M.; Oi, R.; Schaub, B. *Phosphorus and Sulfur* **1983**, *18*, 171.
4. (a) Gosney, I.; Rowley, A. G., in *Organophosphorus Reagents in Organic Chemistry*; Cadogan, J. I. G., Ed.; Academic Press: New York, **1979**; pp 15-253 and references cited therein. (b) Maryanoff, B.; Reitz, A. E. *Chem. Rev.* **1989**, *89*, 863 and references cited therein.
5. Anderson, R. J.; Henrick, C. A. *J. Am. Chem. Soc.* **1975**, *97*, 4327.
6. (a) Froyen, P. *Acta Chem. Scand.* **1972**, *26*, 2163. (b) Aksnes, G.; Khalil, F. Y. *Phosphorus Relat. Group V Elem.* **1973**, *3*, 37, 79, 109.
7. (a) Ward, W. J.; McEwen, W. E. *Phosphorus and Sulfur* **1989**, *41*, 393. (b) Ward, W. J.; McEwen, W. E. *J. Org. Chem.* **1990**, *55*, 493. (c) Loupy, A.;

Tchoubar, *Effets de sels en chimie organique et organometalique*, Ch. 5, Dunod: Paris, **1988**. (d) Tamura, R.; Saeguse, K.; Kakihana, D.; Oda, D. *J. Org. Chem.* **1988**, *53*, 2723.

8. Marí, F.; Lahti, P. M.; McEwen, W. E. *Heteroatom Chem.* **1991**, *2*, 265.
9. Yamataka, H.; Hanafusa, T.; Nagase, S.; Kurakake, T. *Heteroatom Chem.* **1991**, *2*, 465.
10. (a) Maryanoff, B. E.; Reitz, A. B. *Tetrahedron Lett.* **1985**, *38*, 4587. (b) Maryanoff, B. E.; Reitz, A. B.; Mutter, M. S.; Inners, M. S.; Almond, H. R. *J. Am. Chem. Soc.* **1985**, *107*, 1068. (c) Maryanoff, B. E.; Reitz, A. B.; Nortey, S. O.; Jordan, A. D.; Mutter, M. S. *J. Org. Chem.* **1986**, *51*, 3302. (d) Maryanoff, B. E.; Reitz, A. B.; Mutter, M. S.; Inners, M. S.; Almond, H. R.; Whittle, R.; Olofson, R. R. *J. Am. Chem. Soc.* **1986**, *108*, 7664. (e) Maryanoff, B. E.; Reitz, A. B.; Graden, D. W.; Almond, H. R. *Tetrahedron Lett.* **1989**, *30*, 1361.
11. Jones, E.; Trippet, S. *J. Chem. Soc. C* **1966**, 1090.
12. Baughman, L. L., Ph.D. Thesis, 1991, University of Massachusetts at Amherst.
13. The MMX force field embodies many years of work by J. J. Gajewski and K. Gilbert, University of Indiana, Bloomington, IN. The MMX program is available from Serena Software through Professor K. Steliou, University of Montreal, Quebec, Canada.
14. Hammond, G. S. *J. Am. Chem. Soc.* **1955**, *77*, 334.
15. R. R. Holmes, *Pentacoordinated Phosphorus*, Vol. I, ACS Monograph 175, American Chemical Society, Washington, 1980.
16. Marí, F.; Lahti, P. M.; McEwen, W. E. *Heteroatom Chem.* **1990**, *1*, 255.
17. (a) Burkert, U.; Allinger, N. *Molecular Mechanics*, ACS Monograph 177, American Chemical Society, Washington, DC, 1982. (b) Clark, T. *A Handbook of Computational Chemistry*, Wiley, New York, 1985.
18. Kirkpatrick, S.; Gelatt, C. D.; Vecchi, M. P. *Science* **1983**, *220*, 671.
18. Lee, B. K.; Richards, F. M. *J. Mol. Biol.* **1971**, *55*, 400.
19. Fukui, K. *Top Curr. Chem.* **1970**, *15*, 1.
20. (a) Höller, R.; Lischka, H. *J. Am. Chem. Soc.* **1980**, *102*, 4632. (b) Volatron, F.; Eisenstein, O. *J. Am. Chem. Soc.* **1987**, *109*, 1.
21. Carey, F. A.; Sundberg, R. J. *Advanced Organic Chemistry*, 2nd Ed., Plenum, New York, 1984.

RECEIVED December 3, 1991

Chapter 13

Reaction of 2-Amino-4-Substituted Phenols with Aryl Phosphorodichloridates

Formation of a Novel Spirobis(1,3,2-benzoxazaphosphole) System from an Unexpected Elimination of Aryloxy Moieties

C. D. Reddy[1], K. D. Berlin[2], M. ElMasri[2], S. Subramanian[2],
S. V. Mulekar[2], R. S. Reddy[1], S. B. S. Reddy[1], L. Esser[3], X. Ji[3],
and D. van der Helm[3]

[1]Department of Chemistry, Engineering College, Sri Venkateswara
University, Tirupati 517 502, India
[2]Department of Chemistry, Oklahoma State University,
Stillwater, OK 74078
[3]Department of Chemistry, University of Oklahoma, Norman, OK 73019

A series of experiments was performed in which three systems consisting of 2-amino-4-substituted-phenols were allowed to react with aryl phosphorodichloridates in boiling toluene in the presence of triethylamine. Novel spirobis[1,3,2-benzoxazaphosphole] systems were obtained. With 2-amino-4-methylphenol, a pentavalent phosphorus-containing intermediate **1** probably formed and reacted with a third equivalent of the amino-phenol starting reagent to yield **2a**. Analysis of **2a** in DCCl$_3$ via ^{31}P NMR spectroscopy revealed two narrowly separated signals which suggested an equilibrium in solution involving two forms. Condensations using 2-amino-4-t-butylphenol and 2-amino-4-chlorophenol proceeded similarly to give related solids **4** and **5**, respectively. Single crystal X-ray diffraction analysis of crystalline **2a** confirmed the structure as 8,8''-dimethyl-2-(2-amino-4-methylphenoxy)-2λ^5-2,2'(3H,3H')spirobis[1,3,2-benz-oxazaphosphole] possessing one equatorial and two apical P-O bonds and two equatorial P-N bonds.

Dibenzobicyclic phosphoranes containing a spirobis[1,3,2-benzoxazaphosphole] unit are relatively uncommon, are not easily formed, and have been rarely subjected to X-ray analysis (1-3). Related members of **1** have been studied (4), but relatives of **2** are scarce in the literature. Certain heterocyclic phosphoranes possess geometry with some distortion that suggests an intermediate structure close in energy to a trigonal bipyramid (TBP), rectangular pyramid (RP), or square pyramid (SP) (2,3,5). Such distortion can be manifested in low angular deformation, associated with the Berry pseudorotation process (5-7) which affects intramolecular ligand exchange (5,7). Concerning the pseudorotation process from a TBP to a SP arrangement, a correlation may exist between the electronegativity of substituents, the ring structure of phosphoranes, and possibly the electronegativity of the atom bonded to the pivotal

0097–6156/92/0486–0162$06.00/0

ligand (5,7). An assessment of the distribution of bonds is therefore important in a dibenzobicyclic phosphorane with atoms possessing different electronegativities.

Results And Discussion

We have discovered that 2-amino-4-methylphenol condenses with different aryl phosphorodichloridates in boiling toluene/triethylamine to produce (via suspected intermediate **1**) the novel spirobis[1,3,2-benzoxazaphosphole] **2**. One product was obtained from *all* reactions with the assumed elimination of an aryloxy moiety. Apparently the high temperature and increased basicity of the amino group (presumably held out of the ring plane to some extent by the OH group) of the starting material induced conversion of intermediate **1** to **2**.

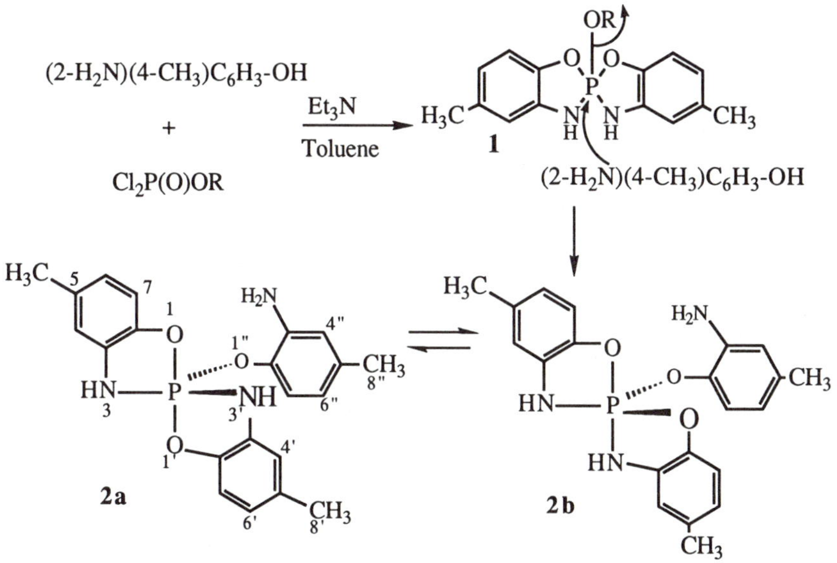

$[R = C_6H_5; 4\text{-}H_3CC_6H_4; 4\text{-}ClC_6H_4; 3,4\text{-}(H_3C)_2C_6H_3]$

Analysis of solid product **2** revealed a very faint color in the ferric chloride test, which was *not* conclusive for the presence of a phenolic hydroxyl group. Absorption (broad 3450-3400 cm^{-1}) occurred in the IR spectrum for N-H (O-H stretching might be obscured). The ^{1}H NMR signals included doublets for $^2J_{PNH}$ (8) but a high ^{1}H signal density at 300 or 400 MHz allowed only the ratio of methyl signals for $H_3C(8,8')$ and $H_3C(8'')$ at δ 2.25 and δ 2.15 respectively, to be determined as close to 2:1. Signals for aromatic protons were somewhat broad (δ 6.26-6.58) but were not convincing in terms of confirming the presence of more than one isomer .

Complex signal patterns in the ^{13}C spectrum with $^2J_{PC}$ couplings and $^3J_{PC}$ couplings were not discernible regarding isomers in solution. Interestingly, the ^{31}P NMR spectrum showed a large signal (2, 9) at -41.39 ppm and a very small signal at -42.8 ppm. Thus, two similar phosphorus atoms were apparently present but in only slightly differing environments and yet widely differing amounts. The ratio of ^{31}P signals was approximately 99:1 (or greater) which was *not* supportive of a coupling. The existence of a slow equilibrium is feasible involving two TBP systems like **2c** or **2d** or an SP system like **2e**. However, assuming that the more electronegative oxygen atoms will occupy the apical positions, that the two ^{31}P NMR signals indicate a similar environment, and that the ferric chloride test is *not* legitimate, we

suggest that **2a** and **2b** exist in an equilibrium, but **2c** cannot be eliminated from consideration. It seems less tenable that an equilibrium could be involved between a closed form, like **2a**, and an "open form" with a tetrasubstituted phosphorus atom since the ^{31}P signals found in our work have very similar shifts. Allcock and Kugel (10) reported a slightly related system (from a different reaction) whose IR spectrum (nujol mull) implied the presence of two isomers.

2-Amino-4-*t*-butylphenol and 2-amino-4-chlorophenol behaved similarly under identical conditions. A solid product was obtained in each case which melted at 205.5-207°C (**4**) and 206-207°C (**5**), respectively. A mixture melting point determination of a mixture of the two materials melted over a range (179-186°C) to confirm the individual identities. Spectral data for both products are in the Experimental.

Because of the ambiguity of the ^{1}H and ^{31}P NMR data for **2a**, a variable temperature study was performed on samples of **2a**, **4**, and **5** in DCCl$_3$. Except for a few very minor changes in chemical shifts, the proton and phosphorus NMR spectra of **2a** were unaffected from +55°C to -55°C. Room temperature ^{1}H NMR

analysis was unrevealing on **4** & **5**, but, surprisingly, both compounds at room temperature gave two sets of ^{31}P NMR signals observable at -40.86 and -42.67 (**4**) ppm and -40.7 and -41.9 ppm (**5**) in ratios of 3:1 and 14.4:1, respectively. For **4**, the ratio of ^{31}P signals was unaltered at -30°C and at -55°C as well. In contrast, the areas of ^{31}P signals for **5** changed drastically to give decreasing ratios of 14.4:1 (RT), 3.1:1 (-30°C), and 1.18:1 (-55°C) as the temperature was lowered. It is tentatively assumed that the isomer in each case with the ^{31}P signal at lower field is structurally related to **2a**, namely **4** and **5** as illustrated. There is no intuitively obvious reason for the differences in ratios among isomers in **2a**, **4**, or **5** nor for the sharp change in ratios of presumably an isomer of **5** as the temperature of the solution is lowered.

An analysis of the possible mechanism for the formation of **2a** (and **4** and **5** as well) must be speculative at this time. Initial attack by the nitrogen atom of the starting aminophenol on phosphorus is reasonable but one cannot eliminate the possibility of an alternative attack by the oxygen atom on phosphorus. Whatever the structure of the initial product may be, a series of pseudorotations is necessary to generate the final structures illustrated. Small ligand exchange energies involving two five-membered rings of the general type present in **2a**, **4**, and **5** have been noted by Wolf and co-workers (11). Related systems with four oxygen atoms and phosphorus in each of two rings have also been observed to undergo rapid ligand exchange (12).

It is worthwhile to discuss briefly the mass spectral analysis for one member, namely **2a**. The M$^{+\cdot}$ is observed at m/z 395 along with principal ions at m/z 273 and 152 via EI mass spectrometry. The latter two large masses are tentatively ascribed to the spirobis[1,3,2-benzoxazaphosphole] and benzoxazaphosphole groups, respectively. There appears to be a stepwise loss of 2-amino-4-methylphenol fragments which supports the presence of three units of 2-amino-4-methylphenol in **2a**. The molecular ion for **5** was only obtainable by FAB (protic) mass spectrometry The EI mass spectra of **5** did furnished some fragments. Generally, the fragments of **2a** and **5** resemble each other except for substituent differences. Similar to **5**, it was only possible to obtain the [M$^{+\cdot}$ + 1] ion for **4** by FAB (protic type) mass spectrometry.

An X-ray analysis confirmed solid **2a** (13). The structure was solved by direct methods using SHELXS-86 (14) with refinement via SHELX76 (15) utilizing a total of 2867 reflections and anisotropic temperature factors for nonhydrogens to an R factor of 0.038 (R$_w$ = 0.036). A perspective view of a single molecule is illustrated (Figure 1) (16) in which A, B, and C represent three ligands [bidentate ligand O(1) through N(3), through O(1') and N(3'), and monodentate through O(1")]. The compound **2a** is a distorted TBP with atoms N(3), N(3'), and

O(1") in equatorial positions while O(1) and O(1') are in apical positions. The distortion of the phosphorus coordination from a TBP is indicated by bond angles at the phosphorus atom. In the equatorial plane are N(3)-P(2)-N(3') = 131.4, N(3)-P(2)-O(1") = 114.1°, and N(3')-P(2)-O(1") = 114.5° which are neither equal to 120° or to each other but do total to 360° [Table I]. This implies the P atom lies in the equatorial plane. The equatorial bond P(2)-O(1") is shorter by 0.03 Å than the two P(2)-N bonds as expected. The apical bonds make an angle of 175.4°, being reduced from 180° because of short interactions (2.711 Å) [Table II] between atoms O(1') and C(7a") which also produce an obtuse angle of 95.6° between apical bonds P(2)-O(1') and P(2)-O(1") and induce the inclination of the apical P(2)-O(1') bond in the direction of the O(1)-P(2)-O(1") plane. The P(2)-O(1) bond is nearly perpendicular to the equatorial plane as evidenced by the angles (89.2°, 89.3°, and 88.9°) made with the three equatorial bonds. Two bidentate ligands and the P atom form two five-membered rings, O(1) through C(7a) and O(1') through C(7a') (Figure 1). Both rings are flat (RMS = 0.03 Å), have apical-equatorial orientation

Table I. Selected Bond Angles and Torsion Angles with e.s.d.'s in Parentheses

Selected Bonds	Bond Angles	Selected Torsion Angles	
O(1)-P(2)-N(3)	89.2 (1)	O(1)-C(7a)-C(3a)-N(3)	-1.7 (2)
P(2)-N(3)-C(3a)	116.0 (2)	O(7a)-C(3a)-N(3)-P(2)	-3.6 (2)
N(3)-C(3a)-C(7a)	108.9 (2)	C(3a)-N(3)-P(2)-O(1)	6.0 (2)
C(3a)-C(7a)-O(1)	112.1 (1)	N(3)-P(2)-O(1)-C(7a)	-6.8 (1)
C(7a)-O(1)-P(2)	113.2 (1)	P(2)-O(1)-C(7a)-C(3a)	6.1 (2)
O(1')-P(2)-N(3')	89.5 (1)	O(1')-C(7a')-C(3a')-N(3')	1.8 (2)
P(2)-N(3')-C(3a')	115.6 (1)	O(7a')-O(3a')-N(3')-P(2)	-6.5 (2)
N(3')-C(3a')-C(7a')	108.8 (2)	C(3a')-N(3')-P(2)-O(1')	7.0 (1)
C(3a')-C(7a')-O(1')	112.6 (1)	N(3')-P(2)-O(1')-C(7a')	-5.8 (1)
C(7a')-O(1')-P(2)	112.9 (1)	P(2)-O(1')-C(7a')-C(3a')	3.2 (2)
C(2")-O(1")-P(2)	126.9 (2)	P(2)-O(1")-C(2")-C(3")	118.9 (2)
O(1)-P(2)-N(3)	89.2 (1)		
O(1)-P(2)-N(3')	89.3 (1)		
O(1)-P(2)-O(1")	88.9 (1)		
O(1')-P(2)-N(3)	88.2 (1)		
O(1')-P(2)-N(3')	89.5 (1)		
O(1')-P(2)-O(1")	95.6 (1)		
O(1)-P(2)-O(1')	175.4 (1)		
N(3)-P(2)-N(3')	131.4 (1)		
N(3)-P(2)-O(1")	114.1 (1)		
N(3')-P(2)-O(1")	114.5 (1)		

and possess almost identical bond angles and distances. Presumably the rings are C_2 half chairs (17) with asymmetry parameters ΔC_2 [O(1)-P(2)] = 1.3 for **A** and ΔC_2 [P(2)-N(3')] = 1.1 for **B** (ΔC_2 is 0.0 for an ideal half-chair conformation of five-membered rings).

In summary, spectral data support the TBP for isomers **2a /2b** in solution while X-ray data confirm TBP **2a** as the solid product. The driving force for the reaction is suggested to arise from ease of displacement of the aryloxy group from the intermediate **1** by the nucleophilic N (or O) atom in a third equivalent of the starting marterial as illustrated. Assuming the entry by N is apical, the initial product is probably converted to **2a** or **2b** by a Berry pseudorotation process (2). Presumably the same mechanism operates in the reaction of 2-amino-4-*t*-butylphenol and 2-amino-4-chlorophenol with the corresponding aryl phosphorodichloridates to give **4** and **5**, respectively. There are much higher concentrations of two isomers in solutions of **4** and **5** compared to **2a** which, in DCCl$_3$, has only one major compound as evidence by one very large ^{31}P signal (18). From the X-ray analysis it is clear that the more electronegative oxygen atoms occupy the apical positions rather than the nitrogen atoms, at least in **2a**. Although the yields are modest, the procedure is simple and thus the method has potential for obtaining a variety of such pentacoordinated phosphorus systems.

Table II. Bond Distances with e.s.d's In Parentheses

Bond	A[a]	B	C
P(2)-O(1)	1.705 (1)	1.708 (1)	1.624 (1)
P(2)-N(3)	1.651 (2)	1.654 (2)	
O(1)-C(7a)[b]	1.371 (2)	1.380 (3)	1.399 (2)
N(3)-C(3a)	1.392 (2)	1.407 (2)	1.396 (3)
C(3a)-C(4)	1.377 (3)	1.376 (4)	1.393 (2)
C(4)-C(5)	1.395 (3)	1.401 (2)	1.383 (3)
C(5)-C(6)	1.387 (3)	1.383 (4)	1.390 (3)
C(6)-C(7)	1.398 (3)	1.399 (4)	1.386 (2)
C(7)-C(7a)	1.374 (2)	1.376 (3)	1.371 (3)
C(7a)-C(3a)	1.388 (3)	1.382 (3)	1.389 (3)
C(5)-C(8)	1.510 (4)	1.510 (4)	1.520 (2)

[a]A, B, and C represent three ligands [bidentate ligand O(1) through N(3), and O(1') through N(3'), and monodentate ligand O(1") through N(3")].
[b]In C, C(7a) is C(2) and C(3a) is C(3) (see Figure 1).

Experimental Section

General Information. The ^{1}H and ^{13}C NMR spectra were recorded at 400 MHz and 100 MHz, respectively, with a Varian XL-400 spectrometer on solutions of the compounds in DCCl$_3$ and with TMS as the internal standard. All ^{13}C NMR spectra were obtained with proton decoupling as were the ^{31}P NMR spectra which were recorded at 161.9 MHz (referenced to external 85% H$_3$PO$_4$). Chemical shifts downfield from the reference are positive and upfield are negative. Some early NMR experiments were done at 300 MHz. All experiments were performed under N$_2$. Off-resonance experiments confirmed the carbon signals.

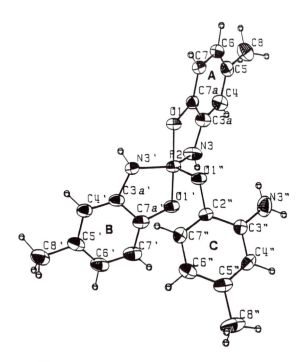

Figure 1. ORTEP[16] drawing of **2a** in which **A**, **B**, and **C** represent three ligands.

8',8"-Dimethyl-2-(2-amino-4-methylphenoxy)-2λ⁵-2,2'(3H,3'H)-spirobis[1,3,2-benzoxazaphosphole] (2a). 2-Amino-4-methylphenol (2.46 g, 0.02 mol) and triethylamine (2.02 g, 0.02 mol) in 60 mL of very dry toluene (dried over sodium and freshly distilled)were heated at reflux in standard equipment. A solution of phenyl phosphorodichloridate (2.11 g, 0.01 mol) in 26 mL of dry toluene was added dropwise. After 6 h at reflux, TLC analysis indicated the reaction was complete. Filtration removed triethylamine hydrochloride, and the remaining solution was evaporated to a solid. After being washed (water), the residual solid was recrystallized (2-propanol) to yield **2a** (0.57 g, 21.6%). Two additional recrystallizations gave an analytical sample with mp 226-228°C. IR (KBr) 3450 (b, N-H), 3400, 1240, 1020, 950, (P-O-C), 940 cm⁻¹. ¹H NMR (DCCl₃) δ 2.15 [s, H_3C(8")], 2.25 [s, H_3C(8,8')], 2.85 [bs, NH_2], 5.58 [d, ²J_{PNH} = 20.4 Hz, NH (3,3')], 6.26-6.58 [m, Ar-H]. ¹³C NMR analysis (DCCl₃ with TMS) ppm 20.9, 21.2 [CH₃, (8, 8', 8")], 109.1, 109.3, 110.32 [C (4, 4', 6")], 110.4 [C(7, 7', 3a")], 116.7, 119.0, 120.7, 120.8, 121.0, 1121.1, 129.6, 130.0, 130.2, 134.4, 134.5, 137.9, 138.0; other aromatic carbons had signals at 143.2, 143.3 [C(3a, 3a',7a")], 151.5, 151.6 [C(7a, 7a', 2")] ppm. ³¹P NMR (DCCl₃) ppm -41.39, -42.8. Mass spectral (EI) analysis: 395 (M⁺·) [2.9%], 273 (M⁺· - C₇H₈NO)⁺· [86%], 168 [(M⁺· - C₇H₈NO) - C₇H₇N]⁺· [5%], 152 [(M⁺· - C₇H₈NO) - C₇H₇NO]⁺· [65%], 137 (C₇H₆OP)⁺·, 122 (C₇H₈NO)⁺· [2%]. Anal. Calcd for C₂₁H₂₂N₃O₃P: C, 63.79; H, 5.61. Found: C, 63.77; H, 5.79.

When the same reaction was run in benzene or THF, only the same product could be isolated. An identical product was obtained when *p*-tolyl phosphorodichloridate or 4-chlorophenyl phosphorodichloridate was employed.

8',8"-Di-*t*-butyl-2-(2-amino-4-*t*-butylphenoxy)2,2'-(3H,3H')spirobis[1,3,2-benzoxazaphosphole] (4). 2-Amino-4-*t*-butylphenol (3.30 g, 0.02 mol) and triethylamine (2.02 g, 0.02 mol) in dry toluene (60 mL) were treated with phenyl phosphorodichloridate (2.11 g, 0.01 mol) in dry toluene (26 mL) in a dropwise manner. After 6 h at reflux, triethylamine hydrochloride was filtered off and the solution was concentrated to a semisolid. Washing with water gave a solid which was allowed to dry. Additional washing with HCCl₃:petroleum ether (1:2) removed a gummy like material and yielded crude **4**. Recrystallization (HCCl₃:petroleum ether, 1:2) of this solid gave pure **4** (0.97 g, 27.5%) which melted at 205.5-207°C. IR (KBr) 3480-3380 (N-H), 2960 C-H), 1240 (P-O-C), 1020, 965, 740 cm⁻¹. ¹H NMR (DCCl₃) δ 1.20 [s, 9 H, 3 H₃C, H(9")], 1.27 [s, 18 H, 6 H₃C, H(9,9')], 3.55 [bs, NH_2], 5.51 [d, ²J_{PH} = 20.1 Hz, NH (3,3')], 6.44-6.81 [m, ArH]. ¹³C NMR (DCCl₃) ppm 30.56 [C(9")], 30.87 [C(9,9')], 33.4 [C(8")], 33.64 [C(8,8')], 106.1 [d, C(7'), ³J_{PC} = 17.41 Hz], 108 [d, C(7,7"), ³J_{PC} =10.0 Hz], 112.5, 114.5, 116.2, 120.0, 128.,9, 129, 136.73 [d, C(3,3'), ²J_{PC} = 5.2 Hz], 137.45 [d, C(3"), ²J_{PC} = 10.9 Hz], 142.4 [d, C(2,2'), ²J_{PC} = 3.0 Hz], 147.0 [C(2"), ²J_{PC} = 2.5 Hz]. ³¹P NMR (DCCl₃) ppm -40.83, -42.64. Mass spectral (FAB-protic) analysis: Calcd [M⁺· + 1]: 522.2886; Found: 522.2855. An attempt to obtain an EI mass spectrum failed under several conditions. Anal. Calcd for C₃₀H₄₀N₃O₃P: C, 69.07; H, 7.73; N, 8.05. C₃₀H₄₀N₃O₃P·0.25 H₂O: C, 68.48; H, 7.66; N. 7.98. Found: C, 68.30 ; H, 7.70; N, 7.98.

Compound **4** was obtained when either 2-chlorophenyl phosphorodichloridate or 2,3-dimethylphenyl phosphorodichloridate was utilized.

8',8"-Dichloro-2-(2-amino-4-chlorophenoxy)-2,2'-(3H,3H')spirobis[1,3,2-benzoxazaphosphole] (5). 2-Amino-4-chlorophenol (2.80 g, 0.02

mol) and triethylamine (2.02 g, 0.02 mol) in dry toluene (60 mL) were treated dropwise with phenyl phosphorodichloridate (2.4 g, 0.01 mol) in dry toluene (26 mL). After 6 h, the reaction was terminated, and triethylamine hydrochloride was filtered off. Evaporation of the solvent gave a solid which was washed (H_2O) and dried. An initial washing with $HCCl_3$:petroleum ether (1:2), followed by two recrystallizations from the same system, gave pure **5** (0.93 g, 30.5%); mp 206-207°C. IR (KBr) 3440 (N-H), 1500, 1230 (P-O-C), 950, 800 cm^{-1}. ^{1}H NMR ($DCCl_3$) δ 3.68 [bs, NH_2], 5.67 [d, $^2J_{NH}$ = 20.5 Hz, NH (3,3')], 6.75-7.4 [b, ArH]. ^{13}C NMR ($DCCl_3$) ppm 109.47, 109.57, 109.67, 115.12, 119.09, 121.69, 121.72, 123.31, 128.25, 128.37, 128.40, 132.67, 132.87, 137.81, 137.92, 141.4, 141.45, 144.17, 144.21. ^{31}P ($DCCl_3$) ppm -40.71, -41.99. Mass spectral (FAB-protic) analysis: Calcd. [M$^{+\cdot}$ + 1]: 454.9; Found: 455. The exact mass could not be obtained. Mass spectral (EI) analysis: 312.9893 (M$^{+\cdot}$ - 141.9866; 100%), 171 (28.3%), and 142 (92.5%). Anal. Calcd for $C_{18}H_{13}Cl_3N_3O_3P$: C, 47.34; H, 2.87. Found: C, 47.26; H, 2.79.

When this procedure was used with 2-chlorophenyl phosphorodichloridate or 2,3-dimethyl phosphorodichloridate, only **5** could be isolated.

Crystal Structure Determination and Refinement of 2a. The data crystal used for X-ray investigation was as small as 0.03 x 0.07 x 0.42 mm. Data were collected at -110(1)°C with an Enraf-Nonius CAD4 automatic counter diffractometer controlled by a Micro VAX II computer and fitted with a low-temperature apparatus. Ni-filtered CuKα (CuKα, λ = 1.54178 Å, μ = 14.04 cm^{-1}) was used. Fifty reflections with 33°>θ>15° and CuKα$_1$ wavelength (λ = 1.54051 Å) were used for lattice constants. A total of 4300 reflections with 1.0° ≤ 2θ ≤ 150.0° in −12 ≤ h ≤ 12, −16 ≤ k ≤ 16, 0 ≤ l ≤ 9 were measured with the ω - 2θ scan technique. The profiles of all the reflections were observed and stored. The maximum scan time for a single reflection was 60 seconds. The scan angle was calculated as (0.08 + 0.20 tanθ)°. The receiving aperture had a variable width calculated as (2.40 + 0.86 tanθ) mm, while its height remained constant at 6 mm. Three intensity control monitors were measured every 7200 seconds of X-ray exposure time and they showed a maximum difference of 0.032 and an e.s.d. of 0.005. No systematic absence was observed. The data set was processed by using a profile analysis method and programs.[19] Lorentz-polarization corrections and numerical absorption correction[20] were applied. Among 3926 unique data, there were 2867 observed ones with I ≥ 2σ(I). The structure was solved by using the direct methods with the program SHELXS-86.[14] The structure was refined with SHELX76[15] utilizing 2867 observed reflections and anisotropic temperature factors for nonhydrogen atoms. The locations for all hydrogen atoms were determined from successive difference Fourier synthesis and refinement isotropically. A final R of 0.038 and R_w of 0.036 were obtainable by the minimization of $\Sigma\omega(|F_o|-|kF_c|)^2$ with ω = $1/\sigma^2$(F). The maximum shift/e.s.d. in the last cycle of the refinement was 0.022 for nonhydrogen atoms and 0.019 for hydrogens. The maximum and minimum electron densities in the final difference Fourier map were 0.21 and -0.31 e/Å^3. The error of fit EOF = $[\Sigma\omega(F_o - F_c)^2/(N - NP)]^{1/2}$ was 1.7, where N was the number of data used (N = 2867) and NP was the number of parameters refined (NP = 341).

Acknowledgment. Support via a grant from Hindustan Lever Research

Foundation, Bombay, India, is gratefully acknowledged. The National Science Foundation (U.S.A.) is also gratefully acknowledged (KDB) for a grants both to purchase (CHE81-06152) the XL-300 NMR unit and to upgrade (DBM-860-3864) the spectrometer and to purchase the XL-400 NMR unit (CHE-8715180). We (KDB) thank the College of A & S of the Oklahoma State University for partial support in the form of salary. Support is gratefully acknowledged from the National Cancer Institute (DvdH) for a grant CA-17562.

Supplementary Material Available: The X-ray crystallographic data for **2a** (17 pages) can be obtained from the one author (DvdH) upon request.

Literature Cited

1. McEwen, W. E.; Berlin, K. D., Eds., *Organophosphorus Stereochemistry, Part II P(V) Compounds,* Dowden, Hutchinson, and Ross: Stroudsburg, PA, 1975, Section 3.

2. Holmes, R. R. *Pentacoordinate Phosphorus, Vol. I and II*, ACS Monograph 175 and 176, American Chemical Society: Washington, D. C., 1980. See also: Meunier, P. F.; Day, R. O.; Devillers, J. R.; Holmes, R. R. *Inorg. Chem.* **1978**, *17*, 3270-3276.

3. Riess, J. G. *Unusual Ligand Modes of Phosphorus Compounds: Phosphoranides*, in *Phosphorus-31 NMR Spectroscopy in Stereochemical Analysis*, Verkade, J. G.; Quin, L. D., Eds., VCH Publishers: Deerfield, FL, 1987.

4. (a) Koizumi, T.; Watanabe, Y.; Yoshii, E. *Tetrahedron Letters* **1974**, 1075-1078. (b) Reddy, C. D.; Reddy, S. S.; Naidu, M. S. R. *Synthesis* **1980**, 12, 1004-1005.

5. Holmes, R. R.; Deiters, J. A. *J. Am. Chem. Soc.* **1977**, *99*, 3318-3326. See also Holmes, R. R., In *Progress in Inorganic Chemistry*, Lippard, S. J., Ed., Wiley: New York, NY, 1987; Vol. 32.

6. Berry, R. S. *J. Chem. Phys.* **1960**, *32*, 933-938.

7. Luckenbach, R. *Dynamic Stereochemistry of Penta-coordinated Phosphorus and Related Elements*, G. Thieme: Stuttgart, Germany, 1973.

8. Malavaud, C.; Barrans, J. *Tetrahedron Letters,* **1975**, 3077-3080. The value of +26 ppm for the ^{31}P signal in a compound related to **2a** may really be -26 ppm via the newer sign convention.

9. For a review of the ^{31}P NMR data in this field, see: Gorenstein, D. G. *Prog. NMR Spectr.* **1983**, *16*, 1-98. See also references 2 and 3.

10. Allcock, H. R.; Kugel, R. L. *J. Am. Chem. Soc* **1969**, *91*, 5452-5456.

11. Klaebe, A.; Cachapuz, A. C.; Brazier, J.-F.; Wolf, R. *J. Chem. Soc., Perkin II* **1974**, 1668-1671.

12. Houalla, D.; Wolf, R.; Gagnaire, D.; Robert, J. B. *Chem. Commun.* **1969**, 443- 444. For a recent review on cyclic oxaphosphoranes, see: Swamy, K. C.; Burton, S. D.; Holmes, J. M.; Day, R. O.; Holmes, R. R. *Phosphorus, Silicon, and Sulfur* **1990**, *53*, 437-455.

13. Crystal data for **2a**: Triclinic, P1(No. 2). Unit cell (at -110°C): a = 9.910(1), b = 13.326(2), c = 7.975(1) Å, α = 81.989(9), β = 106.460(8)°, γ =108.791(8) and V = 954.8 Å^3. Z = 2 for $C_{21}H_{22}PO_3N_3$ (M.W. = 395.40) and D_{calc} = 1.38 g cm^{-3}.

14. Sheldrick, G. M. *SHELXS-86*, Program for crystal struture determination; Institute fur Anorganische Chemie der Universitat: Federal Republic of Germany, 1986.

15. Sheldrick, G. M. *SHELXS76*, Program for crystal structure determination, University of Cambridge, England, 1976.
16. Johnson, C. K. *ORTEP*, Report ORNL-3794; Oak Ridge National Laboratory, Oak Ridge, Tennessee, 1965.
17. Duax, W. L.; Norton, D. A. *Atlas of Steroid Structure*, Vol. 1, Duax, W. L.; Norton, D. A., Eds., Plenum: New York, NY, 1975.
18. For ^{31}P data on related systems in solution, see reference 8 and: Sanchez, M.; Brazier, J. F.; Houalla, D.; Wolf, R. *Bull. Soc. Chim. Fr.* **1967**, 3920.
19. Blessing, R. H. *Cryst. Rev.* **1987**, *1*, 3-58.
20. DeTitta, G. T. *ABSORB*, Medical Foundation of Buffalo, Buffalo, NY, 1982.

RECEIVED December 3, 1991

Chapter 14

Bis(diphenylphosphino)methane in Organometallic Synthesis

New Applications

Alan N. Hughes

Department of Chemistry, Lakehead University, Thunder Bay, Ontario P7B 5E1, Canada

A new, versatile, one-step route to bis(diphenylphosphino)methane (dppm) metal carbonyl complexes involving $NaBH_4$ reduction of a metal salt/dppm mixture under CO has been developed. The reaction outcome is determined by careful control of the conditions. With Co^{II}, the complexes $[Co(\eta^2\text{-dppm})_2(CO)][Co(CO)_4]$, $Co_2(\mu\text{-dppm})_2(CO)_4$, $Co_2(\mu\text{-dppm})_2(\mu\text{-CO})_2(CO)_2$, and $Co_2(\mu\text{-dppm})_2(\mu\text{-}H)(\mu\text{-PPh}_2)(CO)_2$ can all be obtained as the major product. $CoBH_2(\eta^1\text{-dppm})(\eta^2\text{-dppm})(CO)_2$, the first stable metallaborane, is a minor product and other η^1- and η^2-dppm-Co-CO complexes can also be prepared. With Ni^{II}, $Ni(\eta^1\text{-dppm})_2(CO)_2$, $Ni_2(\mu\text{-dppm})_2(\mu\text{-}CO)(CO)_2$ and $[Ni_2(\mu\text{-dppm})_2(\mu\text{-PPh}_2)(CO)_2]X$ can all be prepared as major products. The heterobinuclear $CoRh(\mu\text{-dppm})_2(CO)_3$ is similarly prepared from $Co^{II}/Rh^{III}/dppm/BH_4^-/CO$ reactions and related (e.g., Co-Mn, Ni-Cu) complexes can be prepared by two-step processes. Pd^{IV} gives at least five Pd-dppm-CO complexes.

Bis(diphenylphosphino)methane (dppm) is a versatile ligand in coordination chemistry since it can act as a mono-coordinating, bridging or chelating species with metals. The small bite angle of dppm does not favor chelation and the majority of such complexes reported to date contain mono-coordinating or, more usually, bridging dppm. Indeed, the importance of bridging dppm in the stabilization of homo- and heterobinuclear (and cluster) complexes has received increasing recognition recently (1-4). Binuclear and small cluster complexes are of considerable interest because many reactions which take place with these systems, particularly those which contain metal-metal bonds and/or coordinatively unsaturated metal centers, are analogous (4) to those which take place on metal surfaces in the context of heterogeneous catalysis. Also in the context of catalysis, many metal phosphine, metal carbonyl, and metal carbonyl phosphine complexes are active catalysts for a variety of homogeneous processes (5). Metal carbonyl phosphine complexes are most often prepared by reaction of a metal carbonyl and a phosphine or, less commonly, by treatment of a metal phosphine complex with

0097–6156/92/0486–0173$06.00/0

CO. However, these procedures suffer from the disadvantages that metal carbonyls are generally highly toxic and are sometimes not readily available (e.g., unsubstituted Pd carbonyls are unknown) while the preparation of low oxidation state metal phosphine complexes suitable for reaction with CO is frequently a tedious process.

It is against this background that we have developed in our laboratories a very simple one-step procedure which allows the synthesis, usually in acceptable yields, of a wide variety of M^I and M^0-dppm-CO mono- and (more usually) binuclear complexes directly from metal salts. The procedure has been extended with considerable success to the one-step synthesis of certain heterobinuclear systems and also some phosphido-bridged complexes of Co^I and Ni^I. While this study has so far been largely concerned with Co^{II} and Ni^{II} systems, preliminary work has been carried out with Rh^{III}, Pd^{IV} and Fe^{III}. Some of the work described herein has been reported in preliminary communications but a complete overview of all aspects of the investigations carried out so far is given here in order that the very broad applicability of the procedure can be fully appreciated.

Co^{II}/dppm/NaBH$_4$/CO Reactions

The basic reaction used as the starting point for these studies is the reduction of a CO-purged solution (usually in ethanol/benzene) of a mixture of a metal salt and dppm with ethanolic $NaBH_4$ accompanied by further passage of CO for some minutes after the addition of the $NaBH_4$ has been completed. The outcome of this process is critically dependent upon minor changes in the reaction conditions with the reactant ratios and the rate of addition of the $NaBH_4$ being the most important variables.

Using variations of this procedure with the salts CoX_2 (X = Cl, Br), six products have been isolated and characterized of which four (obtained as major products) are binuclear systems and two (minor, but extremely interesting, products) are mononuclear. Thus, if the reactant ratios of Co^{II}, dppm and $NaBH_4$ are 1:2:2 and the addition time of the $NaBH_4$ is 30 minutes, the ion pair $[Co(\eta^2\text{-}dppm)_2(CO)][Co(CO)_4]$ ($\delta^{31}P$: -6.1; υ_{CO}: 2010, 1950, 1898, 1855, 1835 cm^{-1}), 1, is formed. This product was originally formulated (6) as $Co_2(\eta^2\text{-}dppm)_2(CO)_4$ containing an unsupported Co-Co bond but the structure has since been corrected (7) on the basis of both chemical and crystallographic studies, the latter of which has shown that the cation of 1 is a distorted trigonal bipyramid in which the CO ligand occupies an equatorial position. This cation has also been prepared by others (8) by a less direct route.

If the reactant ratios are kept as noted above but the $NaBH_4$ addition time is decreased to about 10 minutes, the major product (6) of the reaction is $Co_2(\mu\text{-}dppm)_2(CO)_4$, 2 ($\upsilon_{CO}$: 1970, 1955, 1910, 1878 cm^{-1}), which, on stirring in suspension in ethanol, converts smoothly into the isomeric $Co_2(\mu\text{-}dppm)_2(\mu\text{-}CO)_2(CO)_2$, 3 ($\upsilon_{CO}$: 1968, 1950, 1945, 1910, 1905, 1890, 1875, 1865, 1769, 1745 cm^{-1}). Complex 2 can also be obtained from 1 by slow elimination of CO in benzene (heated) or dichloromethane. The structures of 2 and 3 have been shown by X-ray crystallography (9) to be as illustrated below. The complexes 2 and 3 show (6) identical ^{31}P NMR spectra (δ 36.0) implying that the two structures are in rapid equilibrium at room temperature. We have studied this equilibrium in detail (10) and we have shown that, at room temperature, 2 is favored while at ca. -70° C, the equilibrium shifts in favor of 3. In fact, at -70° C, the ^{31}P and isotopically-enriched ^{13}C spectra of the two structures are clearly resolved with **2** showing $\delta^{31}P$ at 44.8 and $\delta^{13}C$ at 213.9 (terminal CO) while **3** shows $\delta^{31}P$ at 37.4 and, in the ^{13}C spectrum, the terminal and bridging CO groups at δ 206.0 and 244.3, respectively. Thermodynamic data (*10*) show that the enthalpy term favors **3** while the entropy term favors **2**. The CH_2 proton (^1H spectrum) and CO carbon

2   **3**   **4**

(^{13}C spectrum) equivalence observed (*10*) in the low-temperature spectra of **2** indicates that polytopal rearrangement about the five-coordinate cobalt atoms still occurs rapidly even at these temperatures.

If the conditions of the basic reaction for CoX_2 are changed slightly (reactant ratios 1:2.8:2.8, $NaBH_4$ addition time 5 minutes), a fourth product, **4** (υ_{CO}: 1990, 1925, 1898 cm^{-1}), is obtained in low (ca. 10%) yield together with **1** and **2**. At room temperature, the ^{31}P NMR spectrum of this complex is quite simple and shows (*7*) that the molecule is fluxional. Thus, P_A and P_B are equivalent giving a broad unresolved signal at δ -3.44, P_C appears as a doublet of triplets at δ 50.35 and P_D as a doublet at δ -28.66 ($J_{AC} = J_{BC} = 24.5$, $J_{CD} = 50.1$ Hz). At -85° C, fluxionality is slowed to the point that all signals are clearly observable with δP_A 12.64 (dd), δP_B -21.0 (dd), δP_C 49.3 (ddd) and δP_D -31.6 (d) (J_{AB} 113.0, J_{AC} 109.4, J_{BC} 58.0, J_{CD} 50.7 Hz).

Before commenting on the fifth and sixth products isolated from these reactions under somewhat different conditions, some discussion of the route(s) by which **1**, **2** (in equilibrium with **3**) and **4** are formed is in order. Such a pathway has been established (*7*) and it is based upon the knowledge (*11*) that while CO can form unstable complexes with Co^{II} and can even slowly reduce Co^{II} to Co^I (*11,12*) under certain circumstances, the reduction of Co^{II} by $NaBH_4$ in the presence of dppm under N_2 is extremely rapid (*13*), and complex formation of CO with Co^I is also known (*12,14*) to be rapid. It seemed likely, then, that under the conditions outlined above for the syntheses of **1**, **2**, and **4**, some reduction of Co^{II} prior to interaction with CO probably occurs. The first identifiable product formed (*13*) from the Co^{II}/dppm/$NaBH_4$/N_2 system is the strongly paramagnetic (5 unpaired electrons) $Co_2X_3(dppm)_2$ (X = Cl, Br) of unknown structure and from this is later derived the tetrahedral $CoX(\eta^1\text{-}dppm)_3$ for which an X-ray structure has been reported (*13*). We have shown that interaction of CO with the Co^I-Co^{II} system $Co_2X_3(dppm)_2$ in the presence of $NaBH_4$ leads, in sequence, to the formation of **4**, **2** and **1**. Thus, treatment of $Co_2X_3(dppm)_2$ with CO gives **4** although this is not immediately obvious from the ^{31}P NMR spectrum of the reaction mixture which has two main signals as broad, unresolved humps at δ -3.5 and + 51.8. However, if the reaction solution is cooled to -68° C, the humps resolve fully and a new doublet appears at around δ -28 to give a spectrum identical, apart from small chemical shift differences, to the low temperature spectrum of pure **4** described above. What apparently happens is that the CO interacts with the high-spin Co^I center of $Co_2X_3(dppm)_2$ to give the low-spin **4** but, at room temperature, the free end of the η^1-dppm unit of **4** is weakly associated with the unreduced paramagnetic Co^{II} center as shown in **5**. This renders this P atom unobservable and broadens the remaining signals considerably. That this explanation is probably

5

6

correct is shown by the fact that if pure **4** (X = Br) is treated in solution with one molar equivalent of $CoBr_2$, the room-temperature ^{31}P spectrum immediately becomes identical to that attributed to **5**. At low temperatures, however, the equilibrium between the associated and non-associated forms of **4** is frozen such that the non-associated form (the major component of the equilibrium) becomes fully observable. Addition of one molar equivalent of $NaBH_4$ to the solution of **5** under CO leads to the formation of the known (9) **6** and this is further reduced by $NaBH_4$/CO to give **2**. This, in turn, disproportionates in the presence of CO to give **1**.

The interaction of CO with $CoX(\eta^1$-dppm$)_3$ (in the absence of $NaBH_4$) is also of interest since it too leads to the formation of **4** and the cation of **1** as well as the synthesis of **7**. Thus, passage of CO into a solution of $CoCl(\eta^1$-dppm$)_3$ gives a mixture of free dppm and the low-spin **7** [^{31}P NMR: simplified AA'XX' pattern with δP_A at 54.8 (dd) and δP_X at -24.8 (dd); J = 41.7 and 29.5 Hz. IR: υ_{CO} at 1988 and 1927 cm^{-1}] in quantitative yield. Complex **7** cannot be isolated from this reaction in pure form since it isomerizes slowly to **4** which, in turn, loses CO to

7   8

give **8** by an intramolecular displacement reaction. However, the sequence is reversible under CO at higher temperatures and careful manipulation of the process allows isolation of pure **7**, **4**, and **8** (X = Br) in 72, 35 and 37% yields, respectively (7). Complexes **7** and **8** are, as will be seen later, useful starting points for the synthesis of heterobimetallics.

Turning now to the fifth complex isolated from the Co^{II}/dppm/NaBH$_4$/CO reaction, this highly unusual compound is obtained (*15*) as a minor (ca. 5%) product, together with **2**, when the reactant ratios are changed to 1:2:2.5, the NaBH$_4$ addition time is 30 minutes and the mixture is stirred for a further four hours under CO after the addition is complete. X-ray crystallographic studies (*15*) have shown the complex to have a distorted version of structure **9** [^{31}P NMR: δP_A 46.9 (d), δP_B 18.1 (d), δP_C 49.5 (d), δP_D -30.6 (d); *J* values: $P_A P_B$ 155, $P_C P_D$ 28 Hz. IR: υ_{CO} 1925, 1865 cm^{-1}; υ_{BH} 2380, 2310 cm^{-1}]. This compound is the first reported example of a stable cobaltaborane containing the Co-BH$_2$ bond and although much work has been carried out in the area of M-BH$_2$ chemistry (*16*), there is only one other example so far reported (*17*). This is the unstable species $(CO)_4CoBH_2$.THF which has been prepared by treatment of $Co_2(CO)_8$ with BH_3.THF at low temperatures and which was characterized by low-temperature ^{11}B NMR and IR studies. It is fairly clear that **9** owes its stability to the bridging of the Co-BH$_2$ unit by dppm but the unusually long Co-B distance (2.227 Å) suggests that the interaction is weak. The dppm may also be a stabilizing agent by virtue of the fact that it increases the effective electron density on the Co center. In this connection, *ab initio* molecular orbital calculations (basis set STO-3G) indicate (*15*) that for the model $[(CO)_4CoBH_3]^-$, the negative end of the dipole is at the Co center.

9   **10**   **11**

The pathway whereby **9** is formed in these reactions remains a mystery. Several attempts to prepare **9** by replacing some or all of the dppm in the basic reaction by dppm.BH$_3$ or H$_3$B.dppm.BH$_3$ were unsuccessful (*18*) as were reactions of these species with $Co_2(CO)_8$. The most promising line of approach, reaction of dppm with the unstable $(CO)_4CoBH_2$.THF, has also proved to be unsuccessful (*18*) but attempts to develop a reliable, high-yield synthesis continue.

The sixth product obtained from these Co^{II} reactions is more appropriately discussed in a later section.

Ni^{II}/dppm/NaBH$_4$ or NaBH$_3$CN/CO Reactions

Three products, two of which are quite novel, have been isolated by us from analogous reactions with Ni^{II}. Thus, if a Ni^{II}/dppm mixture in ethanol/toluene is treated over ten minutes with NaBH$_3$CN (a milder reducing agent than NaBH$_4$) under CO with the reagents in a ratio of 1:3.5:3.5, complex **10** (^{31}P NMR: AA'XX' pattern, δP_A = 26.13, δP_X = -23.74. IR: υ_{CO} 1992, 1930 cm^{-1}) is formed (*19*) in 40% yield. Such a structure has obvious potential for the synthesis of heterobimetallic complexes and, as will be seen shortly, we have had considerable success with such studies.

If, however, the reaction is carried out in the conventional fashion using $NaBH_4$ instead of $NaBH_3CN$ and with the reagent ratios 1:2:3, the CO-bridged binuclear system 11 is formed (20), also in 40% yield. While four other routes to this complex have been reported very recently [decomposition of a Ni tripod complex (21), reduction of $Ni(\eta^1$-dppm$)_2Cl_2$ by Zn under CO (22), treatment of $Ni_2(\mu$-dppm$)_2(\mu$-CNMe)(CNMe$)_2$ with CO_2 under pressure (23), and reaction of $Ni(COD)_2$ with dppm and CO (23)], the route described above seems by far the simplest. Complex 11 is also formed from slow decomposition of 10 in solution at room temperature although the rate of this decomposition can be decreased by addition of free dppm to the solution. The sequence of reactions leading to the formation of 10 and 11 has not yet been established.

The third, and most interesting, product of these Ni^{II} reactions is discussed in the next section.

Co^I and Ni^I Diphenylphosphido-bridged Complexes

Perhaps the most remarkable products of these reactions are phosphido-bridged complexes obtained from both Co^{II} and Ni^{II}. Such complexes are normally prepared [the recent literature concerning phosphido-bridged complexes has been summarized (24)] by any one of several processes such as the addition of secondary phosphines, or alkali metal derivatives thereof, to a variety of metal complexes or the interaction of R_2PX (X = halogen) with certain metal-containing species. Such methods are frequently, however, time consuming since they require the prior synthesis of suitable precursors. They also generally require the use of hazardous and unpleasant-to-handle phosphine derivatives.

The Co^{II} and Ni^{II}/dppm/$NaBH_4$/CO reactions discussed above are normally carried out such that the $NaBH_4$ is added over a period of 10-30 minutes. If, however, the addition time is decreased to one minute or less, the reactions follow an entirely different course and phosphido-bridged complexes are produced (25) in acceptable yields. Thus, with reactant ratios of 1:2:3, the reaction with $CoCl_2$ gives the hydrido-phosphido bridged complex 12 [^{31}P NMR: δP_A 47.6 (d), δP_B 215.6 (quintet), J 54 Hz. ^1H NMR: δCo-H -17.75 (sextet), J values $P_AH = P_BH = 29.2$ Hz. IR: υ_{CO} 1921 cm^{-1}], in 20-30% yield after recrystallization. Small amounts of 2 are also formed and this contaminates later crops of 12 in the recrystallization process. Complex 12 can also be produced, together with the known (26) 13, in

12   13   14

a completely different procedure in which a solution (ethanol/benzene) of a Co^{II}/dppm mixture is first treated with $NaBH_4$ in ethanol over 10 minutes (reactant ratios 1:2.3:2.3) and CO is then passed through the resulting mixture for 90 minutes. Clearly, in this instance, Co^{II} reduction occurs long before any dppm cleavage to give the diphenylphosphido group occurs since similar experiments carried out (*13*) under N_2 rather than CO show no evidence of phosphide formation. Complex **12** can also be obtained (33% yield) by heating **2** under reflux in toluene in the presence of H_2.

Similar reactions using Ni^{II} give (*25*) the phosphido-bridged complex **14** [^{31}P NMR: δP_A 15.6 (d), δP_B 239.8 (quintet), J 57.5 Hz. IR: υ_{CO} 1996 cm^{-1}] although the yield is poor and isolation of the product is difficult because of the presence of numerous other reaction products. However, **14** is most readily, and unexpectedly, obtained (20% yield) when a Co^{II}/dppm mixture is treated first with $NaBH_4$ (reactant ratios 1:2.3:2.3), then $NiCl_2$ is added (originally with the intention of preparing heterobimetallic species, see later discussion) and, finally, CO is passed for 90 minutes. This reaction occurs readily even at $0°$ C. The complex was isolated as the tetraphenylborate salt. Cationic phosphido-bridged species such as **14** appear to be quite rare.

While the mechanisms of these rapid, one-step reactions leading to phosphido-bridged Co^I and Ni^I carbonyl complexes have not yet been elucidated (work is in progress), these BH_4^- induced P-C bond cleavages in dppm take place under some of the mildest conditions yet reported. Most such metal-mediated cleavages in simple tertiary, bis- or polyphosphines take place (*27*) only under forcing conditions although there have been some recent reports of phosphido-bridged complexes being formed under comparatively mild conditions from dppm, and related P-C-P complexes, of di-iron carbonyl (*28,29*) and platinum chloride (*30,31*). What may also be significant is that we have so far been unable to find a Ph_2PCH_2- fragment in any product of the Co^{II} and Ni^{II}/dppm/$NaBH_4$/CO reactions outlined above. In the iron carbonyl (*28,29*) and platinum chloride (*30,31*) induced cleavages just mentioned, both fragments from the cleavage are incorporated into the products. This suggests (although there is, as yet, no positive evidence) that in the Co^{II} and Ni^{II} reactions, a double P-C cleavage occurs. One such double cleavage has recently been observed (*32*) in a di-iron carbonyl system although this occurs in two quite distinct steps. In any event, the procedure discussed above represents by far the simplest and quickest approach to the synthesis of phosphido-bridged cobalt and nickel carbonyls.

Attempts to extend the reaction (phosphido-complex formation) to other transition metal salts have not yet been successful although Rh^{III} shows some promise.

Synthesis of Heterobimetallics

Two approaches to these systems have been the subjects of preliminary investigation: a) direct reduction of mixtures of transition metal salts in the presence of dppm and CO and b) reactions with other metal compounds of some of the M^I and M^0 η^1- and η^2-dppm carbonyl complexes formed in the reactions described above. Dealing first with approach a), Co^{II}/Rh^{III}/dppm/$NaBH_4$/CO mixtures (reactant ratios 1:1:2:5.4) give (*33*) either the Co-Rh system **15** [40% yield; ^{31}P NMR: A_2X_2 pattern, δP_A 31.5 (br,s), δP_B 20.1 (dt), $^2J_{PP}$ 72, $^1J_{RhP}$ 129 Hz. ^{13}C NMR: $\delta Co-\underline{C}O$ 214.5 (br,s), $\delta Rh-\underline{C}O$ 180.7 (dt, J_{PC} 15, J_{RhC} 71 Hz. IR:

υ_{CO} 1965, 1922, 1815 cm^{-1}] or a mixture of **15** and the corresponding known (*34*) homobinuclear Rh-Rh system. The course of the reaction again depends upon the rate of addition of the NaBH$_4$ with rapid addition (ca. 1 minute) favoring the almost exclusive formation of **15**. It is possible that the reaction is mechanistically similar to that which leads to the formation of **2** although it should be noted that another route to **15**, and related heterobimetallic structures in which one of the metals is rhodium, has been developed by others very recently (*35*). In this approach, RhCl(dppm)$_2$ is treated with a metal carbonylate and it is, therefore, also possible that such intermediates are present in the direct Co^{II}/Rh^{III} reaction discussed above. The complex is of particular interest in view of the synergistic effect noted (*36*) for some reactions catalyzed by mixtures of Co$_2$(CO)$_8$ and Rh$_4$(CO)$_{12}$. Also, the general chemical behavior of systems containing two metals with entirely different properties is of considerable interest. The X-ray crystal structure of **15** has been determined (*33*) and this shows clearly that one of the CO ligands on cobalt is semibridging (υ_{CO} 1815 cm^{-1}). However, the two CO groups on cobalt are clearly fluxional and equivalent in solution, even at low temperatures (isotopically enriched ^{13}C NMR).

15   16

Until the mechanism of formation of heterobimetallics, such as **15**, in these one-step reactions is understood better, it is difficult to devize a reliable general synthesis of this type. Consequently, our attempts to synthesize such systems by this method have been somewhat hit and miss. Thus, our as yet limited attempts to produce heterobimetallics by co-reduction of Co^{II} and metal ions other than Rh^{III} have so far been unproductive. Such a system has, however, been produced (*37*) from a related Ni^{II}/Cu^{II} reaction. In this reaction, an Ni^{II}/dppm mixture was treated in the usual manner with NaBH$_3$CN (reactant ratios 1:3.6:4.8)) under CO with an addition time of 10 minutes. The mixture was stirred for two hours after which CuCl$_2$ (1 molar equivalent) was added. From this mixture was obtained **16** (66% yield, ^{31}P NMR: AA'XX' pattern, δNi-$\underline{P}$ 23.0, δCu-$\underline{P}$ -17.1 (v. broad signals). IR: υ_{CO} 2000, 1958, υ_{CN} 2190 cm^{-1}). As will be seen shortly, complex **10** is a probable intermediate in this reaction. An X-ray crystal structure has shown that in the solid state, the molecule has the cradle-like geometry shown in **16**. While this is a heterobimetallic system, it is of less interest than homo- and heterobimetallic systems such as **2, 3, 11** and **15** since the metal-metal bond which is so useful in reactions which mimic those which take place on metal surfaces is absent.

Turning now to approach b), preliminary studies have shown that the Co^{I} species **7** and **8**   and   the Ni^0 complex **10** all generate heterobimetallics in reactions with other metal systems. Dealing first with the Co systems, **7** (X = Cl) reacts smoothly (*7*) with a stoichiometric amount of $Rh_2(CO)_4Cl_2$ in benzene to give **17** [43% yield, ^{31}P NMR: A_2M_2X pattern, δP_A 26.5 (dt), δP_B 46.1 (t), $^2J_{PP}$ 39, $^1J_{RhP}$ 104 Hz. IR: υ_{CO} 1986, 1962, 1850 cm^{-1}] . The same complex has been prepared (*33*) by treatment of **15** with $HgCl_2$ or $CHCl_3$ and, in these reactions, it was isolated as the tetraphenylborate salt.

In a reaction very similar to that developed by Cowie's group (*35*) for the synthesis of **15**, **8** (X = Cl) reacts (*38*) fairly cleanly with [Mn(CO)$_5$]$^-$ to give **18** [40% yield, ^{31}P NMR: δP_A 30.4 (dd, *J* 70.7 and 62.9 Hz), δP_X 83.2 (ddd, *J* 70.7, 62.9, 131.9 Hz)]. An X-ray structure has been determined for **18** and the geometry of the system is highly distorted with the two P atoms at Mn being *trans* while those at Co are *cis*. In the low-temperature ^{13}C NMR spectrum of **18**, all five CO groups are unique and clearly observable. A similar Rh-Mn system has recently been prepared by Cowie's group (*35*).

17   18

With the Ni^0 complex **10**, several interesting reactions have been carried out. Thus, **10** reacts (*19*) with PtCl$_2$(COD) to give **19** (25% yield, ^{31}P NMR: solubility and solution stability too low to obtain analyzable spectra. IR: υ_{CO} 1756 cm^{-1}). A crystal structure determination (*19*) offers strong evidence for a Ni-Pt bond and this implies that the system is best regarded as a Ni^0-Pt^{II} system. Apparently

19   20

similar complexes are obtained (*39*) from reactions of **10** with Pd^{II} and Ni^{II} salts in that analyses, IR spectra and X-ray powder diffraction studies support such formulations. However, it has not yet proved possible to grow crystals of these systems suitable for single crystal X-ray studies.

Complex **10** also reacts (*37*) with [Cu(NCMe)$_4$]ClO$_4$ to give **20** [73% yield, ^{31}P NMR: AA'XX' pattern, δNi-$\underline{P}$ 22.9, δCu-$\underline{P}$ -12.0 (v. broad signals). IR: υ_{CO} 2000, 1918 cm^{-1}]. While the X-ray crystal structure of **20** has not yet been established, it is clearly closely related to **16**. Indeed, **16** is very easily converted into **20** (80% yield) by treatment with NaBH$_3$CN. The two complexes differ, however, in that the Cu atoms have different coordination numbers and **16** is a covalent species while **20** is ionic. Similar reactions of **10** with a variety of other

d^{10} metal species have so far been unproductive except that, at low temperatures, the corresponding Ni^0-Ag^I system appears to be formed (37).

Pd^{IV}/dppm/NaBH$_4$/CO and Fe^{III}/dppm/LiAlH$_4$/CO Reactions

Only preliminary work has been carried out in these areas. Dealing first with palladium, this investigation was motivated by the fact that unsubstituted Pd carbonyls are unknown and, since the most common synthesis of metal carbonyl phosphine complexes is the reaction of a phosphine with a metal carbonyl, there are relatively few Pd carbonyl phosphine complexes in the literature. Clearly, the general method discussed earlier in this account could have considerable potential as a route to such complexes.

Although the investigation is as yet only in its early stages, the method offers a simple route to a wide variety of Pd-dppm-CO complexes with, again, the nature of the major product formed being critically dependent upon the reaction conditions (40). Thus, if a CO-purged K_2PdCl_6/dppm mixture in ethanol/benzene is treated with ethanolic NaBH$_4$ over three minutes in a CO atmosphere (reactant ratios 1:1:2), the principal product (89% yield) is the known (41) and unstable carbonyl-bridged complex **21** which is normally prepared by CO insertion into the Pd-Pd bond of $Pd_2(\mu\text{-dppm})_2Cl_2$. If, however, the reactant ratios are changed to 1:1:4, an unusual complex tentatively formulated as **22** is formed (31% yield). The tentative structural assignment is based upon analytical and spectroscopic data and some aspects of its chemical behavior.

21   22   23

The analytical data for this dark red complex suggest a Pd:P ratio of somewhere between 3:4 and 3:5 indicating a trimeric palladium skeleton which is a common structural unit in palladium chemistry (4). The infrared spectrum (Nujol) is quite complex in the carbonyl region and shows clearly that both terminal (υ_{CO} 1860, 1843, 1810 cm^{-1}) and bridging (υ_{CO} 1765, 1750, 1735 cm^{-1}) CO are present. The ^{31}P NMR spectrum at room temperature shows only an unresolved broad hump at δ 13. At higher temperatures, this sharpens considerably while at lower temperatures (ca. -25° C), the spectrum exhibits three poorly-resolved absorptions at about δ 25, 18 and 0. A solid state ^{31}P MAS spectrum (room temperature) is similar in appearance but is much better resolved with absorptions at δ 23.5, 15.0, 5.5 and 2.6. Clearly, at low temperatures and in the solid state, the molecule contains both bridging and chelating dppm but at room temperature in solution it is a fluxional species. Bearing in mind the preference of Pd to have a 16 electron arrangement, the structure illustrated in **22** is not unreasonable. While it has not yet been possible to grow crystals suitable for X-ray examination, some support for this formulation comes from the chemical behavior of the complex. It is reasonably stable in solution at room temperature under N_2 but, under CO, it disproportionates to give two products which exhibit sharp singlets in the ^{31}P NMR spectrum at δ 11.0 and -9.2, respectively, indicating that the two structures are symmetrical, that the first contains only bridging dppm and that the second contains only chelating dppm. These complexes are readily isolated in pure form and the first of them contains also both bridging (υ_{CO} 1753 cm^{-1}) and terminal (υ_{CO} 1885, 1850, 1835, 1820 cm^{-1}) CO ligands. Microanalyses agree very well for the formulation $Pd_2(dppm)_2(CO)_4$ and the structure **23** is proposed. The second product of this reaction contains only terminal CO (υ_{CO} 1830 cm^{-1}) and analyzes very well for Pd(dppm)(CO). The sum of the evidence suggests that this is the three-coordinate species Pd(η^2-dppm)(CO). Thus, these two products account for all of the Pd atoms and ligands in the red complex for which structure **22** is proposed. Why **22** should disproportionate in the presence of CO but not N_2 is not entirely clear. It is possible that the CO coordinatively saturates the Pd atoms in **22** (to an 18 electron configuration) and that this renders the trimeric unit unstable.

One other product of these reactions ($\delta^{31}P$ 23), possibly derived from **23**, has been detected but it has not yet been isolated.

Consideration of the Fe^{III} reactions has been saved until last partly because the investigation is in its very early stages but also because the approach used incorporates a significant departure from the general procedure outlined in the preceding discussion. It appears that Fe^{III}/dppm/NaBH$_4$/CO reactions do not produce reduced iron-dppm-CO complexes under the normal conditions used for such reactions. We have therefore carried out (*42*) some preliminary studies using the much stronger reducing agent LiAlH$_4$ which, of course necessitates a change from protic to aprotic solvents. Apart from these changes, the procedure folowed is much the same as that outlined for the related reactions discussed herein. For example, treatment of a CO-purged FeCl$_3$/dppm mixture in dry THF with LiAlH$_4$ (reactant ratios 1:2:1) over five minutes gave a complex which is almost certainly **24** (correct analysis, yield ca. 20%, ^{31}P NMR: δ 8.01. IR υ_{CO} 1968, 1956, 1945 cm^{-1}). An X-ray crystal structure determination is pending.

24   **25**

Similar reactions using diphos, $Ph_2CH_2CH_2PPh_2$, produce what is most probably **25** [correct analysis, yield 22%, ^{31}P NMR: δ 60.99 (typical of chelate diphos). IR: υ_{CO} 1930 cm^{-1} (terminal)]. It seems, therefore, that LiAlH$_4$ shows much promise in these reactions with less easily reduced metal ions.

Conclusion

While, because of our interest in bimetallic species, most of the work discussed in this account has concentrated upon reactions which involve dppm, the synthetic procedure outlined clearly represents a novel and extremely simple approach to metal carbonyl phosphine complexes in general. Of particular significance, however, are the one-step reactions which lead to heterobimetallic or phosphido-bridged complexes and the ready access to palladium carbonyl complexes. Further research in this area is likely to be extremely productive.

Acknowledgments

The major contributions of my colleagues Dr. D.G. Holah (Lakehead University) and Dr. R.J. Puddephatt (University of Western Ontario) to this work are gratefully acknowledged as are the crystallographic contributions of Dr. V.R. Magnuson (University of Minnesota-Duluth). Thanks are also due to the Natural Sciences and Engineering Research Council of Canada and the Senate Research Committee of Lakehead University for their generous financial support of this work.

Literature Cited

(1) Puddephatt, R.J. *Chem. Soc. Rev.* **1983**, 99.
(2) Balch, A.L. In *Homogeneous Catalysis with Metal Phosphine Complexes*; Pignolet, L., Ed.; Plenum Press: New York, N.Y., 1983.
(3) Chaudret, B.; Delavaux, B.; Poilblanc, R. *Coord. Chem. Rev.* **1988**, *86*, 191.
(4) Puddephatt, R.J.; Manojlovic-Muir, L.; Muir, K.W. *Polyhedron* **1990**, *9*, 2767.
(5) For a general overview, see Cotton, F.A.; Wilkinson, G. *Advanced Inorganic Chemistry*, 5th ed.; Wiley-Interscience: New York, N.Y., **1988**, chapter 28.
(6) Elliot, D.J.; Holah, D.G.; Hughes, A.N. *Inorg. Chim. Acta* **1988**, *142*, 195.
(7) Elliot, D.J.; Holah, D.G.; Hughes, A.N.; Magnuson, V.R.; Moser, I.M.; Puddephatt, R.J.; Xu, W. *Organometallics*, in press.
(8) Carriedo, C.; Gomez-Sal, P.; Royo, P.; Martinez-Carrera, S.; Garcia-Blanco, S. *J. Organomet. Chem.* **1986**, *301*, 79.
(9) Elliot, D.J.; Holah, D.G.; Hughes, A.N., Magnuson, V.R.; Moser, I.M.; Puddephatt, R.J. in preparation.
(10) Elliot, D.J.; Mirza, H.A.; Puddephatt, R.J.; Holah, D.G.; Hughes, A.N.; Hill, R.H.; Xia, W. *Inorg. Chem.* **1989**, *28*, 3282.
(11) Albertin, G.; Bordignon, E.; Orio, A.A.; Rizzardi, G. *Inorg. Chem.* **1975**, *14*, 944 and references cited therein.
(12) Bressan, M.; Corrain, B.; Rigo, P.; Turco, A. *Inorg. Chem.* **1970**, *9*, 1733.
(13) Elliot, D.J., Holah, D.G.; Hughes, A.N.; Maciaszek, S.; Magnuson, V.R.; Parker, K.O. *Can. J. Chem.* **1988**, *66*, 81.
(14) Klein, H.F.; Karsch, H.H. *Inorg. Chem.* **1975**, *14*, 473.
(15) Elliot, D.J.; Levy, C.J.; Puddephatt, R.J.; Holah, D.G.; Hughes, A.N.; Magnuson, V.R.; Moser, I.M. *Inorg. Chem.* **1990**, *29*, 5014.

(16) See, for example, Fehlner, T.P. *New J. Chem.* **1988**, *12*, 307.

(17) Basil, J. D.; Aradi, A.A.; Bhattacharyya, N.K.; Rath, N.P.; Eigenbrot, C.; Fehlner, T.P. *Inorg. Chem.* **1990**, *29*, 1260.

(18) Holah, D.G.; Hughes, A.N.; Xu, W. unpublished results.

(19) Holah, D.G.; Hughes, A.N.; Magnuson, V.R.; Mirza, H.A.; Parker, K.O. *Organometallics* **1988**, *7*, 1233.

(20) Holah, D.G.; Hughes, A.N.; Mirza, H.A.; Thompson, J.D. *Inorg. Chim. Acta* **1987**, *126*, L7.

(21) Osborn, J.A.; Stanley, G.G.; Bird, P.H. *J. Am. Chem. Soc.* **1988**, *110*, 2117.

(22) Zhang, Z-Z.; Wang, H-K.; Wang, H-G.; Wang, R-J.; Zhao, W-J.; Yang, L-M. *J. Organomet. Chem.* **1988**, *347*, 269.

(23) DeLaet, D.L.; Rosario, R.D.; Fanwick, P.E.; Kubiak, C.P. *J. Am. Chem. Soc.* **1987**, *109*, 754.

(24) Nucciarone, D.; MacLaughlin, S.A.; Taylor, N.J.; Carty, A.J. *Organometallics* **1988**, *7*, 106.

(25) Elliot, D. J.; Holah, D.G.; Hughes, A.N.; Mirza, H.A.; Zawada, E. *J. Chem. Soc., Chem. Commun.* **1990**, 32.

(26) Hanson, B.E.; Fanwick, P.E.; Mancini, J.S. *Inorg. Chem.* **1982**, *21*, 3811.

(27) Garrou, P.E. *Chem. Rev.* **1985**, *85*, 171.

(28) Doherty, N.M.; Hogarth, G.; Knox, S.A.R.; Macpherson, K.A.; Melchior, F.; Orpen, A.G. *J. Chem. Soc., Chem. Commun.* **1986**, 540.

(29) Grist, N.J.; Hogarth, G.; Knox, S.A.R.; Lloyd, B.R.; Morton, D.A.V.; Orpen, A.G. *J.Chem. Soc., Chem. Commun.* **1988**, 673.

(30) Alcock, N.W.; Bergamini, P; Kemp, T.J.; Pringle, P.G. *J.Chem. Soc., Chem. Commun.* **1987**, 235.

(31) Bergamini, P.; Sostero, S.; Traverso, O.; Kemp, T.J.; Pringle, P.G. *J. Chem. Soc., Dalton Trans.* **1989**, 2017.

(32) Hogarth, G.; Knox, S.A.R.; Turner, M.L. *J. Chem. Soc., Chem. Commun.* **1990**, 145.

(33) Elliot, D.J.; Ferguson, G.; Holah, D.G.; Hughes, A.N.; Jennings, M.C.; Magnuson, V.R.; Potter, D.; Puddephatt, R.J. *Organometallics* **1990**, *9*, 1336.

(34) Woodcock, C.; Eisenberg, R. *Inorg. Chem.* **1985**, *24*, 1285.

(35) Antonelli, D.M.; Cowie, M. *Organometallics* **1990**, *9*, 1818.

(36) Horvath, I.T. *Polyhedron* **1988**, *7*, 2345 and references cited therein.

(37) Holah, D.G.; Hughes, A. N.; Magnuson, V.R.; Xu, W., in preparation.

(38) Elliot, D.J.; Holah, D.G.; Hughes, A.N.; Puddephatt, R.J., to be published.

(39) Holah, D.G.; Hughes, A.N.; Mirza, H.A., unpublished results.

(40) Holah, D.G.; Hughes, A.N.; Krysa, E., to be published.

(41) Benner, L.S.; Balch, A.L. *J. Am. Chem. Soc.* **1978**, *100*, 6099.

(42) Holah, D.G.; Hughes, A.N.; Spivak, G., to be published.

RECEIVED December 3, 1991

Chapter 15

Substituted 1,3,2λ⁵-Dioxaphospholanes

New Synthetic Methodologies

Jeffery W. Kelly, Philip L. Robinson, William T. Murray,
Nita Anderson-Eskew, Anne Pautard-Cooper, Isabel Mathieu-Pelta,
and Slayton A. Evans, Jr.

The William Rand Kenan, Jr., Laboratories of Chemistry, University
of North Carolina, Chapel Hill, NC 27599–3290

The synthetic elaboration of 1,3,2λ⁵-dioxaphospholanes is described. Inherent and subtle differences in steric proximity between substituents in transient regioisomeric betaines are shown to influence the enantiomeric enrichment in chiral epoxides. The results of kinetic studies indicate that the control of the reactivity of 1,3,2λ⁵-dioxaphospholanes employing metal ions occurs via cationic association between the apical dioxaphospholanyl oxygen giving rise to a controlled P-O bond cleavage which affords ready access to highly substituted epoxides. Benzoic acid is capable of initiating site-selectivity in the ring opening of 1,3,2λ⁵-dioxaphospholanes leading to a preference for benzoate substitution at the more hindered carbon (*i.e.*, the equivalent of *C*-2 substitution in 1,2-propanediol). Additional mechanistic insights are presented as well as other examples where their exploitation as valuable synthetic intermediates in organic synthesis might be useful.

Popular preparative routes to transient σ-heterophosphoranes involve "redox" reactions between alkyl, aryl, or aminophosphines (and phosphites) and compounds possessing labile heteroatom-heteroatom bonds (*i.e.*, -O-O-, -O-S-, -S-S-, -S-N, O-Cl, -N=N-, etc.) (*1-10*) as well as carbon-halogen bonds (*11-12*). Reactions of trivalent phosphorus compounds with *o*-quinones and α-diketones, the equivalent of [4+2] cycloaddition reactions, also afford valuable synthetic routes to σ-oxyphosphoranes (*13-14*). These "organophosphorus reagents" have emerged as indispensable and highly attractive for effecting mild synthetic transformations, especially substitution and condensation reactions. The principal focus of this report details the preparation and characterization of new 1,3,2λ⁵-dioxaphospholanes with emphasis on their use in the development of versatile synthetic methodologies.

A Proposed Mechanistic Rationale for Dioxaphospholane-Promoted Cyclodehydration of Diols.

While the unique substitution chemistry attending "hydrolytically-sensitive" substituted 2,2,2-triphenyl-1,3,2λ⁵-dioxaphospholanes is featured here, a word concerning the preparations of dioxaphospholanes is also warrented. Synthetic entry into a range of

0097–6156/92/0486–0186$06.00/0

substituted 1,3,2λ⁵-dioxaphospholanes is readily accomplished by bis(transoxyphosphoranylation) of 1,2-diols with diethoxytriphenylphosphorane (DTPP) in "neutral media" (*vide infra*) (*15-16*). Homogeneous DTPP, prepared by the oxidative addition of triphenylphosphine (TPP) with diethylperoxide (*15-16*), is stable indefinitely in anhydrous toluene solvent at 25°C and exhibits no measurable decomposition even at 90°C over short periods of time. However, DTPP slowly decomposes in chloroform (CHCl₃) solvent, presumably by an irreversible Arbusov-like collapse of oxyphosphonium ion pair **A** to triphenylphosphine oxide (TPPO) and diethyl ether (Scheme 1) (*17*). Catalysis, stimulated by the acidic character of chloroform through presumably intermolecular hydrogen bonding between DTPP and CHCl₃, may be instrumental in promoting this decomposition (*16,18*).

When a 1,2-diol reacts with DTPP, the replacement of two equivalents of ethanol by bis(transoxyphosphoranylation) of the diol is rapid at ambient temperature affording dioxaphosphoranes **B** (*19*). Assuming that Berry polytopal isomerization of dioxaphosphoranes **B** as well as equilibration with the regioisomeric oxyphosphonium betaines **C** are also facile (*19*), the chain closure to the cyclic ether by intramolecular displacement (*i.e.*, 3-*exo*-tet) (*20*) of TPPO from betaines **C** is expected to be rate-limiting (Scheme 1).

Regio- and Stereochemical Features of the Dioxaphosphorane-Promoted Cyclodehydration of 1,2-Diols.

Oxyphosphorane-promoted cyclodehydration of an *unsymmetrical* chiral diol can, in principle, give enantiomeric ethers by either of two stereoisomerically-distinct routes (*21*). Separate stepwise decompositions of oxyphosphonium betaines, **F** and **G**, which are presumably interconverted by facile Berry polytopal exchange processes involving dioxaphospholanes **D** and **E**, ultimately afford "enantio-enriched" distributions of chiral cyclic ethers assuming the availability of enantio-pure or enriched diols and $k_{ret} \neq k_{inv}$. Of the two possible regioisomeric betaines, cyclic ether formation from collapse of **F** should be kinetically favored on the basis of (i) the absence of a vicinal, steric proximity between the R and the oxyphosphonium phenyl substituents (as evident in **G**), and (ii) the kinetically-favored intramolecular alkoxide displacement of TPPO involving the least-hindered primary (*C*-1) oxyphosphonium species (Scheme 2). 3-*Exo*-tet decomposition of betaine **F** requires retention of configuration at the stereogenic center of the epoxide where the primary carbinol oxygen is ultimately removed as the phosphoryl oxygen of triphenylphosphine oxide.

Consequently, cyclodehydration of unsymmetrical chiral 1,2-diols should afford predominantly the cyclic chiral ether with *retained configuration* at the *C*-2 stereogenic carbon, and (b) as the steric bulk of the attached R group increases, the percent of regioselection or *stereospecificity of substitution* should also increase.

The enantiomeric excess (*i.e.*, %ee) within the enantio-enriched cyclic ethers is taken as a measure of the regioselectivity attending the oxyphosphorane-promoted cyclodehydration of the chiral diol assuming that *only* 3-*exo*-tet closures of betaines, **F** and **G**, are possible. For example, when (*S*)-(+)-propane-1,2-diol [(*S*)-**1**] was cyclodehydrated with DTPP, the percent regioselection (82%) resulting in propylene oxide (**4**) was attributable to largely retention of stereochemistry at the *C*-2 stereocenter. Assuming that the specific rate constants, k_{ret} and k_{inv}, characterize the rate-limiting processes and that the equilibration of betaines **F** and **G** (through oxyphospholanes **D** and **E**) are rapid, then the percent of regioselection can be viewed as the difference in the rates of collapse of betaine **F** versus betaine **G** (*i.e.*, $k_{ret} = 4.56\ k_{inv}$). Perhaps surprisingly, the DTPP-mediated cyclodehydration of (*S*)-(+)-phenylethane-1,2-diol [(*S*)-**2**] gave completely racemized styrene oxide (**5**) suggesting that k_{ret} is

Scheme 1

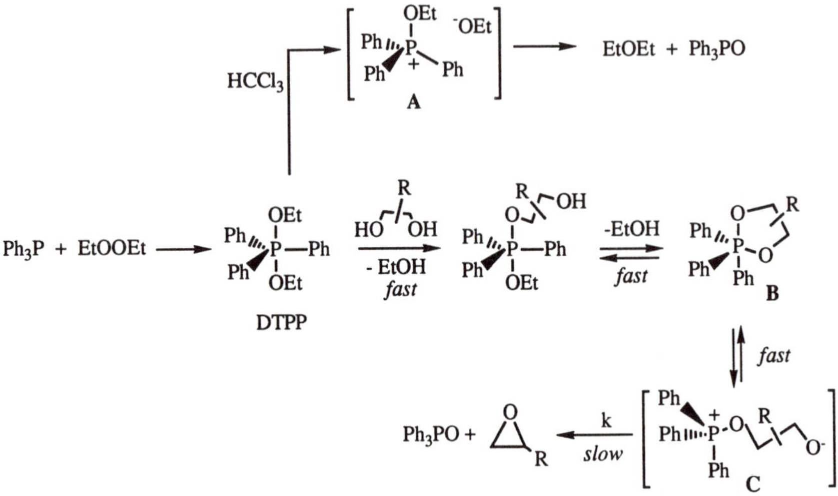

Scheme 2

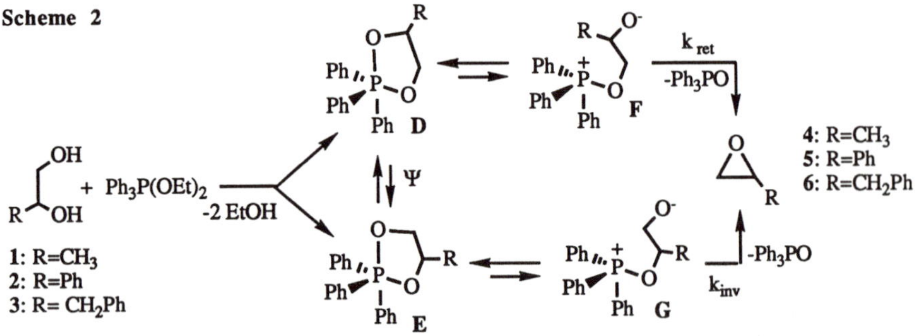

1: R=CH₃
2: R=Ph
3: R= CH₂Ph

4: R=CH₃
5: R=Ph
6: R=CH₂Ph

coincidentially identical to k_{inv} *(16,21)*. Other possibilities that might adequately rationalize the formation of (R,S)-**5** were suggested including PhCH-O bond cleavage in **E** which would compromise the stereochemical integrity of the C-2 stereocenter. In addition, reversible ring opening of epoxide **5** under the reaction conditions was also suggested. However, our additional experiments did not provide corroborative evidence in support of these suggestions *(16,21)*.

If the assumption that the severity of the steric interactions between the R and the Ph_3P^+-OC- substitutents impacts the extent of regioclosure is valid, then a substituent more sterically-demanding than methyl should exert a profound directive influence on the course of the regioclosure of betaines **F** and **G** to produce enantio-enriched epoxides. This expectation was realized from the results of the reaction between (S)-(+)-benzylethane-1,2-diol [(S)-**3**] and DTPP where epoxide **6** is obtained with 94% retention of configuration at the C-2 stereocenter *(22)* (Scheme 2).

The reaction of $(4R)$-(-)-pentane-1,4-diol (**7**) with DTPP gave (R)-(-)-2-methyltetrahydrofuran (**8**) as the predominant enantiomer reflecting largely retention (80.5%) of stereochemistry at C-2 *(16,21)*. Generally, the free energies of activation (*i.e.*, $\Delta G^{\ddagger}$) for m-*exo*-tet closure of chains to three- and five-membered rings *(23)* are often similar; however, the similarity in the percent regioselection for cyclodehydration of (S)-**1** and (R)-**7** (equation 1) is probably coincidental.

Finally, the *stereospecificity* of the phosphorane-promoted cyclodehydration is adequately demonstrated in the reaction of d,l-2,3-butanediol (**9**) with DTPP in CD_2Cl_2 (35°C, 10 h). The stepwise nature of the cyclodehydration process gives exclusively *cis*-2,3-epoxybutane (**10**; >99%) by ^{13}C NMR analysis [δ 12.9 (CH_3) and 52.4 (CHO)] *(16)*. This latter result is consonant with the previous findings of Denney *et al.* *(24)* where a mixture containing 88% d,l- and 12% *meso*-4,5-dimethyl-2,2,2-triethoxy-1,3,2λ⁵-dioxaphospholanes (**11**) gave a mixture of 85% *cis*- and 15% *trans*-2,3-butene oxides (**10**), respectively, during thermolysis (117°C, 42 h) (equations 2 and 3).

Polymeric Dioxaphospholanes.

We envisioned that a cross-linked, polymer-supported dioxaphosphorane (*i.e.*, diethoxydiphenylpolystyrylphosphorane; DDPP) *(25)* might allow for expeditious product isolation while simultaneously incorporating the characteristically mild cyclodehydrating properties of DTPP. As a bonus, it seemed reasonable to expect the steric bulk and rigidity of the polymeric backbone to favorably influence the level of regioselective release of the phosphine oxide, and perhaps provide a more efficient method for enhancing the enantio-enrichment within chiral cyclic ethers.

Polymeric DDPP is easily prepared by oxidative addition of diethyl peroxide to commercially-available diphenylpolystyrylphosphine which is cross-linked with 2% divinylbenzene *(26)* (anhydrous toluene solvent under a nitrogen atmosphere) *(25)*. As expected, the ^{31}P NMR resonance for the phosphorus atom in DDPP (δ -55.3 ppm) is broader in bandwidth, but similar in resonance frequency to the phosphorus atom in DTPP (δ -55.0 ppm) and it seems reasonable to suggest that the trigonal bipyramidal conformer having both P-ethoxy groups in the apical array is probably preferred in DDPP as has been described for DTPP *(16,27)* (Scheme 3).

In general, the cyclodehydrative potential of DDPP is particularly impressive. For example, high conversions of minimally-substituted 1,2-, 1,4-, and 1,5-diols to the epoxides (70-92%), tetrahydrofurans (95-99%), and tetrahydropyrans (85%), respectively, are noteworthy *(25)*.

More importantly, the stereospecific conversion of *meso*-1,2-diphenylethane-1,2-diol (**12**) to *trans*-stilbene oxide (**13**; >99%) with DDPP is consistent with a process

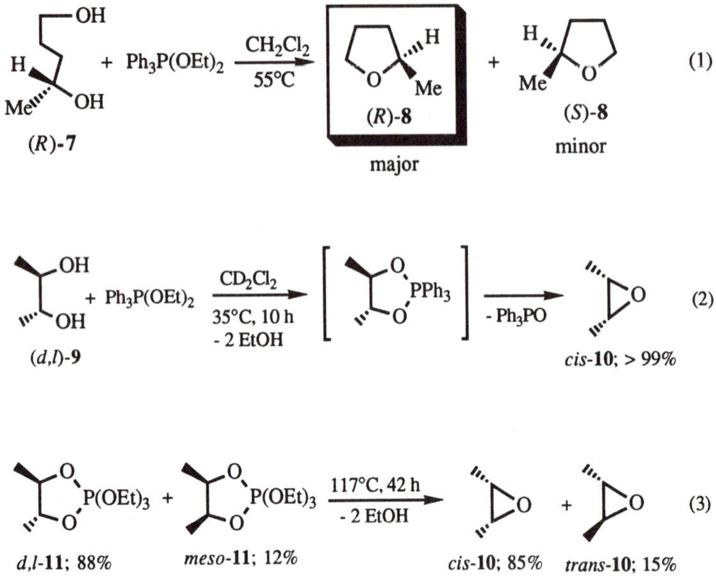

OH
H
Me
OH
(R)-7
+ Ph3P(OEt)2
CH2Cl2
55°C
H
O
Me
(R)-8
major
+
H
Me
O
(S)-8
minor
(1)
OH
OH
(d,l)-9
+ Ph3P(OEt)2
CD2Cl2
35°C, 10 h
- 2 EtOH
O
PPh3
O
- Ph3PO
O
cis-10; > 99%
(2)
O
P(OEt)3
O
+
O
P(OEt)3
O
d,l-11; 88%
meso-11; 12%
117°C, 42 h
- 2 EtOH
O
+
O
cis-10; 85%
trans-10; 15%
(3)

Scheme 3

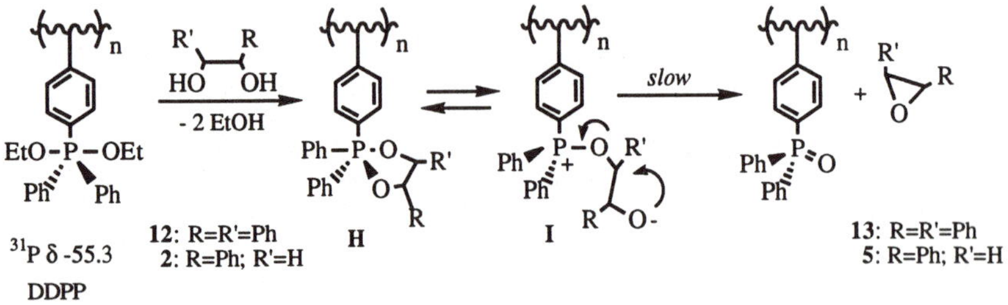

n
EtO-P-OEt
Ph
Ph
31P δ -55.3
DDPP
R'
R
HO
OH
- 2 EtOH
12: R=R'=Ph
2: R=Ph; R'=H
Ph-P
O
R'
Ph
O
R
H
Ph
P-O
R'
+
Ph
R
O-
I
slow
Ph
P
O
Ph
+
R'
R
O
13: R=R'=Ph
5: R=Ph; R'=H

involving an energetically favorable 3-*exo*-tet cyclization with the expected expulsion of TPPO and the accompanying inversion of stereochemistry at the displacement terminus (*i.e.*, betaine I). Regioselective cyclodehydration of (*S*)-2 with DDPP is evident from an examination of the enantiomeric mixture of styrene oxides rich in the (*S*)-(+)-1,2-epoxyethane enantiomer [(*S*)-5] (68%ee). This %ee translates into 85% retention of configuration at the C-2 stereocenter in styrene oxide. By contrast, DTPP converts (*S*)-2 to racemized (*R*,*S*)-styrene oxide (*vide supra*) (*16,21*).

Lewis Acid-Catalyzed Decomposition of 1,3,2λ⁵-Dioxaphospholanes.

It is well-documented that substituted 1,3,2λ⁵-dioxaphospholanes are viable precursors to the substituted epoxides (*15-16*). However, while mono- and disubstituted 1,3,2λ⁵-dioxaphospholanes derived from the corresponding diols readily collapse to epoxides (35-50°C), highly substituted 1,3,2λ⁵-dioxaphospholanes from tri- and tetrasubstituted 1,2-diols are thermodynamically more stable (no reaction <60°C) which is undoubtedly a response to the stabilizing influence of the geminal-dialkyl groups juxtaposed to the ring oxygens. Consequently, these dioxaphospholanes do not, in the absence of high thermal activation, readily collapse to the epoxides and phosphine oxide (*28*). In fact, high temperatures (>100°C) initiate fragmentation of substituted dioxaphospholanes in a nonselective manner affording ketones, allylic alcohols, epoxides, and hydroxy ethers in relative proportions that depend on the severity of the reaction conditions (*29*). We suggested (*29*) that the a cationic metal ion (*i.e.*, Li⁺) might coordinate with the basic apical oxygens and weaken the phosphorus-oxygen (P-O) bonds in the 1,3,2λ⁵-dioxaphospholanes. This action would facilitate formation of the requisite betaines, and ultimately promote their facile collapse specifically to the epoxides and thereby minimize the formation of unwanted side-products (*30*) (Scheme 4).

A basic understanding of the impact of metal ion catalysis on the decomposition of dioxaphospholanes is best gained from the results of kinetic studies performed with 4-methyl-2,2,2-triphenyl-1,3,2λ⁵-dioxaphospholane (14) in tetrahydrofuran (THF)/benzene-d_6 solution. Thermal decomposition of dioxaphospholane 14 gave $k_{obsd} = 5.7 \times 10^{-5}$ s⁻¹ at 52°C. However, at 31°C dioxaphospholane 14 rapidly decomposed to propylene oxide (k = 2.0 x 10⁻⁴ M⁻¹s⁻¹) in the presence of lithium bromide (LiBr). In the presence of tetrabutylammonium bromide (*n*-Bu₄N⁺Br⁻), dioxaphospholane 14 was essentially unaffected at 27°C. These results indicate that there is no significant decomposition of 14 in the presence of *n*-Bu₄N⁺Br⁻ which implies that Br⁻ (*i.e.*, hexavalent phosphorane 15) is probably not an important contributor to the catalytic potential of LiBr. It seems certain that the catalytic influence of LiBr is associated with Li⁺'s ability to weaken the apical P-O bond of 1,3,2λ⁵-dioxaphospholane 14 by cationic coordination (See Scheme 4).

A response to a more fundamental question (*29,30*) concerning the structure of "LiBr" and exactly how it affects the rate of 1,3,2λ⁵-dioxaphospholane 14 decomposition was required, particularly since LiBr exists in organic solvents in high concentrations as LiBr aggregates. In fact, Goralski and Chabenal (*31*) have shown that LiBr is most probably dimerized in THF solvent, while lithium iodide (LiI) and lithium perchlorate (LiClO₄) are less aggregated and monomeric in THF solvent. Consequently, knowledge of the importance of ion pair/monomer/dimer compositions of the lithium salts on the rate of decomposition of 1,3,2λ⁵-dioxaphospholane 14 seemed appropriate. The rate constant for decomposition of 14 in the presence of LiBr (31°C) is $k_{LiBr} = 2.0 \times 10^{-4}$ M⁻¹ s⁻¹.

With LiI, the rate constant increases to $k_{LiI} = 2.7 \times 10^{-4}$ M^{-1} s^{-1} and for $LiClO_4$, a further rate enhancement was observed: $k_{LiClO4} = 4.3 \times 10^{-4}$ M^{-1} s^{-1}. These findings are best interpreted in terms of the extent of aggregation as well as the effective charge on Li^+ in the lithium salts. In other words, formation of LiBr dimers in THF solvent reduces the "relative concentration" of free Li^+ cations available for catalysis as compared to monomeric LiI or $LiClO_4$. Lithium perchlorate is monomeric in THF solvent (*31*) and it increases the rate of decomposition of $1,3,2\lambda^5$-dioxaphospholane **14** considerably when compared to both LiI and LiBr. Intuitively, $LiClO_4$ should be less "covalently" bound than LiI considering the ionic size and stability of ClO_4^-. The subsequent rate enhancement thus reflects a larger positive charge density about Li^+ in $LiClO_4$ or substantially more $Li^+//ClO_4^-$, resulting in a greater Lewis acidity and a higher potential for binding to the ethereal oxygens.

The rate of zinc chloride $(ZnCl_2)$-promoted decomposition of $1,3,2\lambda^5$-dioxaphospholane **14** is more impressive than that observed for the lithium salts. In fact, the addition of one equivalent of $ZnCl_2$ to a 10-fold excess of dioxaphospholane **14** at 25°C initiates rapid (*i.e.*, pseudo-first order rate constant, $k_1 = 3.1 \times 10^{-2}$ M^{-1} s^{-1}) and quantitative opening of **14** to the zinc-coordinated betaine intermediates, **J** and **K**, whose identities were easily confirmed by ^{31}P NMR spectroscopy (δ 63.5 and 62.0 ppm, respectively) (*29*). These results clearly show that the ability of Lewis acids to accelerate dioxaphospholane decomposition is directly related to their coordinating potential with the ethereal oxyens. Subsequent warming of **J** and **K** allows for the direct assessment of the rate constant for the 3-*exo*-tet displacement of TPPO (k_2). At 0°C, k_2 for decomposition is 1.2×10^{-4} s^{-1} and $\Delta G^{\ddagger} = 20.8$ kcal/mol (Scheme 5).

The reaction of *cis*-1,2-cyclohexanediol (**16**) with DTPP affords bicyclic $1,3,2\lambda^5$-dioxaphospholane **17** which thermally decomposes to cyclohexanone (**18**) via a 1,2-hydride migration process (*e.g.*, **L** in Scheme 6) (*16*). The reaction of dioxaphospholane **17** with LiBr in THF solvent is characterized by a pseudo-first order rate constant, $k_{LiBr} = 9.2 \times 10^{-5}$ M^{-1} s^{-1} at 52°C (*29*). Moreover, the reaction of phosphorane **17** with 1 equivalent of $ZnCl_2$ gives betaine intermediate **M** (^{31}P NMR δ 61.6 ppm) and its long-term stability allows for a direct measure of $\Delta G^{\ddagger}$ for hydride migration and its subsequent collapse to ketone **18**. Decomposition of betaine **M** has a rate constant, $k_M = 1.2 \times 10^{-4}$ s^{-1} at 25°C and $\Delta G^{\ddagger} = 22.9$ kcal/mol (Scheme 6).

Comparison of the $\Delta G^{\ddagger}$ data for the collapse of betaines **J** and **K** (associated with the 3-*exo*-tet alkoxide displacement of $-O-^+PPh_3$ to afford epoxides in Scheme 5) and **M** indicates that the hydride migration is *ca*. 2.0 kcal/mol higher in energy (*29*).

Explorations of The Synthetic Utility of $1,3,2\lambda^5$-Dioxaphospholanes.

Anhydropyranosides. There are few synthetic routes to anhydropyranosides in the absence of strategically placed protecting groups. The DTPP-promoted bis(transoxaphosphoranylation) of methyl α-D-glucopyranoside (**19**) represents a highly efficient stereoselective route to isomeric anhydropyranosides (*30,32-33*) (Scheme 7).

Scheme 4

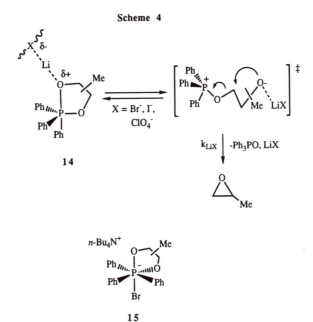

Scheme 5

Scheme 6

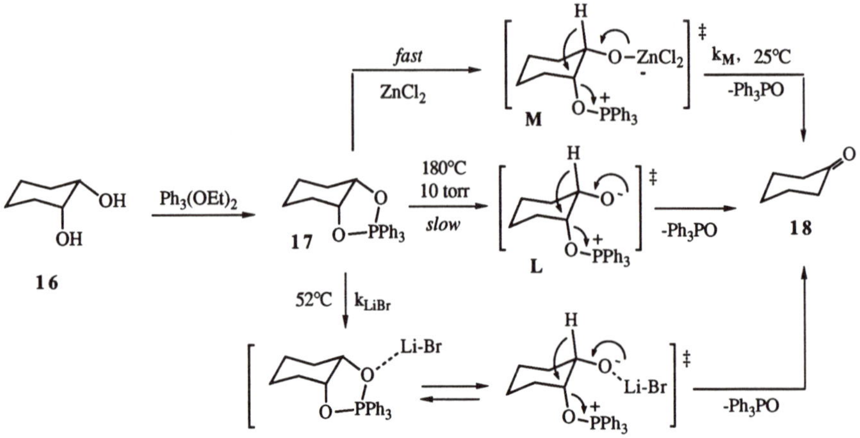

Scheme 7

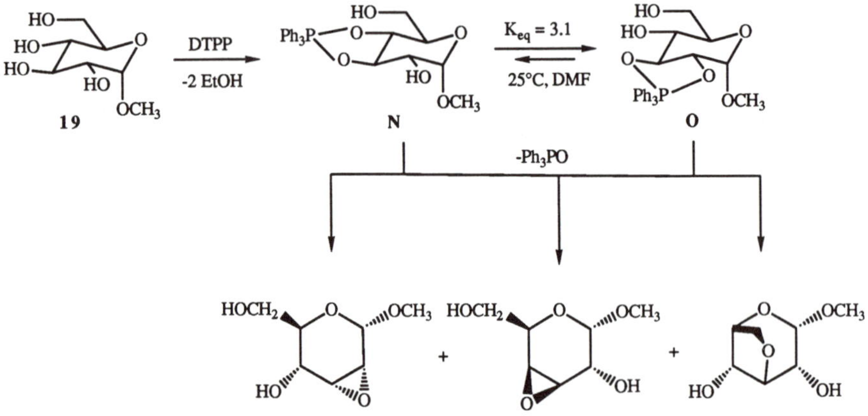

The bis(transoxaphosphoranylation) of glucopyranoside **19** with DTPP affords two intermediate 1,3,2λ^5-dioxaphospholanes, **N** (^{31}P NMR δ -36.1) and **O** (-37.7 ppm) where **O** is kinetically favored over **N** in a ratio of 5:1. Their facile, thermal equilibration (*ca.* 25°C; ^{31}P NMR) in *N*,*N*-dimethylformamide solvent enhances the population of **N** although **O** still predominates (*i.e.*, $K_{eq} = 3.1$; $\Delta G° = -0.66$ kcal mol^{-1}) (*33*).

A *ca.* 1:1 mixture of two epoxides, methyl 2,3-anhydro-α-D-allopyranoside (**20**; 45%) and methyl 3,4-anhydro-α-D-galactopyranoside (**21**; 40%), as well as some methyl 3,6-anhydro-α-D-glucopyranoside (**22**; 15%) are formed in DMF or toluene solvent from the thermal decomposition (65-90°C, 6 h) of the rapidly equilibrating 1,3,2λ^5-dioxaphospholanes, **N** and **O** (*32-33*) (Scheme 7).

Mechanistically, the requisite betaine intermediates [from dioxaphospholanes **N** and **O**] adopt either the chair ($^4C_1 \rightarrow {}^1C_4$) or twist-boat ($^1S_5$, $^0S_2$) conformations so that the C-O$^-$ and -O-$^+$PPh$_3$ groups can assume the requisite "pre-transition state" antiperiplanar arrangement for suitable displacement of Ph$_3$PO. From the results of molecular modeling studies (*i.e.*, MacroModel II) on both the chair (*e.g.*, $^1C_4$) and twist-boat betaine (*e.g.*, $^1S_5$, $^0S_2$) conformers, the small energy differences between them suggest that the twist-boat ($^1S_5$ and $^0S_2$) betaine conformers may also be energetically-favored intermediates and crucial to the efficacy of the cyclodehydration process (Scheme 8). It also seems conceivable that even glucopyranoside **22** arises via a transition state that encourages a 5-*exo*-tet displacement of *C*-6-O-$^+$PPh$_3$ by the pseudo-axial *C*-3 oxyanion [*e.g.*, betaine **P** ($^0S_4$) in equation 4].

Finally, we invented a highly regioselective reaction promoting the cyclodehydrative collapse of specifically dioxaphosphorane **O** that is presumably controlled by Li$^+$ ion chelation (*32*). When LiBr was added to a DMF solution containing 1,3,2λ^5-dioxaphospholanes **N** and **O**, a facile decomposition (50°C) occurred, affording methyl allopyranoside **20** as the exclusive product (>99% by ^{13}C NMR). The rate constant for the *uncatalyzed* decomposition (in the absence of LiBr) is $k_{uncat} = 1.5 \times 10^{-4}$ s^{-1} (65°C) while the *catalyzed* process has a rate constant, $k_{LiBr} = 2.7 \times 10^{-4}$ M^{-1} s^{-1} (50°C) [also, $k_{LiI} = 4.9 \times 10^{-4}$ M^{-1} s^{-1} (25°C)].

The exclusive formation of **20** and the rate dependence on LiX are best rationalized by assuming that rapid equilibration of oxaphospholanes **N** and **O** via the intermediate betaines is initiated through Li$^+$ ion coordination to the basic anomeric oxygen, and subsequently relayed to the apical P-O of 1,3,2λ^5-dioxaphospholane **O** as depicted in chelate **Q**. In this way, Li$^+$ ion ligation to the apical P-O oxygen encourages P-O cleavage to afford the Li$^+$-chelated *C*-2 oxyanion species **R** which could easily control the equilibrium distribution of the requisite betaines to favor **S** and perhaps **T**. Thus, subsequent collapse of betaine **T** could account for the high regiospecificity observed in the formation of anhydropyranoside **20** (Scheme 9). This latter reaction is remarkable and establishes the *regiospecific* cyclodehydration of α-D-glucopyranoside **19** to pyranoside **20** with DTPP/LiBr in a single synthetic event (*32-33*).

Regioselective Substitution Chemistry. The 1,3,2λ^5-dioxaphospholanes are valuable precursors in other novel synthetic transformations. For example, the benzoylation of 1,2-propanediol employing triphenylphosphine/benzoyl peroxide is

Scheme 8

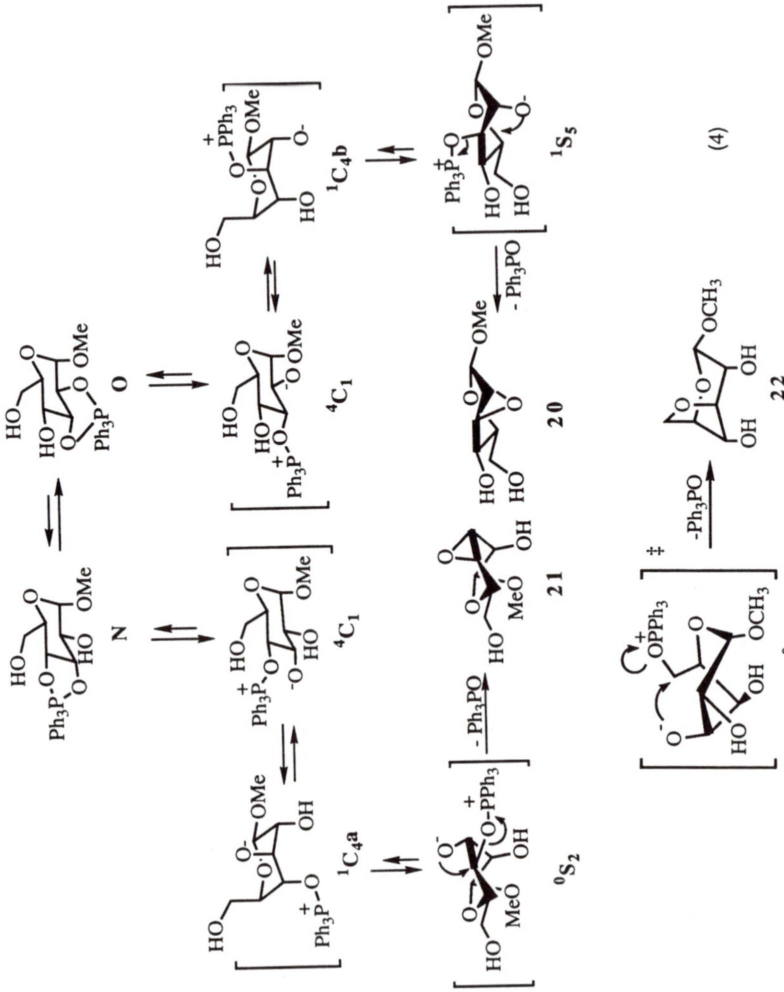

$$(4)$$

highly chemoselective favoring the *C*-2 benzoate over the *C*-1 regioisomer (*34*) (Scheme 10). Dioxaphospholane **14** (**14'**) was identified as the quintessential intermediate in this reaction and it undergoes a highly chemoselective ring opening initiated by hydrogen bonding and ultimately proton transfer involving benzoic acid to give the regioisomeric oxyphosphonium ions **U** and **V**. Presumably, proton transfer occurs at the least hindered oxaphospholanyl oxygen to effect ring opening to mainly **U**. Displacement of TPPO by benzoate anion from regioisomeric oxyphosphonium ions **U** and **V** affords *C*-2 benzoate **23** and the *C*-1 regioisomer **24** in a 7.3:1.0 ratio, respectively. Independent synthesis of dioxaphospholane **14** [reaction of diol **1** with DTPP and removal of ethanol (*19,34*)] followed by reaction with benzoic acid in CH$_2$Cl$_2$ or THF solvent at -78°C gave a 7:1 ratio of the *C*-2:*C*-1 benzoates. With (*S*)-(+)-1,2-propanediol, *C*-2 benzoate **23** was formed with essentially complete inversion of configuration (>93%ee) at the secondary carbinol stereocenter (*34,35*).

Finally, we previously reported that 1,3,2λ⁵-dioxaphospholane **14**, prepared *in situ* from the reaction of diol **1** with diisopropyl azodicarboxylate (DIAD), reacts with benzoic acid at -78°C to afford the two regioisomeric *C*-1 and *C*-2 oxyphosphonium ions in tetrahydrofuran (THF) solvent. Based on the low temperature ^{31}P NMR data, we concluded that these regioisomeric oxyphosphonium ions, **U** and **V**, undergo rapid equilibration with dioxaphospholane **14** in benzoic acid. However, the results of our most recent experiments (Mathieu-Pelta, I., The University of North Carolina at Chapel Hill, unpublished data) involving dioxaphospholane **14** [prepared from the bis(transoxyphosphoranylation) of diol **1** with DTPP] do not corroborate this slow equilibration between oxyphosphonium ions **U** and **V** and dioxaphospholane **14** in benzoic acid in THF solvent. In fact, at -78°C the addition of benzoic acid to dioxaphospholane **14** in THF solvent gives the regioisomeric benzoates directly and the isomeric oxyphosphonium ions **U** and **V** are not observed by ^{31}P NMR spectroscopy. However, in *dichloromethane solvent* ions **U** and **V** *are* actually observable and *do* undergo a temperature dependent, dynamic exchange with dioxaphospholane **14** and benzoic acid (Mathieu-Pelta, I., The University of North Carolina at Chapel Hill, unpublished data).

An extension of this novel reaction was envisioned through the activation of phosphorane **14** by proton transfer from a relatively strong Brønsted acid, followed by the release of TPPO by an efficient, and more versatile nucleophile (*36*). Specifically, dioxaphospholane **14** was allowed to react with *p*-toluenesulfonic acid (*p*-TsOH) in the presence of *sparingly-soluble* sodium azide (NaN$_3$) to afford both the *C*-2 and *C*-1 oxyphosphonium ions, **W** and **X**, respectively, in THF solvent (^{31}P NMR). At ambient temperature (*ca*. 25°C), the sulfonate anion (*p*-TsO⁻) displaces TPPO from presumably oxyphosphonium ion **W** affording >95% yield (^{13}C NMR) of (*R*)-2-(4-methylbenzenesulfonyloxy)-1-propanol (**25**) with high stereospecificity (>92%ee) at the *C*-2 carbinol center (Scheme 11). None of the regioisomeric *C*-1 tosylate was observed by ^{1}H and ^{13}C NMR analyses and there was no evidence of azide incorporation at *C*-1 or *C*-2. While we have no firm evidence concerning the nature of the requisite reactive intermediate, it does seem clear that a facile equilibration between oxyphosphonium ions **W** and **X** must occur. Nevertheless, subsequent heating of the reaction mixture (80°C, 8 h) containing **25** and NaN$_3$ in THF solvent afforded (*S*)-2-azidopropanol (**26**) in 87%ee and 88% yield. The configurational identity of (*S*)-(+)-**26** was confirmed by comparing the sign and magnitude of its optical rotation with that of the (*R*)-(-)-**26** antipode (*37*). These findings require that tosylate **25** form with nearly complete inversion of configuration from **14**.

Scheme 9

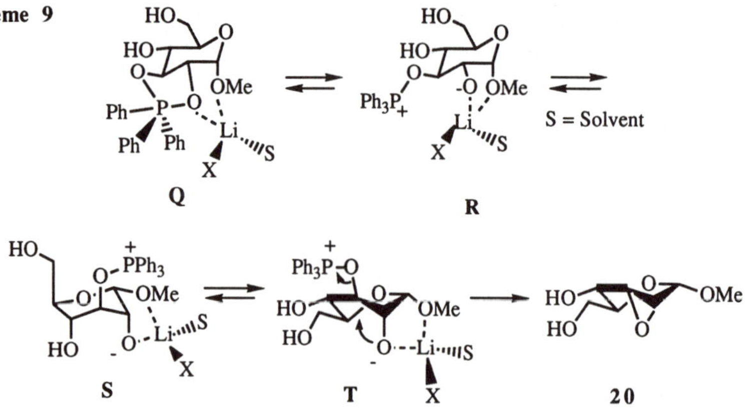

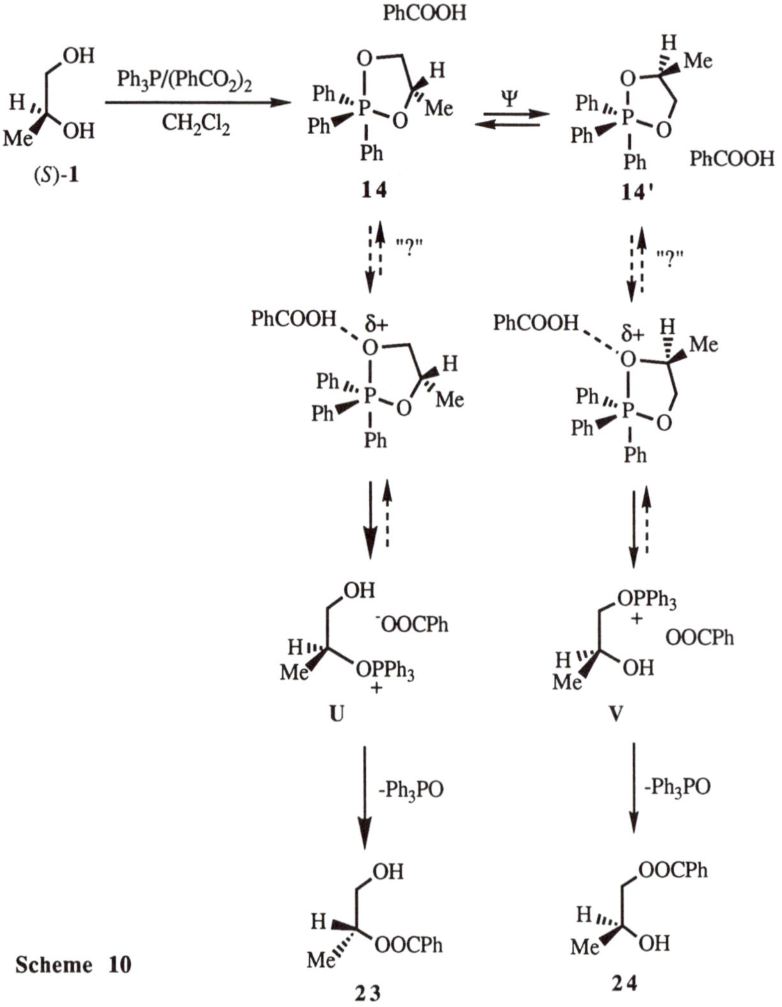

Scheme 10

Scheme 11

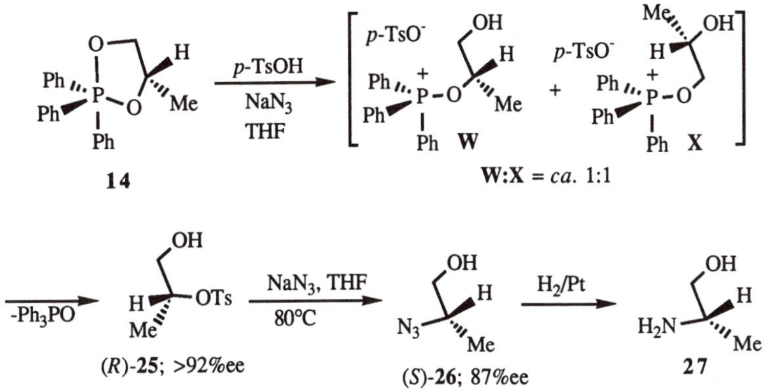

W:X = *ca.* 1:1

14

(R)-25; >92%ee

(S)-26; 87%ee

27

Conclusions

It is clear that dioxaphosphoranes exhibit significant potential as valuable synthetic intermediates for a host of novel synthetic transformations. While they do not serve as traditional protecting auxiliaries for 1,2-diols, they do respond to metal ion catalysis, and under the appropriate reaction conditions, acid-promoted nucleophilic displacements control the regiochemical and configurational integrity of the more hindered stererogenic carbon of the intermediate $1,3,2\lambda^5$-dioxaphospholanes.

Acknowledgment is made to the National Science Foundation and Rhône-Poulenc, Agrochimie for support of this research as well as for a fellowship from Rhône-Poulenc to IMP. We are also grateful to ATOCHEM of North America, Elf Aquitaine for generous supplies of triphenylphosphine.

Literature Cited

(1)   Denney, D. B.; Denney, D. Z.; Hall, C. D.; Marsi, K. L. *J. Am. Chem. Soc.*, **1972**,*94*, 245.
(2)   Baumstark, A. L.; McCloskey, C. J.; Williams, T. E.; Chrisope, D. R. *J. Org. Chem.*, **1980**, *46*, 3593.
(3)   Denney, D. B.; Denney, D. Z.; Garridovic, D. M. *Phosphorus Sulfur*, **1981**, *11*, 1.
(4)   Harpp, D. N.; Adams, J.; Gleason, J. G.; Mullins, D.; Steliou, K. *Tetrahedron Lett.*, **1978**, 3989.
(5)   Harpp, D. N.; Gramata, A. *J. Org. Chem.*, **1980**, *45*, 271.
(6)   Mukaiyama, T.; Ueki, M.; Maruyama, H.; Matsueda, R. *J. Am. Chem. Soc.*, **1968**, *90*, 4490.
(7)   Denney, D. B.; Hanifin, Jr., J. W. *Tetrahedron Lett.*, **1983**, 2177.
(8)   Barry, C. N.; Evans, Jr., S. A. *J. Org. Chem.*, **1983**, *48*, 2825.
(9)   Mitsunobu, O. *Synthesis*, **1981**, 2.
(10) Markovskii, L. N.; Kolesnik, N. P.; Shermolovich, Yu. G. *Russ. Chem. Revs.*, **1987**, *56*, 894.
(11) Appel, R.; Halstenberg, M. in *"Organophosphorus Reagents in Organic Synthesis"*; Cadogan, J. I. G. (Ed.); Academic Press: New York, 1979; pp 378-424.
(12) Murray, W. T.; Kelly, J. W.; Evans, Jr., S. A. *J. Org. Chem.*, **1987**, *52*, 525.
(13) Kutyrev, A. A.; Moskva, V. V. *Russ. Chem. Revs.*, **1987**, *56*, 1028.
(14) Lowther, N.; Beer, P. D.; Hall, C. D. *Phosphorus Sulfur*, **1988**, *35*, 133.
(15) Chang, B. C.; Conrad, W. E.; Denney, D. B.; Denney, D. Z.; Edelman, R.; Powell, R. L.; White, D. W. *J. Am. Chem. Soc.* **1971**, *93*, 4004.
(16) Robinson , P. L.; Barry, C. N.; Kelly, J. W.; Evans, Jr., S. A. *J. Am. Chem. Soc.* **1985**, *107*, 5210.
(17) Bass, S. W.; Barry, C. N.; Robinson, P. L.; Evans, Jr., S. A. *ACS Symp. Ser.* **1981**, No. *171*, 165.
(18) Robinson, P. L.; PhD Dissertation, The University of North Carolina-Chapel Hill, NC, 1985.
(19) Kelly, J. W.; Evans, Jr., S. A. *J. Am. Chem. Soc.*, **1986**, *108*, 7681.
(20) Baldwin, J. E. *J. Chem. Soc. Chem. Commun.*, **1976**, 734.
(21) Robinson, P. L.; Barry, C. N.; Bass, S. W.; Jarvis, S. E.; Evans, Jr., S. A. *J. Org. Chem.*, **1983**, *48*, 5396.
(22) Kelly, J. W. PhD Dissertation, The University of North Carolina-Chapel Hill, NC, 1986.
(23) Stirling, C. J. M. *J. Chem. Educ.*, **1973**, *50*, 844.
(24) Denney, D. B.; Jones, D. H. *J. Am. Chem. Soc.*, **1969**, *91*, 5821.

(25) Kelly, J. W.; Robinson, P. L.; Evans, Jr., S. A. *J. Org. Chem.*, **1985**, *50*, 5007.

(26) Amos, R. A.; Emblidge, R. W.; Havens, N. *J. Org. Chem.*, **1983**, *48*, 3598.

(27) Denney, D. B.; Denney, D. Z.; Wilson, L. A. *Tetrahedron Lett.*, **1968**, 85.

(28) Murray, W. T.; Evans, Jr., S. A. *New J. Chem.*, **1989**, *13*, 329.

(29) Murray, W. T.; Evans, Jr., S. A. *J. Org. Chem.*, **1989**, *54*, 2440.

(30) Murray, W. T.; Pautard-Cooper, A.; Eskew, N. A.; Evans, Jr., S. A. *Phosphorus, Sulfur,* **1990**, *49/50*, 101.

(31) Goralski, P.; Chabernal, M. *Inorg. Chem.*, **1987**, *26*, 2169.

(32) Eskew, N. A.; Evans, Jr., S. A. *J. Chem. Soc. Chem. Commun.*, **1990**, 706.

(33) Eskew, N. A.; Evans, Jr., S. A. *Heteroatom Chem.*, **1990**, *1*, 307.

(34) Pautard-Cooper, A.; Evans, Jr., S. A. *J. Org. Chem.*, **1988**, *53*, 2300.

(35) Pautard-Cooper, A.; Evans, Jr., S. A. *J. Org. Chem.*, **1989**, *54*, 2485.

(36) Pautard-Cooper, A.; Evans, Jr., S. A. *Tetrahedron*, **1991**, *47*, 1608.

(37) Vander Werf, C. A.; Heisler, R. Y.; McEwen, W. E. *J. Am. Chem. Soc.*, **1954**, *76*, 1231.

RECEIVED December 26, 1991

Chapter 16

DNA—The Ultimate Phosphorus Polymer

^{31}P One- and Two-Dimensional NMR Spectroscopy of Oligonucleotides and Protein–DNA Complexes

David G. Gorenstein, Christine Karslake, Jill Nelson Granger, Yesun Cho, and Martial E. Piotto

Department of Chemistry, Purdue University, West Lafayette, IN 47907

^{31}P NMR spectroscopy serves as an important probe of the conformation and dynamics of nucleic acids, particularly the sugar phosphate backbone, which is not well defined by 2D ^{1}H/^{1}H NOESY data. The ^{31}P NMR spectra of various 14-base pair *lac* operators free and bound to both wild-type and mutant *lac* repressor headpiece proteins (N-terminal 56-residue fragments) were analyzed to provide information on the backbone conformation in the complexes. It is proposed that specific, tight-binding operator-protein complexes retain the inherent phosphate ester conformational flexibility of the operator itself, whereas the phosphate esters are conformationally restricted in the weak-binding operator-protein complexes. This retention of backbone torsional freedom in tight complexes is entropically favorable and provides a new mechanism for protein discrimination of different operator binding sites. It demonstrates the potential importance of phosphate geometry and flexibility on protein recognition and binding. The ^{1}H and ^{31}P resonances of the 3′-thymidine phosphorodithioate decamer $d(CGCTpS_2^- TpS_2^- AAGCG)$ were assigned by two-dimensional NMR and the solution structure determined. Other dithiophosphate antisense oligonucleotides are described.

^{31}P NMR has developed as a powerful probe of the structure and dynamics of nucleic acids in solution. With the development of synthetic schemes in which defined RNA and DNA fragments can be routinely produced in multi-milligram quantities and the concomitant development of sequence-specific 2D ^{1}H/^{1}H and ^{1}H/^{31}P NMR methodologies (1-9) for assignment of individual ^{31}P signals, analysis of the ^{31}P NMR spectra of modest size oligonucleotides (12-20 base pairs) is now possible.

It is now widely appreciated that significant local conformational heterogeneity exists in the structures of nucleic acids (10). While X-ray crystallography has

0097–6156/92/0486–0202$06.00/0

provided much of our understanding of largely DNA duplex structural variations, high resolution NMR has also begun to provide detailed three-dimensional structural information on duplex, hairpin loop and other oligonucleotide structures.

Traditionally the phosphate ester moiety is often overlooked and viewed as a passive element of nucleic acid structure. Base pairing and other nucleic acid base oriented structural features have been assumed to control nucleic acid structure and conformation. More recently (and as discussed in this paper) we have begun to appreciate the important role played by the phosphate ester (Figure 1) in defining the structure and dynamics of nucleic acids as well as nucleic acid-drug or protein complexes.

Thus, most attention on understanding the binding specificity between amino acid sequences of DNA transcriptional regulating proteins and DNA sequences has centered on hydrogen-bonding to the acceptor/donor groups on the Watson-Crick base-pairs in the major groove (cf. 11). Even with the extensive X-ray and NMR studies on this problem, we still do not understand this "second genetic code" of protein-DNA recognition. Perhaps one reason for the inability to dissect the basis for this specificity is the emphasis on base-pair interactions alone. Localized, sequence-specific conformational variations in DNA are quite likely another important component of a protein's recognition of specific sites on the DNA.

Remarkably, in every high resolution X-ray crystal structure of a repressor-operator complex, the majority of the contacts are to the *phosphates* (12-14)! Indeed, the recent crystal structure of the *trp* repressor demonstrated that every one of the direct protein contacts were mediated through interactions with the phosphate backbone (14). Ionic interactions involving the phosphate backbone have been implicated as being important factors in the recognition of the *lac* operator as well (15, 16). It is not known whether any of these ionic interactions provide a specific recognition mechanism for these repressors. As described in this paper, we suggest that the conformation and, in particular, the flexibility of the phosphates may be an important component of protein-DNA recognition.

One of the main reasons for assigning ^{31}P resonances of oligonucleotides is to obtain information on the conformation of the phosphodiester backbone (Figure 2) (17, 18). The internucleotide linkage is defined by six torsional angles from one phosphate atom to the next along the DNA backbone. Theoretical studies have shown that the conformation of two of the six torsional angles (α: O3'-P-O5'-C5' and ζ: C3'-O3'-P-O5') appear to be most important in determining ^{31}P chemical shifts (18, 19). While 2D ^{1}H/^{1}H nuclear Overhauser effect spectroscopy (NOESY) has largely been used to provide the structural information to derive 3D structures of nucleic acids in solution, the distances derived from NOESY spectra give no direct information on the phosphate ester conformation and NOESY-distance derived structures have been believed to be effectively disordered in this part of the structure. However, as discussed in this paper, ^{31}P chemical shifts and ^{1}H-^{31}P coupling constants can provide valuable information on the phosphate ester backbone conformation and the role the phosphate plays in defining nucleic acid structure and recognition.

Our laboratory has been able to show experimentally [as originally predicted

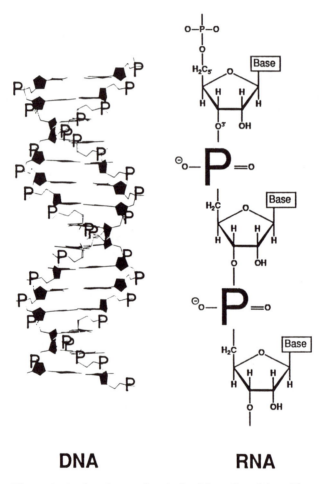

DNA   **RNA**

Figure 1: A phosphorus chemist's vision of nucleic acids.

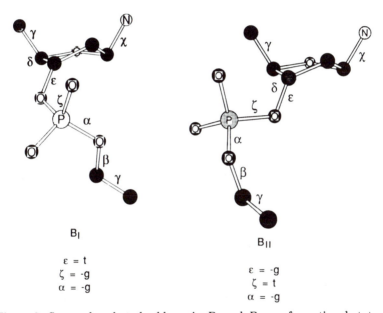

B_I

$\epsilon = t$
$\zeta = -g$
$\alpha = -g$

B_{II}

$\epsilon = -g$
$\zeta = t$
$\alpha = -g$

Figure 2: Sugar phosphate backbone in B_I and B_{II} conformational states. $\alpha, \beta, \epsilon, \zeta$: g^-, t, t, g^- (A); g^-, t, g^-, t(B). Torsional angles gauche(-) (g^- or -60°); trans (180°).

theoretically (20)] that phosphorus chemical shifts indeed provide direct information on the conformation of the phosphate ester (21, 22). The phosphates in a gauche⁻ (g⁻: dihedral angle, –60°) conformation about one of the P–O ester bonds have been demonstrated to have a ³¹P chemical shift ca. 1.6 ppm upfield of the phosphates in a trans (t: dihedral angle 180°) conformation (19, 20). Local helical parameters such as helix twist and base-pair roll appear to affect the ³¹P chemical shift of nucleic acids and their complexes by stretching or contracting the sugar phosphate backbone which in turn alters the time-averaged conformation of the phosphate ester. Therefore, sequence-specific information about the DNA is potentially contained within the geometry of the phosphates and may be obtained from the ³¹P NMR spectrum. In this paper we describe our ³¹P NMR studies of regular DNA duplexes and duplex protein complexes as well as various dithiophosphate DNA analogues.

Assignment and Interpretation of ³¹P NMR Spectra of Oligonucleotides

A major limitation in the use of ³¹P NMR in providing information on the backbone conformation has been the difficulty in assigning the signals. Conventional 2D ³¹P – ¹H heteronuclear correlation (HETCOR) NMR spectroscopy has been applied with limited success to the assignment of the ³¹P NMR spectra of oligonucleotides (7-9). Assignment of the phosphorus residues by regiospecific labelling with ¹⁸O and ¹⁷O proved quite useful (3, 23, 24).

More recently, reverse detection HETCOR (25) and long range, constant time HETCOR (26) methods have been successfully applied to the ³¹P assignment problem. Because of these methods, assignment of the ³¹P resonances of modest-size oligonucleotide duplexes is now quite feasible (19, 27). Thus the assignment problems that initially held back the use of ³¹P NMR have now been solved.

In order to understand the origin of the variation in ³¹P chemical shifts of nucleic acids, it has proven to be critical to be able to measure the backbone torsional angles. Fortunately the $J_{H3'-P}$ coupling constants in oligonucleotide duplexes can often be readily determined from a selective 2D J-resolved spectra. We have noted a "Christmas-tree" shaped pattern with monotonically decreasing coupling constants for the signals to higher field. This is usually observed in most duplex oligonucleotides (11, 19, 21) and strongly supports a correlation between ³¹P chemical shifts and $J_{H3'-P}$ coupling constants.

The observed three-bond coupling constants may be analyzed with a proton-phosphorus Karplus relationship to determine the C4'-C3'-O-P torsional angle ε (28). As shown by Dickerson (10) there is a strong correlation (R = –0.92) between torsional angles C4'-C3'-O3'-P (ε) and C3'-O3'-P-O5' (ζ) in the crystal structures of various duplexes. Thus both torsional angles ε and ζ can often be calculated from the measured P-H3' coupling constant.

The Karplus relationship provides solutions for four different torsional angles for each value of the same coupling constant (29). However, it appears that for regular duplexes, the coupling constants, which can vary only between 1 and 11 Hz, all correspond to ε torsional angles on a single limb nearest the crystallographically observed average for nucleic acids (ε = -169±25°). Assuming a linear correlation between the coupling constants and ^{31}P chemical shifts (effectively a linear fit to the data on the main limb of the Karplus curve) the correlation coefficient is -0.92 for the combined data from many oligonucleotide duplexes (11). It is important to recognize that ^{31}P chemical shifts are perturbed by factors other than torsional angle changes alone (19). However, knowledge of *both* ^{31}P chemical shift and $J_{H3'-P}$ coupling constant can be very helpful in the proper interpretation of phosphate ester conformation. Thus, for "normal" B-DNA geometry, there is an excellent correlation between the phosphate resonances and the observed phosphate ester torsional angles.

The possible basis for the sequence-specific variation in the ^{31}P chemical shifts of oligonucleotides can be understood in terms of sugar phosphate backbone distortions involved in duplex geometry changes. Briefly, as the local helical geometry changes, the length of the sugar phosphate backbone must change, which in turn requires changes in the sugar phosphate backbone angles.

These backbone changes can be accomplished by the phosphate switching from the "B_I" ($\zeta = g^-$, $\alpha = g^-$) to the "B_{II}" conformation ($\zeta = t$, $\alpha = g^-$; Figure 2). The B_I and B_{II} states represent ground-state minima with the P-O and C-O torsional angles in the staggered conformations (19). Partially or fully eclipsed conformations which are not energy minima are only accessible through libration in each of the staggered states or through transient passage during the rapid picosecond-time scale jumps between the two ground states.

Our laboratory (21) has analyzed the variation in the ^{31}P chemical shifts and $J_{H3'-P}$ coupling constants in terms of fractional populations of the two thermodynamically stable B_I and B_{II} states. We have determined that the ^{31}P chemical shift and $J_{H3'-P}$ coupling constant of a phosphate in a purely B_I conformational state is estimated to be ca. -4.6 ppm and 1.3 Hz respectively. Similarly the ^{31}P chemical shift and $J_{H3'-P}$ coupling constants of a phosphate in a purely B_{II} conformational state should be ca. -3.0 ppm and 10 Hz respectively. The dispersion in the ^{31}P chemical shifts of oligonucleotides is attributable to different ratios of populations of the B_I and B_{II} states for each phosphate in the sequence. The phosphate makes rapid jumps between these two states (30). Thus the measured coupling constant (and ^{31}P chemical shift) provides a measure of the populations of these two conformational states.

lac Repressor Headpiece-Operator ^{31}P NMR Spectra

The regulation of the expression of the *lac* genes has served as the archetypal example of a negatively controlled operon in prokaryotes (31). While site-specific mutagenesis and mutant operators have proven to provide important information on the role of individual amino acids and base pairs in the recognition process (32) it is still not understood at a detailed molecular level. X-ray crystallographic structural refinements, extensive mutagenesis experiments, as well as model-building, suggest that certain side chain residues recognize the individual bases of nucleic acids, while others "recognize" the backbone phosphates, possibly through sequence-specific variations in the DNA conformation.

It is possible to duplicate the basic *lac* operator-*lac* repressor protein interaction by using the smaller *lac* repressor headpiece N-terminal domain fragment (33, 34). Recent NMR derived structures of repressor headpiece bound to *lac* operator DNA fragments have begun to provide details confirming the sequence-specific interactions of a recognition α-helix binding within the major groove of the operator DNA. In this section, the ^{31}P NMR spectra of various 14-base pair *lac* operators free and bound to both wild-type and a Y7I (tyrosine 7 to isoleucine) mutant *lac* repressor headpiece proteins are shown to provide information on the backbone conformation in the complexes.

The ^{31}P spectra of both wild-type *lac* operator 14-mer d(TGTGAGCGCTCA-CA)$_2$ (O1) as well as a number of base-pair mutants, d(T*A*TGAGCGCTCA*T*A)$_2$ (O2) and d(TGTG*T*GCGC*A*CACA)$_2$ (O3; complementary sites of mutation in the palindromic operators are italicized) have been assigned (22). These symmetrical base sequences are about two-thirds the length of the 21 base-pair wild-type sequence, and the 14-mers are believed to contain most of the important recognition sites for the *lac* repressor protein.

The ^{31}P spectral changes upon binding the N-terminal 56-residue headpiece (HP) to the 14-mer operators (O) demonstrate that all of the phosphate resonances remain in fast chemical exchange during the entire course of the titration because only one set of peaks is observed at all DNA/protein ratios (16). In the wild-type O1 operator binding to the wild-type headpiece, the ^{31}P signals of the G5 phosphate shifts 0.20 ppm downfield while the G7 phosphate shifts 0.16 ppm upfield during the titration. The A8 phosphate shifts 0.13 ppm upfield during the titration. The ^{31}P signals of phosphates 9 and 11 shift 0.15 ppm upfield during the titration. The ^{31}P signals of the remaining phosphates 4, 10 and 12–16 show either no or small perturbations (< 0.1 ppm) upon titration with headpiece.

Remarkably the changes in the ^{31}P chemical shifts of the wild-type symmetrical O1 operator upon binding Y7I mutant headpiece were strikingly similar to the changes observed for the wild-type headpiece-operator complex (16, 36). This might be expected since the Y7I mutant repressor protein binds a 322 bp DNA fragment containing the wild-type non-palindromic operator with only a three-fold poorer binding constant than the wild-type repressor protein. However 2D ^{1}H NMR studies show that the mutation significantly disrupts the overall structure and stability of the recognition helix of the headpiece (residues 17-25; see ref. 37). In the mutant, loss of the Y7-Y17 aromatic sidechain interaction, proposed to exist in the wild-type 56-residue headpiece, presumably selectively destabilizes the helix. The presence of nearly all of the tertiary structure cross-peaks (38) in the mutant protein indicates that the overall folding has not been dramatically altered.

Analysis of ^{31}P Chemical Shift Perturbations in Protein-DNA Complexes

The perturbations in ^{31}P chemical shifts in forming the O-HP complex can arise from several sources. Electrostatics and local shielding effects by the bound protein certainly can play a role (19). However, as described earlier, the phosphate ester conformation plays a dominant role in the ^{31}P chemical shift differences in small DNA fragments. Sequence-specific variations in the conformation of the DNA sugar phosphate backbone thus can possibly explain the sequence-specific recognition of DNA, as mediated through direct contacts and electrostatic complementarity between the phosphates and the protein.

Assuming then that the ^{31}P chemical shifts also represent the relative populations of the B_I and B_{II} states in the HP-O complexes, we can calculate the fractional populations by assuming a simple two-state model. A plot of ^{31}P chemical shifts (and B_I/B_{II}) populations vs. sequence for several of the operator complexes is shown in Figure 3. Note as expected for the very similar perturbation in ^{31}P chemical shifts in the O1 wild type and Y7I mutant headpiece complexes, the pattern of sequence-specific variation of ^{31}P chemical shifts in the two complexes is very similar (Figure 3A). However, in the *lac* O2 mutant operator-headpiece complex several of the phosphates are significantly perturbed from the phosphate ester populations in the O1-HP complex. Phosphates such as p6, p8 and p11 in the O2 complex are apparently constrained to a B_I conformation (Figure 3B; p15 is also shifted to a more B_I-type state).

These results have suggested that discrimination between the operators may be based upon the degree to which the repressor protein restricts phosphate ester conformational freedom in the complex. We suggest that specific, tightly bound complexes *retain the inherent phosphate ester conformational flexibility of the operator itself*, whereas more weakly bound operator-protein complexes restrict the phosphate ester conformational freedom in the complex relative to the free DNA.

This requirement for *retention* of backbone torsional freedom in strongly bound complexes (which is entropically favorable) provides a new mechanism for protein discrimination of different operator binding sites. Thus upon binding, various rotational degrees of freedom must be lost if the repressor were to bind to only one of the phosphate conformations. The dissociation constant will reflect this internal entropic disadvantage, which may be as large as 8 eu (2.4 kcal/mol at 25° C) per lost degree of torsional freedom (39). Binding of the repressor in such a way as to restrict the intrinsic conformational freedom of 20 or more of the operator phosphates could, in principle, contribute to a sizable entropic disadvantage.

This analysis suggests another explanation both for the entropy differences in the binding of various wild-type and mutant operator for repressor as well as an entirely new appreciation for the difficulty in which a protein must recognize the structural features *and dynamics* of operator DNA. Thus in the tightly bound, specific complexes the protein must not only provide a binding surface that matches the sequence-specific variation in the phosphate conformation of the operator but also allows for retention of the phosphate conformational freedom in the complex. This can only be possible if the protein-DNA interface is flexible enough and if there is a "coupling" of the motion of the amino acid residues in the binding site with the phosphate ester motion. In contrast in a mutant operator that does not bind as tightly to the repressor, the subtle structural and dynamical requirements for providing the necessary coupling or flexibility is presumably lost so as to preclude free rotation of at least some of the phosphate esters in the protein-DNA complex. Recall that in the O2-HP complex the ^{31}P data suggest that several of the phosphates appear to be restricted to a B_I state.

In mutant operators presumably the structural and dynamical coupling between the protein and DNA is disrupted such that the backbone conformational freedom in the complex is greatly restricted through either steric or electrostatic interactions. This constrains one or more of the phosphodiester bonds as shown by the O2 mutant operator headpiece complexes. This is reflected in significant entropic penalties, resulting in poorer binding to the repressor. Perhaps understanding the "second genetic" code will require that we more fully appreciate the role of phosphate ester flexibility in defining protein-DNA interactions.

^{31}P NMR of Phosphoryl-modified Analogues of Oligonucleotides

The role of the phosphate ester has also proven important in the design of various "antisense" oligonucleotide agents (40). Oligodeoxyribonucleotides, oligodeoxynucleoside methylphosphonate, phosphorothioate, phosphoroamidate and phosphorodithioate analogues among others have been studied (cf. 40). These complementary, "antisense" oligonucleotide agents are believed to operate by forming an DNA-RNA hybrid complex which is degraded by RNase H or alternatively blocks ribosome processing of the message (40).

While unmodified oligodeoxynucleotides are susceptible to nuclease digestion, the phosphate-modified oligodeoxynucleoside methylphosphonate, phosphorothioate and phosphorodithioate analogues are generally nuclease resistant, and have recently received great attention (40). Our laboratory has been interested in investigating the structural perturbations involved in the dithio substitution and how this modification affects the overall molecular structure, stability and properties of these dithiophosphate agents. Synthesis and purification of dithiophosphate oligonucleotide analogs may be accomplished using thiophosphoramidite chemistry (41-43). The large downfield ^{31}P shift (ca. 110 ppm) of the dithiophosphate groups relative to the phosphate moiety (ca. -4 ppm) readily identifies the substitution. As shown in Figure 4 the ^{31}P spectrum of d(CGCTpS$_2^-$TpS$_2^-$AAGCG) [10-mer(pS^{2-})$_2$] (a site of dithio substitution is designated with the symbols pS^{2-}) multiple dithiophosphoryl substitutions show the expected reduction in the number of phosphoryl-oligonucleotide signals with replacement by new dithiophosphoryl signals (44). A comparison of peak intensity at 110.67 ppm and 109.47 ppm vs. peak intensity at ca. -4 ppm shows the expected ratio of 2:7 (2 dithiophosphates to 7 phosphates). Moreover, the absence of additional ^{31}P signals (including signals around 56 ppm for monothiophosphates) confirm the purity and conformational homogeneity of the sample. A PAC, Constant time ^{1}H/^{31}P heteronuclear correlated spectrum has been used to assign the resonances indicated (44).

Other dithio substituted oligonucleotides including CpS^{2-}GpS^{2-}CpS^{2-}TpS^{2-}-TpS^{2-}ApS^{2-}ApS^{2-}GpS^{2-} CpS^{2-}G [10-mer(pS^{2-})$_9$], CGCTpS^{2-}TpS^{2-}ApS^{2-}-ApS^{2-}GCG [10-mer(pS^{2-})$_4$], ApS^{2-}GCT, and ApS^{2-}GpS^{2-}CpS^{2-}T have also been synthesized in our laboratory using the same manual solid phase synthetic methodology. The 10-mer(pS^{2-})$_9$ with all of the phosphates substituted for dithiophosphate groups has also been synthesized by both manual and automated routes. The ^{31}P NMR spectrum shows that all of the phosphate region signals have been

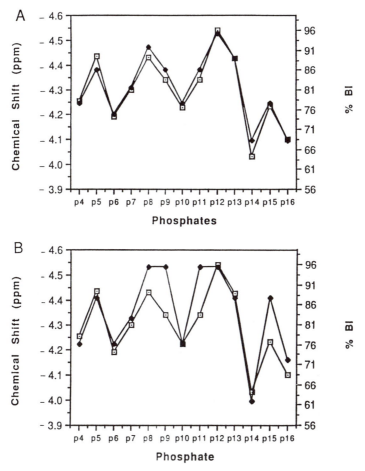

Figure 3: ^{31}P chemical shifts (♦) vs. sequence for A) wild-type O1 operator-Y7I mutant headpiece and B) mutant O2 operator-wild type headpiece. The ^{31}P chemical shifts (□) vs. sequence for the wild-type O1 operator-wild type headpiece is shown for comparison.

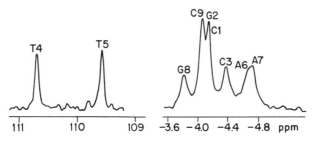

Figure 4: ^{31}P NMR spectra and phosphate assignments of dithiophosphate decamer d(CGCTpS$_2^-$TpS$_2^-$AAGCG).

eliminated and only dithiophosphate signals can be seen (Figure 5). Importantly there are no monothiophosphate impurities which are often observed in sample preparations.

Interestingly, analysis of the 2D spectra of the dithiophosphate analogue indicated that it was not in a duplex B-DNA conformation. The proton spectrum of the 10-mer(pS^{2-})$_2$ dithiophosphate analogue in this low salt buffer was assigned through analysis of two-dimensional TOCSY and NOESY spectra following the sequential assignment methodology (44). In the 300 ms NOESY spectrum, base to H1' connectivity was observed along the entire backbone from C1 to G10 with interruption only for the absence of a T4 H1' – T5 H6 cross peak. A striking feature of the base to H1' NOESY spectrum is the observed connectivity between A6 H8 and T4 H1'. This type of connectivity does not exist in regular A or B DNA where the distance between bases i and i+2 is too large. This connectivity clearly indicates an unusual structure which would place A6 H8 in close proximity to T4 H1'.

These observations are consistent with a structure in which the base of T5 loops out in a possible single strand hairpin loop or bulged duplex. The NOESY crosspeaks suggest that the T5 base stacks out of the structure, whereas the A6 pair stacks inside. Figure 6 is a model of the NOESY-distance restrained structure derived from these spectra. Thus, in contrast to the parent palindromic decamer sequence (30) which has been shown to exist entirely in the duplex B-DNA conformation under comparable conditions (100 mM KCl, 10 mM strand concentration), the dithiophosphate analogue forms a hairpin, even at twice the strand concentration. However, at higher salt concentrations (200 mM), four ^{31}P signals for the dithiophosphoryl groups of the dithiophosphate analogue are observed (Figure 7). The additional pair of upfield ^{31}P dithiophosphate signals is assigned to the normal duplex form of the dithiophosphate analogue. It is known that palindromic sequences will generally form hairpin loop structures at low salt and low strand concentrations. At 200 mM salt we observe a 1:2 mixture of duplex and hairpin loop. This has been confirmed by analysis of the ^{1}H 2D NMR spectra of the mixture, which clearly shows the expected connectivity and chemical shifts for both duplex and hairpin (44). As would be expected the unusual A6H8 to T4H1 crosspeak was not found for the duplex form.

The destabilization of the duplex form for the dithiophosphate is possibly attributable to unfavorable dithiophosphate electrostatic repulsion in the duplex form. In the hairpin the thymidine dithiophosphates do not interact with an adjacent phosphate group across the major and minor grooves. The P-S bond length is ca. 0.5 Å longer than the phosphoryl P-O bond length and the larger van der Waals radius for sulfur is presumably responsible for the increased electrostatic destabilization of the duplex form. As shown in Figure 6, the hairpin structure also separates the large dithiophosphates (van der Waals surfaces for the sulfurs are indicated) along the strand to a much greater degree than is possible in the duplex form.

The ^{31}P NMR spectrum of the 10-mer(pS^{2-})$_4$ contains eight different resonances in the dithiophosphate region corresponding to both hairpin and duplex

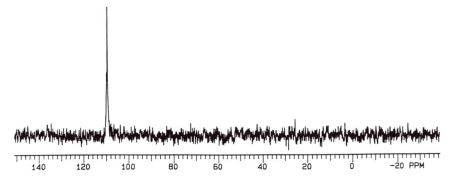

Figure 5: ^{31}P spectrum of oligonucleotide d($CpS^{2-}GpS^{2-}CpS^{2-}TpS^{2-}TpS^{2-}$-$ApS^{2-}ApS^{2-}GpS^{2-}CpS^{2-}G$). Unpurified oligonucleotide was prepared by automated synthesis.

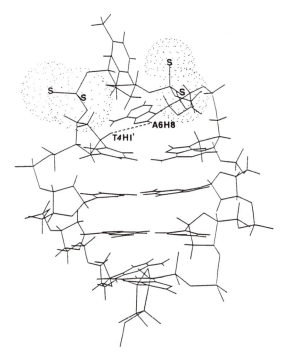

Figure 6: Model of d($CGCTpS^{2-}TpS^{2-}AAGCG$) derived from NOESY-distance restrained, molecular dynamics calculations.

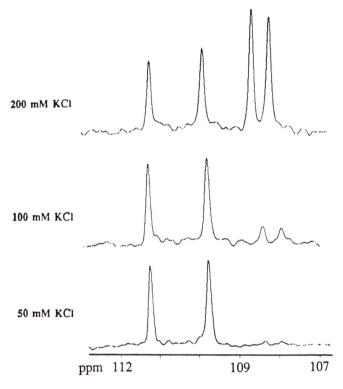

Figure 7: Dithiophosphate region of the ^{31}P spectrum of d(CGCTpS^{2-}TpS^{2-}-AAGCG) at various KCl concentrations, 22° C.

forms. Based upon integration, the ratio of hairpin and duplex form is 4:1 (see Figure 8). Although we have not quantified the amount of 10-mer(pS^{2-})$_4$ strand concentration, it appears that adding two additional dithiophosphates has not significantly further destabilized the duplex form.

These results demonstrate the potential importance of electrostatic interactions in the relative stabilization of duplex and hairpin DNA. It also raises potential implications for design of monothiophosphate and dithiophosphate antisense analogues.

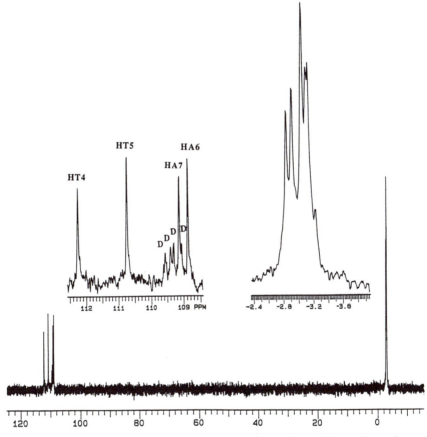

Figure 8: ^{31}P spectrum of oligonucleotide d($CGCTpS^{2-}TpS^{2-}ApS^{2-}ApS^{2-}$-GCG) in D_2O.

Acknowledgments

Supported by NIH (AI27744), the Purdue University Biochemical Magnetic Resonance Laboratory which is supported by the NIH designated AIDS Research Center at Purdue (AI727713) and the NSF Biological Facilities Center on Biomolecular NMR, Structure and Design at Purdue (grants BBS 8614177 and 8714258 from the Division of Biological Instrumentation).

References

1 Feigon, J.; Leupin, W.; Denny, W. A.; Kearns, D. R. *Biochemistry* **1983**, *22*, 5930-5942; 5943-5951.

2 Hare, D. R.; Wemmer, D. E.; Chou, S. H.; Drobny, G.; Reid, B. *J. Mol. Biol.* **1983**, *171*, 319.

3 Shah, D. O.; Lai, K.; Gorenstein, D. G. *J. Am. Chem. Soc.* **1984**, *106*, 4302-4303.

4 Frechet, D.; Cheng, D. M.; Kan, L. - S.; Ts'o, P. O. P. *Biochemistry* **1983**, *22*, 5194-5200.

5 Broido, M. A.; Zon, G.; James, T. L. *Biochem. Biophys. Res. Commun.* **1984**, *119*, 663-670.

6 Scheek, R. M.; Boelens, R.; Russo, N.; vanBoom, J. H.; Kaptein, R. *Biochemistry* **1984**, *23*, 1371-1376.

7 Schroeder, S. A.; Fu, J. M.; Jones, C. R.; Gorenstein, D. G. *Biochemistry* **1987**, *26*, 3812-3821

8 Pardi, A.; Walker, R.; Rapoport, H.; Wider, G.; Wüthrich *J. Am. Chem. Soc.* **1983**, *105*, 1652.

9 Lai, K.; Shah, D. O.; Derose, E.; Gorenstein, D. G. *Biochem. Biophys. Res. Commun.* **1984**, *121*, 1021.

10 Dickerson, R. E.; Drew, H. R. *J. Mol. Biol.* **1981**, *149*, 761-786.

11 Gorenstein, D. G.; Schroeder, S. A.; Fu, J. M.; Metz, J. T.; Roongta, V. A.; Jones, C. R. *Biochemistry* **1988**, *27*, 7223-7237.

12 Wolberger, C.; Dong, Y.; Ptashne, M.; Harrison, S. C. *Nature* **1988**, *335*, 789-795.

13 Jordan, S. R.; Pabo, C. O. *Science* **1988**, *242*, 893- 899.

14 Otwinowski, Z.; Schevitz, R. W.; Zhang, R. G.; Lawson, C. L.; Joachimiak, A.; Marmorstein, R. Q.; Luisi, B. F.; Sigler, P. B. *Nature* **1988**, *335*, 321.

15 deHaseth, P. L.; Lohman, T. M.; Record Jr, M. T. *Biochemistry* **1977**, *16*, 4783-4790.

16 Karslake, C.; Schroeder, S.; Wang, P. L.; Gorenstein, D. G. *Biochemistry* **1990**, *29*, 6578-6584.

17 Gorenstein, D. G. *In* "Phosphorus-31 NMR: Principles and Applications"; Gorenstein, D. G., Ed.; Academic Press: Orlando, 1984; pp 7-36.

18 Gorenstein, D. G.; Findlay, J. B.; Momii, R. K.; Luxon, B. A.; Kar, D. *Biochemistry* **1976**, *15*, 3796-3803.

19 Gorenstein, D. G. *In* "DNA Structures, Methods in Enzymology"; Lilley, D. M. and Dahlberg, J. E., Eds.; Academic Press, Inc.: Orlando, FL, 1991.

20 Gorenstein, D. G.; Kar, D. *Biochem. Biophys. Res. Commun.* **1975**, *65*, 1073-1080.

21 Roongta, V. A.; Powers, R.; Nikonowicz, E. P.; Jones, C. R.; Gorenstein, D. G. *Biochemistry* **1990**, *29*, 5245-5258.

22 Schroeder, S. A.; Roongta, V.; Fu, J. M.; Jones, C. R.; Gorenstein, D. G. *Biochemistry* **1989**, *28*, 8292-8303.

23 Petersheim, M.; Mehdi, S.; Gerlt, J. A. *J. Am. Chem. Soc.* **1984**, *106*, 439-440.

24 Joseph, A. P.; Bolton, P. H. *J. Am. Chem. Soc.* **1984**, *106*, 437-439.

25 Sklenár, V.; Miyashiro, H.; Zon, G.; Miles, H. T.; Bax, A. *FEBS Lett.* **1986**, *208*, 94-98.

26 Fu, J. M.; Schroeder, S. A.; Jones, C. R.; Santini, R.; Gorenstein, D. G. *J. Magn. Reson.* **1988**, *77*, 577-582.

27 Ott, J.; Eckstein, F. *Biochemistry* **1985**, *24*, 2530- 2535.

28 Lankhorst, P. P.; Haasnoot, C. A. G.; Erkelens, C.; Altona, C. *J. Biomol. Struct. Dynam.* **1984**, *1*, 1387-1405.

29 Nikonowicz, E. P.; Gorenstein, D. G. *Biochemistry* **1990**, *29*, 8845-8858.

30 Powers, R.; Jones, C. R.; Gorenstein, D. G. *J. Biomol. Struct. Dyn.* **1990**, *8*, 253-294.

31 Gilbert, W.; Gralla, J.; Major, J.; Maxam, A. "Protein-Ligand Interactions"; de Gruyter: Berlin, 1975; pp 270-288.

32 Lehming, N.; Sartorius, J.; Oehler, S.; von Wilcken-Bergmann, B.; Müller-Hill, B. *Proc. Natl. Acad. Sci., U.S.A.* **1988**, *85*, 7947-7951.

33 Wade-Jardetzky, N.; Bray, R. P.; Conover, W. W.; Jardetzky, O.; Geisler, N.; Weber, K. *J. Mol. Biol.* **1979**, *128*, 259-264.

34 Zuiderweg, E.; Scheek, R.; Boelens, R.; Gunsteren, W.; Kaptein, R. *Biochemie* **1985**, *67*, 707-715.

35 Betz, J.; Sasmor, H.; Buck, F.; Insley, M.; Caruthers, M. *Gene* **1986**, *50*, 123-132.

36 Karslake, C.; Botuyan, M. V.; Gorenstein, D. G., submitted.

37 Karslake, C.; Wisniowski, P.; Spangler, B. D.; Moulin, A. - C.; Wang, P. L.; Gorenstein, D. G. *J. Am. Chem. Soc.* **1991**, *113*, 4003-4005.

38 Wisniowski, P.; Karslake, C.; Piotto, M. E.; Spangler, B.; Moulin, A. - C.; Nikonowicz, E. P.; Kaluarachchi, K.; Gorenstein, D. G. *J. Biomolec. Struct. Stereodynamics*, in press.

39 Page, M. I.; Jencks, W. P. *Proc. Natl. Acad. Sci. USA* **1971**, *68*, 1678-1683.

40 Uhlmann, E.; Peyman, A. *Chem. Rev.* **1990**, *90*, 543-584.

41 Brill, W. K. D.; Nielsen, J.; Caruthers, M. H. *Tetrahedron Lett.* **1988**, *29*, 5517-5520.

42 Farschtschi, N.; Gorenstein, D. G. *Tetrahedron Lett.* **1988**, *29*, 6843-6846.

43 Brill, W. K. D.; Tang, J. - Y.; Ma, Y. - X.; Caruthers, M. H. *J. Am. Chem. Soc.* **1989**, *111*, 2321-2322.

44 Piotto, M. E.; Granger, J. N.; Cho, Y.; Farschtschi, N.; Gorenstein, D. G. *Tetrahedron* **1991**, *47*, 2449-2461.

RECEIVED December 3, 1991

Chapter 17

Phosphorus Flame Retardants for a Changing World

Alan M. Aaronson

Akzo Chemicals Inc., 1 Livingstone Avenue, Dobbs Ferry, NY 10522–3401

The growing use of plastics and other flammable
synthetic materials in construction, home furnishings,
and various industrial applications is creating opportunities for polymer compatible, thermally stable,
environmentally friendly, and cost-efficient flame retardants. New legislation emphasizing test methods that
provide input for mathematical firemodeling is changing the focus of interest in preferred flame retardant
action from simple ignition resistance to fire propagation and rate-of-heat release. These factors together
with concerns about combustion product toxicity and
corrosivity, long-term worker exposure, and manufacturing waste minimization are creating product parameters uniquely met by various phosphorus compounds.
The best phosphorus compounds for both new and
existing applications will be significantly different from
those compounds in the marketplace today. This
paper will describe some new compounds and the driving forces defining the product parameters that will
challenge phosphorus chemists for the next decade.

Flame retardants are an important commercial end use for organophosphorus compounds. Many opportunities exist for organophosphorus

0097–6156/92/0486–0218$06.00/0
© 1992 American Chemical Society

compounds to play an increasingly greater role in this end use in the future. I will point out some of the factors in our changing world, which to some may appear to be problems, but which I believe present excellent opportunities for organophosphorus compounds.

Phosphorus compounds, as flame retardants, have been around for a very long time but it is only in the last several decades that various events have occurred creating a situation in which organophosphorus compounds are fast becoming the most important and most versatile class of flame retardants. This new era in flame retardancy was ushered in by the tremendous growth of new synthetic polymers as replacements for traditional materials in all aspects of our daily life, from construction to home furnishings. A major turning point was the 1974 Federal Trade Commission complaint against 25 companies, testing agencies and standards organizations. In this complaint, the Federal Trade Commission, spurred on by the increasing incidence of fire deaths involving synthetic materials, and by one particularly horrible case, accused essentially an entire industry and its voluntary consensus standards of misrepresenting the fire safety of materials. The straw that broke the camel's back was a case in which a homeowner insulated his garage with exposed rigid polyurethane foam which he was assured was flame retarded and safe for use because the foam was "self extinguishing" by a commonly applied, small-scale, standard test method (ASTM 1692). The fire that eventually occurred in the garage was not small-scale, and the foam was far from "self-extinguishing".

In the resulting consent agreement, the FTC required that ambiguous and misleading terminology be eliminated, that test methods be developed to more accurately reflect the hazards of materials used in actual applications and that sufficient warning accompany any such test results. In addition, the FTC required that parties to the consent agreement fund a five million dollar, five-year research program to develop better and more-scientifically-sound test methods to measure the fire performance of materials, in particular cellular or foamed plastics. This research program, administered by the Products Research Committee, set the tone for considerable work in both industry, academia and at the National Bureau of Standards which has since changed its name to the National Institute of Standards and Technology or NIST.

Not long after the FTC complaint, the various model building-code-writing organizations responded to the growing use of plastics in construction by writing separate sections in their codes defining acceptable applications and establishing fire test performance criteria for plastic materials. These building codes were subsequently written into law throughout the country.

In 1977, The Consumer Product Safety Commission banned the sale of children's sleepwear treated with trisdibromopropyl phosphate, better known simply as "Tris". This ban was imposed after a National Cancer Institute study linked Tris to kidney cancer.

Trisdibromopropyl phosphate is no longer an item of commerce.

In the area of household furnishings, the Consumer Products Safety Commission encouraged the development by the furniture industry of a voluntary flammability standard designed to provide a minimum of cigarette-ignition resistance. Various local jurisdictions, most notably the State of California, the City of Boston and the Port Authority of New York and New Jersey established their own standards going beyond cigarette-ignition resistance to much more severe open flame testing, especially for furniture used in high-risk occupancies.

In just 17 years, the entire field of flame retardancy has taken a giant leap forward and the technology to provide FR synthetic materials has made remarkable progress. Today, flame retardant grades of synthetic polymers have found applications in areas considered impossible a few years ago. One example is the use of polyurethane foam in upholstered furniture for high-risk occupancies such as hotels, prisons and hospitals. Polyurethane foam, which was once described in the press as "solid gasoline" is now formulated to withstand severe fire test exposure without propagating the fire.

This remarkable progress was made possible by advances in the development of fire-test methodology.

Many of these advances are directly related to Products Research Committee sponsored projects.

Fire-test method development has followed two separate but complementary paths. One path, theoretically oriented, is characterized by the measuring of scientifically-meaningful fire properties, such as mass loss and rate-of-heat release. This approach also includes the development of mathematical models incorporating these properties to predict propagation and flame spread. A new lab-scale apparatus, the "cone calorimeter" developed at NIST is an example of the hardware now available to measure these fire properties.

The second, more empirical path, is marked by the development of larger scale test methods that are meant to simulate potential "real world" or actual exposure conditions. Examples of such test methods are room fire tests in which actual pieces of furniture are placed in 8'x8'x12' room-size enclosures and exposed to fire sources ranging from multiple sheets of newspaper or waste baskets in full conflagration to even larger sources such as 20 lb. and 25 lb. wood cribs.

The measurement of the rate-of-heat release during the combustion of materials is done by measuring the depletion of the oxygen in the combustion atmosphere under varying heat flux exposures. This has become the dominant technique applied in the development of new small-scale tests. The apparatus developed at the National Institute of Standards and Technology known as the "cone calorimeter", after its cone-shaped radiant heat source, has been so universally well received by fire researchers, especially in the government regulatory agencies of Europe, that it is probably the only test apparatus that will be agreed upon by all parties during the harmonization activities to establish European fire test standards. Using this apparatus, it is possible to measure not only the rate-of-heat released, but also a) the rate-of-smoke generation, b) some measure of the toxic potential of the smoke generated and c) the corrosivity of the smoke generated.

Another test method that has acquired worldwide acceptance as a reliable benchmarking procedure is the Underwriters Laboratory subject 94. This method primarily impacts those materials used in electrical applications. The UL94 method, though it has been around for many years, has not remained static. The criteria for acceptance have become more severe, and in many instances where materials could previously gain acceptance by melting and dripping away from the ignition source (UL94V2), the stricter criteria of UL94V0 will not only prohibit ignition of the specimen but will not allow burning drips either. Thus, formulation techniques to promote melting and polymeric breakdown to low viscosity liquids will eventually become less important than techniques to promote char formation as a barrier to ignition and consumption of material.

These advances in test methodology have brought about a movement away from the "beat-the-test" scenarios of the previous era of very small-scale tests, a practice which contributed to the events which brought about the 1974 FTC complaint.

Instead, attention is now focused on controlling the burning behavior of materials by a) inhibiting ignition in the first place, b) slowing flame propagation, c) reducing the rate of energy (heat) released, and d) limiting the amount of material consumed.

This has created opportunities for organophosphorus flame retardants because you can achieve these effects best by the way organophosphorus flame retardants work.

To be more specific, organophosphorus flame retardants work most effectively in the solid or condensed phase by inhibiting ignition and promoting char formation. Usually, this is most effective in oxygenated polymers like wood, cotton, polyurethanes, polyesters, etc. where char formation is an effective means to limit further burning. A good char layer is not only difficult to ignite, but it also insulates underlying virgin polymer minimizing thermal degradation.

In more hydrocarbon-like polymers that melt, drip and depolymerize to highly-flammable and volatile fragments, such as polyolefins and styrenics, the traditional way to pass FR tests is to poison the flame by quenching radical reactions in the gaseous combustion zone (with halogens from brominated or chlorinated flame retardants) or by diluting combustible fragments with inert gases or by removing heat with molecules like water (from hydrated inorganics such as alumina trihydrate or magnesium hydroxide).

These mechanisms, however, are not as effective in limiting the rate-of-mass loss, energy release or total mass consumption as is the formation of a monolithic, insulating char barrier. Those organophosphorus compounds most commonly used today, the chloroalkyl phosphates and the aryl phosphates, or even the inorganic ammonium polyphosphates and elemental red phosphorus are not very effective char formers in these polymer systems.

To achieve effective char formation in hydrocarbon-like polymers, we are beginning to see combinations of phosphorus compounds with other molecules, and we are beginning to see phosphorus containing molecules combining structural elements that will induce char formation. Several recent examples of the blend approach which is really an extension of FR technology

borrowed from the coatings field are formulations of ammonium polyphosphate with nitrogenous polymers and carbonifics like pentaerythritol as in
recent patents granted to Hoechst and Enichem.

More interestingly, phosphorus-containing molecules which also contain
char-inducing moieties are exemplified by melamine amyl phosphate
(Figure 1). In another case,(Figure 2) the carbonific, pentaerythritol, is built
into the molecule in the form of the spirostructure linking two phosphoric
acid moieties. This intumescent approach, long useful in coatings, is just
beginning to take hold in thermoplastics. Market growth is slow because
these new materials are clear departures from the old highly-filled flame
retarded forms of the same polymers, and it takes time for designers and
engineers to incorporate these new materials into applications.

Shifting our attention to environmental concerns, the Dutch Ministry of
the environment is considering legislation to reduce a number of different
waste streams by the year 2000. Under the intense lobbying efforts of the
green movement, the Dutch Ministry is looking to grossly reduce, if not
prohibit the use of flame retardants, especially organic bromine compounds,
antimony oxide and chloroalkyl phosphates. The acute and chronic toxicity of
these materials are not the only issues. Important other considerations are
the characteristics of these additives in wastes. For example, how these flame
retardant additives leach from their substrates with time, what happens to
them on incineration of the plastics, and the ecotoxicity of these materials in
streams, soil, ground water, etc.

Environmental concerns are not just limited to the flame retardant
molecules themselves or their decomposition products. The implementation
of RCRA and the Clean Air Act have also taken their toll of the manufacturing processes for flame retardants. In the last year, one major flame retardant, tetrakis (2-chloroethoxy)ethylene bisphosphate (Figure 3) was withdrawn
from the market because of environmental problems stemming from its
manufacturing process. Similar manufacturing concerns may threaten other
chloroalkyl phosphates if current government scrutiny of their chronic toxicity
data doesn't do them in first.

The Falklands war also contributed to the loss of confidence in halogenated materials. Analysis of some of the naval experiences during the
Falklands war showed that corrosive atmospheres generated during shipboard
fires did considerable damage to the sensitive electronic guidance and communication systems. These corrosive gases were blamed on halogenated
materials. As a result, the U. S. Navy has subsequently banned the use of
halogenated flame retardants in wire and cable on U.S. Navy ships.

It is generally recognized that approximately 80% of fire deaths are the
result of smoke inhalation, smoke being the airborne solid and liquid particu-
lates, and gases evolved when a material undergoes pyrolysis or combustion
(ASTM E176 (1979)). Thus, the combustion products released from burning
plastics has attracted great interest as the uses of plastics in construction and
furnishings have grown. In particular, certain flame retardant additive
systems, those based on halogens, are suspected of increasing the amount of
toxic and corrosive gases generated by burning plastics. In addition to

presenting a threat to the lives of people caught in fires, there is a concern that these FR products present a threat to the environment when plastic waste is incinerated and a threat to the health of workers exposed to the fumes when molding or extruding these FR plastics. While all halogenated compounds can and will generate hydrogen halides in a fire, certain aromatic bromine compounds are suspected of also generating brominated benzo-dioxins and benzofurans during combustion. Some are also suspected of containing dioxins and benzofurans formed during their manufacture.

As a result of these suspicions, the West German Chemical Industry Association arrived at a consensus to voluntarily stop using decabromodi-phenyl ether and octabromodiphenyl ether in new product development. Many large companies in the U.S. are also looking for non-halogenated flame retardant alternatives.

Since organophosphorus FR's primarily work in the condensed or solid phase, development of less volatile molecules (either higher MW or reactive) will most likely reduce the risk of toxic phosphorus containing species in the smoke from burning materials such as plastics, fabrics and other combustible substrates.The effective inhibition of combustion, though, may cause an increase in the relative amount of carbon monoxide formed. But because less material is consumed, the total amount of toxicants and, therefore, the toxic hazard produced should be less than in uninhibited or poorly-inhibited combustion.

To respond to these challenges, we are seeing considerable new product development activity. The thrust of new flame retardant product development is aimed at a new generation of products that a) will not release halogenated gas, b) will reduce visible smoke, c) will achieve the same density as non-FR polymers and d) will meet more-stringent FR test standards (such as UL94VO vs. UL94V2). Some polymers are processed at temperatures of 250°C and higher. Many of the current flame retardants are not suitable for use under these conditions because they are too volatile, or not sufficiently thermally stable. New flame retardants will also demonstrate greater compat-ibility with polymers, have greater thermal stability, and they will be more permanent.

One way to increase the permanence of a flame retardant is to reduce its vapor pressure by increasing its molecular weight. In one new commercial product, the molecular weight of triphenyl phosphate was more than doubled by making the hexabrominated derivative, tris(2,4-dibromophenyl)phosphate (Figure 4). Three other commercial examples achieve higher molecular weight while retaining high phosphorus content by adding additional struc-turally-similar moieties.

In the first example, resorcinol is used to couple two diphenyl phosphate moieties, creating a molecule which is almost a dimer of triphenyl phosphate (Figure 5). The commercially-available product also contains higher oligo-mers.

In the second example, two bicyclic phosphate molecules are coupled together by an Arbuzov reaction with dimethyl methylphosphonate (Figure 6). This creates a very non-volatile material from three relatively volatile but thermally-stable molecules.

Figure 1 Melamine Amyl phosphate (1)

Figure 2 Dimelaminium pentate (2)

$$(ClCH_2CH_2O)_2P(O)OCH_2CH_2OP(O)(OCH_2CH_2Cl)_2$$

TETRAKIS(2-CHLOROETHOXY) ETHYLENE

BISPHOSPHATE

Figure 3 Tetrakis(2-chloroethoxy)ethylene
bisphosphate (3)

$$3Br_2C_6H_3OH+POCl_3 \xrightarrow{MgCl_2} (Br_2C_6H_3O)_3P=O+3HCL$$

2,4-DIBROMOPHENOL

Figure 4 Tris (2,4-dibromophenyl) (4)

$$HOC_6H_4OH+POCl_3 \longrightarrow Cl_2P(O)OC_6H_4OP(O)Cl_2$$

M-RESORCINOL

$$Cl_2P(O)OC_6H_4OP(O)Cl_2+4C_6H_5OH \longrightarrow$$

$$(C_6H_5O)_2P(O)OC_6H_4OP(O)(OC_6H_5)_2$$

Figure 5 Resorcinol bis(diphenyl phosphate) (5)

Figure 6 Cyclic phosphonate esters (6)

In the third example, two low molecular weight alkylphosphine molecules are coupled together with glyoxal more than doubling the molecular weight of the phosphorus starting molecule (Figure 7). The polarity of the phosphonyl groups and hydrogen bonding by hydroxyl groups also contribute to reducing volatility.

Volatility can also be reduced by making a metal salt such as the aluminum salt of a methylphosphonic acid, as claimed in another recent U.S. patent (Figure 8).

Another way to increase the permanence of a flame retardant is to react it into the polymer substrate as the polymer is being manufactured. General examples of this have been practiced in polyester fiber for several years. These compounds have been around for a few years and have achieved significant commercial success (Figure 9).

Two new examples in the patent literature describe the formation of linear copolyesters. The first example is the phenyl analog of the methyl compound of the previous figure (Figure 10). Even though this molecule has a lower phosphorus content than its methyl analog, this variation is significant because of the worldwide concern about the proliferation of methylphosphorus molecules and their building blocks.

The second example is interesting because the hydroxymethyl monomer is oligomerized thus reducing its volatility before blending with polyester resins (Figure 11). It is also reported to copolymerize with the polyester on blending.

In the past, the approach to reducing textile flammability was by application of FR treatments to the finished fabric or of the final fabricated product. Though several of these posttreatment products (Figure 12) continue to enjoy commercial success, the new reactive FR's seem to indicate a significant change in emphasis from posttreatment of fabrics to manufacturing FR fibers. This is not unrelated to the facts that a) the demand for FR fibers is increasing and b) the textile fiber business has become more global in nature. Competition from low-cost third world fiber producers is forcing the producers in the developed countries of the U.S., Europe and Japan to differentiate their products by adding unique properties such as flame retardancy to gain added value.

I have briefly covered some of those factors I see having a strong influence on the future development of organophosphorus flame retardants. In the future, the design of new FR molecules will be based on a moving target. Gone are the days when all that had to be considered was relative FR efficiency and effect on substrate properties. Today, and in the future, in concert with the chemical industry's dedication to Responsible Care^{TM}, we must consider everything from a flame retardants manufacture, through its distribution, application and service life until its eventual disposal or recycling.

$$C_2H_5CHPH_2 + O=CHCH=O \longrightarrow$$
$$\qquad\;\; |$$
$$\qquad CH_3$$

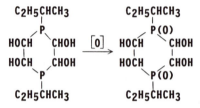

Figure 7   1,4-diphosphacyclohexanes (7)

$$3(CH_3O)_2P(O)CH_3 + AlCl_3 \longrightarrow (CH_3OP(O)O)_3Al + 3CH_3Cl$$
$$\qquad\qquad\qquad\qquad\qquad\qquad\quad |$$
$$\qquad\qquad\qquad\qquad\qquad\qquad CH_3$$

Figure   8   Aluminum tris(methyl methyl-
             phosphonate) (8)

$$CH_3P(O)C_2H_4COOH \qquad\qquad\qquad COOH$$
$$\quad |\qquad\qquad\qquad\qquad\qquad\qquad\qquad |$$
$$\quad OC_2H_4OH \qquad\qquad\qquad P(O)CH_2CHCH_2COOH$$

Figure   9   Reactive FR's for polyester fibers (9)

$$C_6H_5P(O)C_2H_4COOH$$
$$\quad |$$
$$\quad OC_2H_4OH$$

Figure 10   Reactive FR for polyester fibers (10)

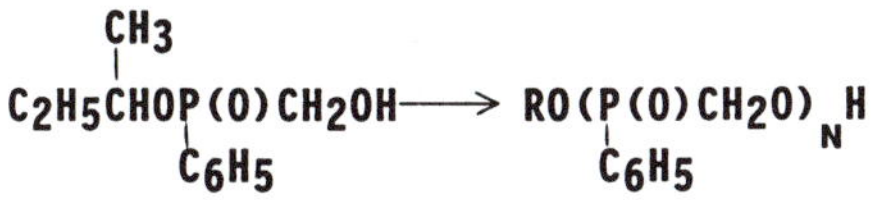

$$\underset{\underset{C_6H_5}{|}}{C_2H_5CHOP(O)CH_2OH} \longrightarrow RO(\underset{\underset{C_6H_5}{|}}{P(O)CH_2O})_NH$$

with CH_3 above C_2H_5CHO

Figure 11 Oligomeric FR for thermoplastic
polyesters (11)

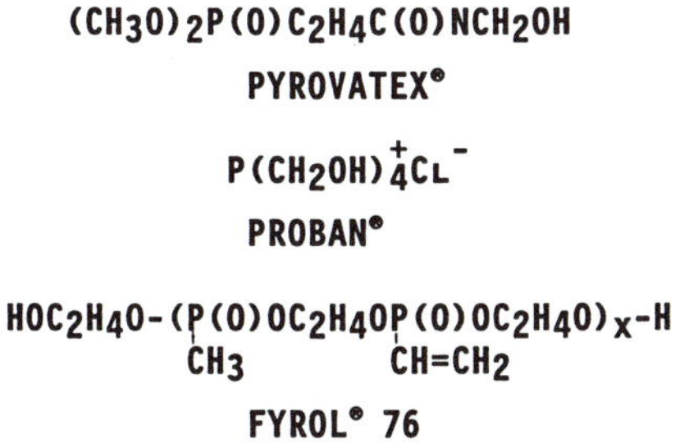

$$(CH_3O)_2P(O)C_2H_4C(O)NCH_2OH$$

PYROVATEX®

$$P(CH_2OH)_4^+CL^-$$

PROBAN®

$$HOC_2H_4O-(\underset{\underset{CH_3}{|}}{P(O)OC_2H_4O}\underset{\underset{CH=CH_2}{|}}{P(O)OC_2H_4O})_x-H$$

FYROL® 76

Figure 12 Commercial organophosphorus FR's used
to treat textiles (12)

Literature Cited

1. Grossman, R. F.; McKane, F. W. Jr.,
    U.S. Patent 5,047,458 1991

2. Halpern, Y.; Hall, D. D. Jr.,
    U.S. Patent 4,342,682, 1982

3. Turley, R. J., U.S. Patent 3,707,586, 1972

4. Gunkel, L. T.; Crosby, J.,
    U.S. Patent 4,897,502, 1990

5. Abolins, V.; Betts, J. E.; Holub, F. F.; Lee,
    G. F., Jr., U.S. Patent 4,808,647 1989

6. Anderson, J. J.; Camacho, V. G.; I.
    Kinney, R. E., U.S. Patent 3,849,368 1974

7. Robertson, A. J.; Gallivan, J. B.,
    U.S. Patent 4,968,416 1990

8. Richardson, J.; Dellar, R. J.,
    U.S. Patent 4,972,011 1990

9. Bollert, U.; Lohmar, E.; Ohorodnik, A.,
    U.S. Patent 4,033,936 1977;
    and Endo, S.; Kashihara, T..; Osako, A.;
    Shizuki, T.; Ikegami, T.,
    U.S. Patent 4,127,590 1978

10. Hazen, J. R., U.S. Patent 4,769,182 1988

11. Gianluigi, L., U.S. Patent 4,981,945 1991

12. Pyrovatex® registered trademark
    Ciba Geigy; Proban® registered trademark
    Albright and Wilson; Fyrol®76 registered
    trademark Akzo

RECEIVED November 12, 1991

Chapter 18

Aryl Group Interchange in Triarylphosphines and Its Role in Catalyst Stability in Homogeneous Transition-Metal-Catalyzed Processes

A. G. Abatjoglou, E. Billig, D. R. Bryant, and J. R. Nelson

Technical Center, Union Carbide Chemicals and Plastics Company, South Charleston, WV 25303

Abstract: Group 8 transition metals catalyze the intermolecular interchange of the aryl groups of triarylphosphines. Reaction rates vary among metals and complexes of a metal. In the rhodium-catalyzed olefin hydrogenation, aryl group interchange is more facile than the competing aryl-alkyl exchange. Kinetic rate data with substituted arylphosphines show little effect of substitution on the rates of aryl interchange.

Background. Soluble transition-metal phosphine complexes are employed as homogeneous catalysts in industrial processes such as olefin hydroformylation, hydrogenation, oligomerization, with significant economic and process advantages (*1*). Long-term catalyst stability is very important in the commercial processes because of the high costs of catalysts and the adverse effects of changing catalyst properties on process performance.

It has been known for many years that transition metals catalyze reactions of coordinated phosphines (*2*). Known reactions of phosphines as ligands include carbon-hydrogen bond cleavage (cyclometalation), as well as direct carbon-phosphorus bond cleavage. Such metal-catalyzed reactions of phosphines lead to formation of new metal complexes which can affect catalyst properties. A known example is the reaction of triphenylphosphine to propyldiphenylphosphine during the rhodium-catalyzed propylene hydrogenation or hydroformylation (*3*).

During a study of this reaction, we observed a surprisingly facile rhodium-catalyzed interchange of aryl groups of tertiary phosphines (*4*). Thus we observed that during rhodium-catalyzed propylene

0097–6156/92/0486–0229$06.00/0
© 1992 American Chemical Society

hydrogenation using a mixture of triphenylphosphine and tris-para-tolylphosphine, seven new tertiary phosphines were formed via rhodium catalyzed interchange of phenyl, tolyl and propyl groups (Scheme 1).

$$(C_6H_5)_3P + (p\text{-}CH_3C_6H_4)_3P \overset{Rh}{\rightleftharpoons} (C_6H_5)_2(p\text{-}CH_3C_6H_4)P + (C_6H_5)(p\text{-}CH_3C_6H_4)_2P$$

$$\Big\Updownarrow \begin{array}{c} Rh \\ CH_2=CHCH_3 / H_2 \end{array}$$

$$(C_6H_5)_2(C_3H_7)P + (p\text{-}CH_3C_6H_4)_2(C_3H_7)P + (C_6H_5)(p\text{-}CH_3C_6H_4)(C_3H_7)P$$

$$\Big\Updownarrow \begin{array}{c} Rh \\ CH_2=CHCH_3 / H_2 \end{array}$$

$$(C_6H_5)(C_3H_7)_2P + (p\text{-}CH_3C_6H_4)(C_3H_7)_2P$$

Scheme 1. Rhodium catalyzed aryl-aryl and aryl-propyl interchange between tertiary phosphines during propylene hydrogenation. (Conditions: 130°C, 2.5 hours; 1000 ppm Rh as Rh(CO)$_2$acetylacetonate; 2.5 wt. percent each triphenylphosphine and tris-para-tolylphosphine; 100 psia H$_2$:propylene; toluene solvent.)

The mass spectrogram of this reaction mixture (Figure 1) shows that virtually complete equilibrium between reactants and products was achieved. Under olefin hydroformylation conditions, aryl-alkyl exchange is much slower, and details for this reaction have been published (*3b*).

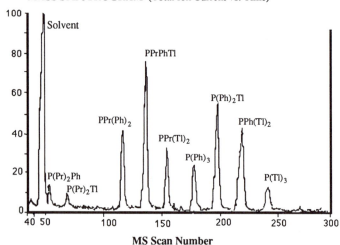

MASS SPECTROGRAM (Total Ion Current vs. Time)

MS Scan Number

Figure 1: Gas chromatography-mass spectrometry of the products for the reaction in Scheme 1. Ph = C$_6$H$_5$; Tl = para-CH$_3$C$_6$H$_4$; Pr = C$_3$H$_5$

This paper deals with work aimed at developing an understanding of the aryl-aryl interchange reaction between tertiary phosphines. This reaction has significant implications on the long-term integrity of certain arylphosphine ligands in catalytic processes.

Relative rates of aryl-aryl versus aryl-alkyl interchange in rhodium-catalyzed olefin hydrogenation. To determine the relative rates of aryl-aryl versus aryl-alkyl interchange during rhodium-catalyzed olefin hydrogenation, we employed monodeuterated triphenylphosphine [(para-deuterophenyl)diphenylphosphine, TPP-d1] as ligand. The formation of the phenyl interchange products, TPP-d0, -d1, -d2 and -d3, was monitored by mass spectrometry, and consumption of TPP and formation of propyldiphenylphosphine (PDPP) by gas chromatography. The mass spectral data in Figure 2 show that the conversion of the starting TPP-d1 ligand to the equilibrium mixture is complete within 5 hours of continuous propylene hydrogenation.

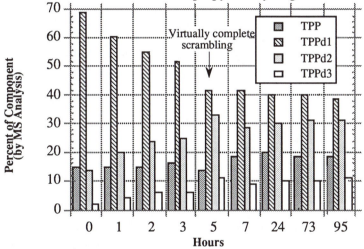

Figure 2. Relative amounts of TPP-d0,-d1,-d2,-d3, formed by aryl interchange between molecules of TPP-d1 (determined by mass spectral analysis) during continuous propylene hydrogenation. Conditions: 105°C; 300 ppm Rh; 10 wt% TPP-d1; 100 psia hydrogen:propylene.

In this period very little propyldiphenyl phosphine is formed as determined by GC analysis. Over longer reaction time, the PDPP formed from TPP by phenyl-propyl exchange can be measured. From the data in (Figure 3) the amount of PDPP formed in 5 hours can be calculated by interpolation. Based on these results we conclude that the relative rate of phenyl-phenyl is 100 times higher than the parallel phenyl-propyl exchange.

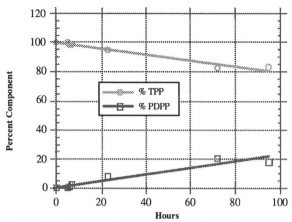

Figure 3. Formation of propyldiphenylphosphine from triphenylphosphine via phenyl-propyl interchange during continuous propylene hydrogenation (Conditions: 300 ppm Rh; 10 wt% TPP; 100 psia hydrogen:propylene).

The ease by which aryl interchange occurs varies with reaction conditions. In general, reaction conditions that preserve catalyst activity, such as olefin hydroformylation or hydrogenation, allow the interchange reaction to proceed and chemical equilibrium between reactants and products can be reached. Conditions that cause catalyst deactivation by formation of CO- and phosphido-bridged clusters retard aryl interchange and the reaction can completely stop before chemical equilibrium is reached. The results detailed below have been obtained using batch laboratory experiments under hydrogen or propylene: hydrogen pressure.

Kinetic rate model for aryl interchange between triarylphosphines. The phosphines which can be formed by interchange of aryls between two triarylphosphines or between molecules of a non-symmetrically substituted triarylphosphine are shown in Scheme 2.

A kinetic rate model for aryl interchange was developed based on the following assumptions. The carbon-phosphorus bonds of the different triarylphosphines cleave with equal ease, and aryl interchange proceeds by a reversible second order reaction. For example, the reaction of a TPP and a TRI molecule always yields one MONO and one DI molecule, and there are nine distinct ways for the forward reaction to occur (any of the three aryls of one molecule can replace or be replaced by any of the three aryls on the other). Of the many possible ways for the molecules to react however, there are three which are unique. These and the respective chemical equilibrium constants are shown in Scheme 3.

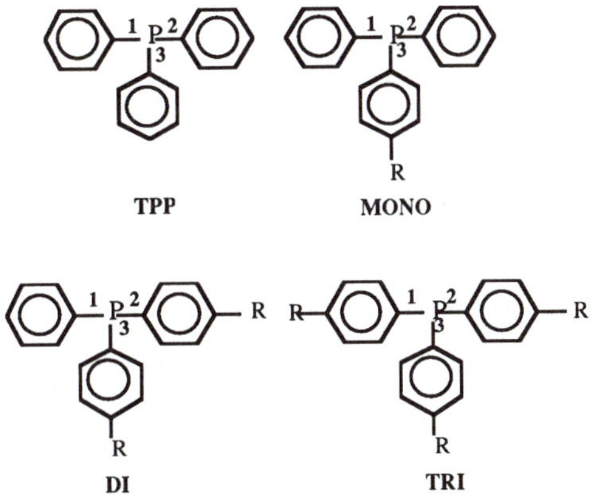

Scheme 2. Possible products of interchange of aryls between two different triaryl phosphines or between molecules of a non-symmetrically substituted triarylphosphine.

$$2\ MONO \xrightleftharpoons{K_1} TPP + DI \qquad (1)$$

$$2\ DI \xrightleftharpoons{K_2} MONO + TRI \qquad (2)$$

$$MONO + DI \xrightleftharpoons{K_3} TPP + TRI \qquad (3)$$

$$K_1 = \frac{(TPP)\ (DI)}{(MONO)^2}$$

$$K_2 = \frac{(MONO)\ (TRI)}{(DI)^2}$$

$$K_3 = \frac{(TPP)\ (TRI)}{(MONO)(DI)} = K_1\ K_2$$

Scheme 3. Unique reactions for the interchange of aryls between two different triaryl phosphines or between molecules of a non-symmetrically substituted triarylphosphine.

The rate expressions for reactions (1), (2), (3) are given are given in Scheme 4, and k_f and k_r are the respective forward and reverse reaction rate constants.

$$\text{Rate}_1 = k_{1f}(\text{MONO})^2 - k_{1r}(\text{TPP})(\text{DI})$$

$$\text{Rate}_2 = k_{2f}(\text{DI})^2 - k_{2r}(\text{MONO})(\text{TRI})$$

$$\text{Rate}_3 = k_{3f}(\text{MONO})(\text{DI}) - k_{3r}(\text{TPP})(\text{TRI})$$

Scheme 4. Second order rate expressions for the interchange of aryls between two different triarylphosphines or between molecules of a non-symmetrically substituted triarylphosphine.

The chemical equilibrium constants are equal to the ratio of the forward and reverse reaction rate constants, as described below:

$$K_1 = k_{1f} / k_{1r}$$

$$K_2 = k_{2f} / k_{2r}$$

$$K_3 = k_{3f} / k_{3r}$$

The relative values of the rate constants can be deduced as follows. Based on reaction probabilities, for example, the relative rate constant k_{3r} is nine times higher than k_{3f} (there are nine ways by which one molecule of TPP and one of TRI can react to produce one molecule of MONO and one of DI, and only one way for the reverse reaction to take place). The same probability analysis is used to calculate the other relative rate constants k_{1f}, k_{1r}, k_{2f}, k_{2r}. The six relative rate constants are displayed below:

k_{1f}	k_{1r}	k_{2f}	k_{2r}	k_{3f}	k_{3r}
2	6	2	6	1	9

The three chemical equilibrium constants can then be calculated.

K_1	K_2	K_3
1/3	1/3	1/9

Notice that K_3 indeed equals K_1 times K_2.

Based on these assumptions experimental rate data can be fitted by calculating only one rate constant. The remaining five are fixed using the relative rate constants listed above. An example of a comparison of actual and predicted data for the aryl interchange between molecules of (para-tolyl)diphenylphosphine is shown in Figure 4.

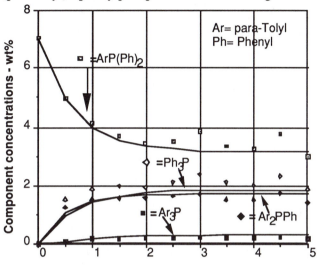

Reaction time (hours)

Figure 4. Aryl group interchange between molecules of (para-tolyl) diphenylphosphine.(Conditions: 7 wt % (paratolyl)diphenylphosphine; 1000 ppm Rh in toluene. 120°C; 60 psia propylene: hydrogen. Solid lines are the computed).

Equally good agreement between experimental and predicted data has been obtained for the aryl interchange starting with DI compounds, as well as the interchange between mixtures of (TPP and TRI) or (MONO and DI).

The effect of temperature on the aryl interchange between TPP and TRI was studied between 100 and 140°C, to make sure that these equilibrium constants developed from probability analysis apply at all temperatures, and to estimate an activation energy for these reactions. The technique of using probability analysis to calculate chemical equilibrium constants, presumes that these "constants" are independent of temperature. That is, the various aryl interchange reactions are presumed to be thermally neutral, and the breaking of the various carbon-phosphorus bonds requires the same amount of energy.

Analysis of the data demonstrated that these equilibrium constants did indeed apply at all temperatures. The activation energy for these reactions was estimated to be 27±1 Kcal/mole, indicating a substantial effect of temperature upon reaction rates.

In another study, the effect of aryl substitution on the rate of interchange between TPP and the TRI ligands below revealed that there is small if any electronic substituent effect of the rate of interchange.

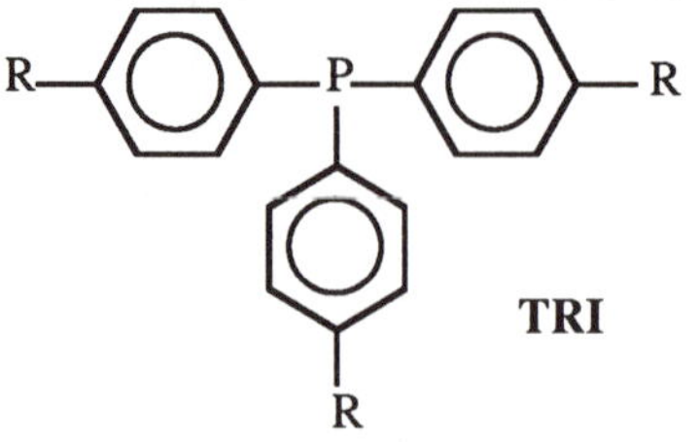

TRI

R= -CH$_3$, -Cl, -F, CH $\underset{3}{O}$-

One reaction variable which affects the rate or aryl interchange is phosphine concentration. We found that in the case of para-tolyl diphenylphosphine the doubling and quadrupling of the concentration, results in redistribution rate constants which are about 1/2 and 1/4 respectively of the reference values. A possible explanation for the effect is that the ligand concentration L determines the concentration of the rhodium which is catalytically active for interchange, according to the equilibrium in Scheme 5

$$Rh_{active} + L \xrightleftharpoons{K} Rh\text{-}L$$

$$Rh_{total} = Rh_{active} + Rh\text{-}L$$

$$Rh_{active} = \frac{Rh_{total}}{1 + (K)(L)} \qquad \text{for } (K)(L) \gg 1$$

$$Rh_{active} = \frac{Rh_{total}}{(K)(L)}$$

Scheme 5. Effect of ligand concentration on the active form of rhodium which is capable of aryl interchange reactions.

Assuming that the product (K)(L) is much greater than one, the active rhodium concentration is inversely proportional to ligand concentration which agrees with the observed results. The possibility of having rhodium complexes which have different catalytic reactivity for aryl interchange is supported experimentally. We have found that different types of rhodium complexes yield different rates of aryl interchange between triphenylphosphine and tris-p-tolylphosphine (Table 1). Mononuclear complexes such as Rh(acac)(CO)Ph$_3$P and RhCl(Ph$_3$P)$_3$, show high reactivity, whereas the more coordinatively and structurally stable clusters Rh$_6$(CO)$_{16}$ and Rh$_4$(μ-CO)$_2$(μ-Ph$_2$P)$_4$-(CO)$_3$(Ph$_3$P)$_7$ are less active.

Table 1. Rates of Aryl Group Interchange between triphenylphosphine and tris-para-tolylphosphine catalyzed by different rhodium complexes. Conditions: 120 °C; 1000 ppm of Rh; 100 psi of H$_2$; phosphine:rhodium mole ratio= 20

Rhodium complex catalyst	Relative Rate
Rh(acac)(CO)(Ph$_3$P)	25
RhCl(Ph$_3$P)$_3$	7.5
RhCl(CO)(Ph$_3$P)$_2$	1
Rh$_4$(μ-CO)$_2$(μ-Ph$_2$P)$_4$(Co)$_3$(PhP$_3$)	a
Rh$_6$(CO)$_{16}$	a

aRate too low to measure, but some interchange was unambiguously observed.

The low reactivity of the rhodium clusters is surprising if one assumes that aryl group interchange can occur between vicinal metal centers as is proposed by Kaneda et al. (5). It is possible however that the rhodium clusters used here are rather coordinatively and structurally stable and do not promote aryl interchange at the employed temperatures. At the higher temperatures however, cluster breakdown to reactive mononuclear species and/or coordinatively unsaturated clusters is likely.

Catalysis of aryl group interchange in triarylphosphines is not unique to rhodium. Other group 8 transition metals show varying degrees of activity (Table 2).

Table 2: Relative rates of aryl group interchange between triphenylphosphine and tris-para-tolylphosphine catalyzed by group 8 transition metals. Reaction conditions: 1000 ppm of metal catalyst; phosphine:metal mole ratio = 20; 100 psi of H$_2$:C$_3$H$_6$ (1:1)

Catalyst Precursor	Temp., °C	Relative Rate
Co$_2$(CO)$_8$	130	25.5
Os$_3$(CO)$_{12}$	130	1.8
Ni(CO)$_2$(Ph$_3$P)$_2$	170	0.6
Pd(Ph$_3$P)$_4$	170	0.4
Ru$_3$(CO)$_{12}$	170	7.5
Rh$_6$(CO)$_{16}$	170	1.0

Of the metals screened, cobalt is the most active. The activities of the ruthenium and the rhodium clusters are very low at 130 °C. At 170 °C the ruthenium is significantly more active than the rhodium cluster catalyst.

Possible mechanism of aryl interchange between triaryl-phosphines. It is well documented that a major pathway for transition metal-catalyzed decomposition of triarylphosphines involves initial aryl-phosphorus bond cleavage by the metal (2). One possible mechanism for aryl interchange between triarylphosphines may therefore involve a sequence of reversible reactions such as those in (Scheme 6). Thus, an active rhodium catalyst complex oxidatively cleaves different aryl-phosphorus bonds causing eventual formation of phosphido-bridged rhodium dimers or oligomers which can break apart with interchange of aryl groups.

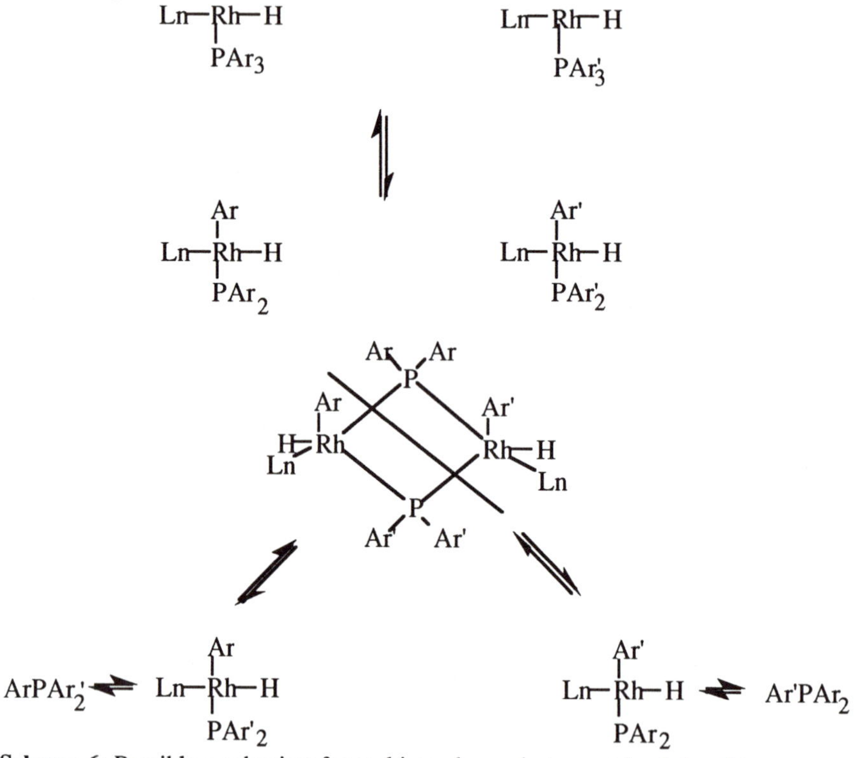

Scheme 6: Possible mechanism for aryl interchange between triarylphosphines

It is apparent that aryl-aryl interchange is greatly facilitated under propylene hydrogenation conditions. The reaction is also observed, albeit at much lower rates, under olefin hydroformylation conditions, and it is likely to occur in other group 8 transition metal catalyzed process employing arylphosphine ligands, and it should be taken into consideration in the long-term viability of certain arylphosphine ligands in catalytic processes.

Acknowledgments: The authors thank Union Carbide Chemicals and Plastics Co., Inc. for permission to publish this work. We gratefully acknowledge Dr. P. C. Price for mass spectral data acquisition and evaluation.

References:

(*1*)   Parshall, G. W. *"Homogeneous Catalysis: The Applications and Chemistry of Catalysis by Soluble Transition Metal Complexes"*; Wiley: New York, **1980**.

(*2*)   Garrou P. E. *Chemical Reviews*, **1985**, 85, 171

(*3*)   a. Gregorio, G.; Montrasi, G.; Tampieri, M.; Cavalieri d'Oro, P;; Pagani, G.; Andreetta, A. *Chim. Ind. (Milan)* 1980, 62, 389.
b. Abatjoglou, A. G.; Billig, E.; Bryant, D. R. *Organometallics,* **1984**, 3, 923.

(*4*)   Abatjoglou, A. G.; Bryant, D. R. *Organometallics,* **1984**, 3, 932.

(*5*)   Kaneda, K.; Sano, K.; Teranishi, S. *Chem. Lett.* **1979**, 821.

RECEIVED January 2, 1992

Chapter 19

Crystalline Chloro-Trisodium Orthophosphate

C. Y. Shen and David R. Gard

Monsanto Company, 800 North Lindbergh Boulevard, St. Louis, MO 63167

Chemistry of the crystalline chloro-trisodium orthophosphate is elucidated. This widely used cleaning and sanitation ingredient can be produced by a melt crystallization approach. A hot phosphatesolution with a Na/P mole ratio of about 2.8 is mixed with a hypochlorite solution. The resulting solution is vacuum cooled to produce crystalline product with a size distribution of about -20+200 mesh, suitable for formulating into automatic dish washing compounds.

Crystalline chloro-trisodium orthophosphate (hence after, this name is abbreviated as Cl-TSP, commonly used in the trade) has versatile properties in saponification, emulsification, peptization, dispersion, and sanitation (1). It is widely used in formulations, such as hard surface cleaners, scouring cleansers, industrial and institutional cleaners, and automatic dish washing compounds. In the seventies, its growth rate is about 4 to 5 percent per year and in 1978 its annual consumption rate reached about 200 million pounds per year (2). Cl-TSP with a high level of water hydration is easy to cake. Storage and transport of Cl-TSP is difficult, especially in hot summer days. For this reason, the use of Cl-TSP in commercial formulation is being replaced by chlorinated cyanuric salts. If a simple preparation procedure of producing Cl-TSP can be developed to produce Cl-TSP at the site of formulators, the use of Cl-TSP may be revived because of low costs, safety, and simple equipment requirement.

0097–6156/92/0486–0240$06.00/0
© 1992 American Chemical Society

According to Bell (3), Cl-TSP has the empirical formula $4(Na_3PO_4 \cdot XH_2O) \cdot NaY$, where X=11-12 and Y is OCl. This empirical formula, however, is not in agreement with the composition given by Mathias (4) and Alder (5). The earlier inventors showed that Cl-TSP has a Na/P mole ratio to be less than three. The Bell proposed formula has been well accepted and a term, chlorinated trisodium phosphate has often been used, as depicted by reaction 1 below. Unfortunately, the implied chemical reaction, chlorination of crystalline trisodium orthophosphate, does not produce the desired product, Cl-TSP. Some clarification of this preparation is needed.

$$2 \, [4(Na_3PO_4 \cdot 12H_2O) \cdot NaOH] + Cl_2 \rightarrow$$
$$4(Na_3PO_4 \cdot 12H_2O) \cdot NaOCl + 4(Na_3PO_4 \cdot 12H_2O) \cdot NaCl + H_2O \tag{1}$$

CHEMISTRY OF Cl-TSP

There are probably more than 50 US patents related to Cl-TSP. Additives, from surfactants to silicates, were used to produce crystals with more stable active chlorine. Examples of these patents are references (6-13) which apparently followed Bell's proposed formula.

Analyses of carefully crystallized Cl-TSP crystals from a liquor prepared from a phosphate liquor with a Na/P ratio between 2.7-3.0 and sufficient amounts of available chlorine from a bleach solution (Sunny Sol 150, Jones Chemicals, Inc., Calendonia, New York, 14423) show the composition of Cl-TSP crystals to be as follows:

Equivalent Composition of Cl-TSP	Weight %
Na_3PO_4	35.6
Na_2HPO_4	5.5
Cl_2	4.2
Na_2O combined with Cl_2	3.7
Water	51.0
Total	100.0

Excluding Na_2O combined with Cl_2, the Na_2O to P_2O_5 mole ratio of Cl-TSP was found to be 2.85. This substantiated the fact that Cl-TSP cannot be produced by reacting crystalline TSP with chlorine as depicted by equation 1.

The Cl-TSP crystals were studied using the x-ray diffraction equipment. The lattice constants and space group were determined using the Syntex P_2 Autodiffractometer. Results are in agreement with earlier publications (6). The x-ray patterns of commercial products with a wide range of moisture and available chlorine levels are nearly all the same as the crystalline TSP.

It is of interest to note when crystalline TSP with a theoretical amount of water of about 56% will lose its characteristic x-ray pattern when it is dehydrated to contain about 54% water. The Cl-TSP crystals can tolerate a wide level of water. The x-ray pattern was unchanged when a Cl-TSP with 52% water was dried at 25°C and 50% RH to about 49%. This property appears to facilitate the production of Cl-TSP. For references, the characteristics of Cl-TSP lattice constants are given below:

$$a = b = 11.908 \ (4) \ \text{Å}$$
$$c = 12.660 \ (4) \ \text{Å}$$
$$\alpha = \beta = 90.00°$$
$$\gamma = 120.00°$$

Crystal class = Trigonal, Space Groups = $\overline{P}_3 Cl^-(D^4_{3d})$
Calculated density = 1.589

2θ	I/I₀	2θ	I/I₀
8.7	75	22.3	20
15.0	75	22.9	20
16.5	50	24.0	20
20.6	45	26.0	80

PROCESS DEVELOPMENT

One of the purposes of this work is to find a simple way for producing Cl-TSP in an operation which could be easily set up in the users' plants. Three approaches were considered and evaluated.

Dry Mix Approach. In this approach, solid Na_2HPO_4 or its hydrates are mixed with an appropriate amount of solutions containing NaOH and Cl_2 at temperatures below 60°C to avoid losses of available chlorine. The product is not sufficiently crystalline thus showing high tendencies to cake and to lose available chlorine.

Crystallization from Solution. Although Cl-TSP crystals could be produced from a solution as mentioned before, this approach is not commercially viable. Some hypochlorite converts to chlorate which causes a disposal problem. The crystalline sizes are large and not suitable for commercial formulations.

Crystallization from a Melt. It is known that Cl-TSP can be melted in its water of hydration. Laboratory tests show that when 103.68 weight unit of a solution of sodium phosphate with Na/P = 2.8 at 90°C is mixed with 33.05 weight unit of hypochlorite solution containing 14.5% available chlorine and 3% excess Na_2O, it will form a Cl-TSP melt which can be vacuum cooled to produce about 100.68 parts crystals with a size distribution essentially -20+200 meshes meeting users' specifications. This product can be further cooled in 25°C dry air to give the desired water content and flow properties. Since the process produces no waste and the product does show the desired properties, it is considered to be the choice approach (14).

The sodium phosphate solution with a Na/P mole ratio of 2.7-2.8 can be produced by many possible ways such as:

a. Reacting 50% NaOH and 77-85% H_3PO_4 in a simple steel mixing tank. The heat of reaction could be used to evaporate the excess water. Figure 1 shows the solubility characteristic of a solution with Na/P=2.8.

b. Melting commercially available crystalline TSP 4[$Na_3PO_4 \cdot 12H_2O$] and disodium phosphate, Na_2HPO_4. Bags of crystalline TSP and DSP are more easily available.

c. Reacting monosodium phosphate $NaH_2PO_4 \cdot 2H_2O$ with caustic solutions.

OPTIMIZATION OF MELT-CRYSTALLIZATION

Test Procedure. The hypochlorite solution was prepared by absorbing a known weight of chlorine in known weight of hydroxide solution at a temperature of less than 20°C by cooling in an ice bath. For example, per 100g of hypochlorite solution, the following composition was used:

Combined Na_2O = 12.72g	- Na_2O/H_2O composition is
Free Na_2O = 2.30g	equivalent to 38.83 parts of
H_2O = 70.40g	50% NaOH in 46.62 parts of
Cl_2 = 14.55g	water.

The amount of available chlorine was determined by titration with standard sodium sulfite solution. There was no loss of any available chlorine in this preparation.

The phosphate solution was prepared from reacting analytical grade 85% H_3PO_4 and 50% NaOH in a heavy duty mixer. After the temperature of the phosphate solution was cooled to the desired level, the specific amounts of hypochlorite solution at a temperature of 20°C were added. The mixture was mixed for a predetermined time

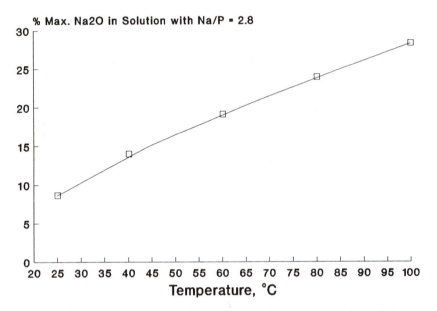

Figure 1. Solution Points of Sodium Phosphate with a
Na/P Mole Ratio of 2.8.

interval before vacuum cooling was applied. The mixture was cooled in about 5 minutes to 45°C and essentially free-flowing damp crystals with an aimed water content of 56.0% were obtained. These crystals were aged under 25°C and 50% RH air for a period of 27 and 47 hours and analyzed.

Test Variables. Examining the melt crystallization process, there are three variables that should be optimized. The variables and the ranges used for this study are:

Available Cl_2 in NaOCl solution:
Let A_1 = 14.5% Cl_2 with 3.0% excess Na_2O
    A_2 = 13.5% Cl_2 with 0.8% excess Na_2O

Temperature of phosphate solution when hypochlorite solution is added:
Let B_1 = High Temperature = 90°C
    B_2 = Low Temperature = 80°C

Time period of mixing before vacuum is applied:
Let C_1 = 5 minutes.
    C_2 = 1 minute.

The results are summarized in Table I. The initial products were aimed to have the same composition in weight percent Na_2O = 24.46, P_2O_5 = 16.04, ave. Cl_2 = 4.50, and H_2O = 56.0%. The large variation of water in the fresh product was unexpected. These experiments could lead to the following tentative conclusions. It is desirable to use the more concentrated hypochlorite solution with an excess of caustic. The temperature of the phosphate solution when the hypochlorite was added, and the length of mixing time before the vacuum was applied are relatively not significant, thus giving the operation more leeway to prepare the desired product.

SUMMARY

A relatively simple process to produce the versatile Cl-TSP from melt crystallization approach was demonstrated. The melt composition can be obtained by mixing a sodium phosphate solution with a Na/P mole ratio of about 2.7-2.8 and a hypochlorite solution containing 14.5% available chlorine and 3.0% excess Na_2O. The mixture is vacuum cooled to result in essentially -20 +200 mesh crystals which can be used in various cleaning products such as automatic dish washing compounds to avoid transportation and storage of the troublesome Cl-TSP crystals.

Table I. Effects of Preparation Conditions on
Available Chlorine Stabilities of Cl-TSP

Sample No.	Initial Sample, Wt.%		After 27 Hrs Ageing / Drying, Wt.%			After 47 Hrs Ageing / Drying, Wt.%		
	Ave. Cl_2	H_2O	Ave. Cl_2	H_2O	Loss	Ave. Cl_2	H_2O	Loss
$A_1B_1C_1$	4.10	54.13	4.10	53.00	0	4.03	52.92	1.71
$A_1B_1C_2$	4.27	53.65	4.19	52.67	1.87	4.17	52.61	2.34
$A_1B_2C_1$	4.24	53.73	4.14	52.48	2.36	4.14	52.42	2.36
$A_1B_2C_2$	4.11	54.77	3.94	52.61	4.14	3.83	52.54	6.81
$A_2B_1C_1$	4.11	57.32	3.31	53.28	19.46	3.28	52.34	20.20
$A_2B_1C_2$	4.11	57.82	3.64	53.20	11.43	3.56	51.54	13.38
$A_2B_2C_1$	4.33	55.24	3.58	52.68	17.32	3.57	52.40	17.55
$A_2B_2C_2$	4.09	56.97	3.33	53.37	18.58	3.31	52.28	19.07

ACKNOWLEDGMENTS

We are in debt to S. H. Ramsey for his help in various tests and determinations, and to B. R. Stults for the X-ray diffraction studies of crystalline TSP and Cl-TSP.

LITERATURE CITED

1. Dibello, P.M., et al., Soap/Cosmetics/Chem. Spec. Aug. 1974, 46.
2. Ayers, J.H., CEH Marketing Research Report, SRI Menlo Park, CA., April 1978.
3. Bell, R.N., Ind. Eng. Chem. 1945, 41, 2901.
4. Mathias, L.D., U.S. Pat. No. 1,555,474, Sept. 29, 1925.
5. Alder, H., U.S. Pat. No. 1,965,304, July 3, 1934.
6. Clark, G.L. and Gross, S.T., Z. Krist. 1937, 98, 107.
7. Hull, H.H., U.S. Pat. No. 2,324,302, July 13, 1943.
8. Miller, D.E., U.S. Pat. No. 2,536,453, Jan. 2, 1951.
9. Shere, L.; Carrera, R.T., U.S. Pat. No. 3,281,364 Oct. 25, 1966.
10. Stamm, J.K., U.S. Pat. No. 3,364,147, Jan. 16, 1968.
11. Taylor, J.A., U.S. Pat. No. 3,342,737, Sept. 19, 1967.
12. Toy, A.D.F.; Bell, R.N., U.S. Pat. No. 3,656,890, April 18, 1972.
13. Vickers, R.H., U.S. Pat. No. 3,525,583, Aug. 25, 1975.
14. Shen, C.Y., U.S. Pat. No. 4,402,926, Sept. 6, 1983.

RECEIVED November 12, 1991

Chapter 20

Development of Functional Organophosphorus Compounds

Danielle A. Bright[1], Fred Jaffe[1], and Edward N. Walsh[2]

[1]Akzo Chemicals Inc., 1 Livingstone Avenue, Dobbs Ferry, NY 10522–3401
[2]Consultant, 33 Concord Drive, New York, NY 10956

Throughout the years many functional organophosphorus compounds have been prepared for incorporation into polymer systems. In order to prevent loss of these phosphorus-containing additives from the polymer, either through migration or vaporization, one approach has been to provide the phosphorus-containing additive with functional groups which allow it to become chemically bound into the polymer system. Also, modifications of the additives were often required to prevent the phosphorus-containing additive from either interfering with the polymer-forming reactions or destabilizing the polymer. This paper describes some of the approaches recently taken at this laboratory to address these requirements. Also reported is a novel use of elemental phosphorus as a reducing agent.

In the course of preparing organophosphorus compounds as flame retardant additives for plastics, a number of requirements were established.

First, the additive must be permanantly incorporated. The additive should not migrate from the polymer, it should not be so volatile that it escapes the polymer through vaporization, and it should not be readily extracted under normal use conditions.

Second, it should be neutral and non-reacting, in the sense that the additive may not be a salt or an acid that can interfere with the components of the polymer-forming reaction system. Thus, all groups on the phosphorus atom must be bound to the phosphorus atom through a C-P, C-O-P or N-P bond.

Third, the additive should have a high phosphorus content, above 10%, and preferably 15-20%.

0097–6156/92/0486–0248$07.50/0
© 1992 American Chemical Society

Fourth, the additive must be low cost and its incorporation into a polymer system must not require costly processing.

One approach to meet these requirements for permanent incorporation is to introduce functionality into the organophosphorus compound. Thus, the additive will be able to react to become bound into the polymer, either in the polymer backbone or as a pendant group on the polymer chain and thereby to become permanently fixed. This paper will illustrate some of the approaches taken at the Dobbs Ferry, New York laboratory to prepare such functional organophosphorus compounds.

One method of incorporating functionality into an organophosphorus compound, a method used early in to our research for the preparation of flame retardant additives for rigid polyurethane foams, was to prepare organophosphorus compounds containing reactive hydroxy groups. These groups can react with the isocyanate groups of the polyisocyanates used in the polymer formulation and thereby the organophosphorus additive can be part of the polyol system. Upon conducting the urethane forming reaction, these hydroxy-containing phosphorus additives become permanently bound as part of the polyurethane matrix, to form phosphorus-containing urethanes.

Two of the first commercially successful compounds of this type are Vircol 82, developed by Virginia-Carolina Corp, now part of Albright and Wilson Co. (1), and Fyrol 6, developed by Victor Chemical Works, now part of Akzo Chemicals Inc. (2). Simplified portrayals of the syntheses and compositions of these two products are illustrated in Fig. 1.

Although the composition of Vircol 82 is described as a pyrophosphate, because of the complexity of the reaction of P_4O_{10} with butanol, there are many other species present. Also, later in this paper, when we discuss the reaction of ethylene oxide with the pyrophosphate bond, there will be presented evidence addressing the level of pyrophosphate content believed present in Vircol 82. Fyrol 6, although produced commercially through processes other than those described in this slide, continues to be a commercially useful flame retardant for rigid urethane foams.

As we proceeded to try to discover other, perhaps more effective, functional additives, several different approaches were considered to prepare such reactive organophosphorus additives. We will now discuss some of the alternate approaches, not all of which became commercial.

One of these approaches, conducted by R. L Muntz and E. N. Walsh (3), was to prepare derivatives of phosphorus-containing aldehydes and ketones. Although phosphorus-containing aldehydes and ketones are themselves functional additives, they also offer a number of routes to introduce additional functionality into phosphorus-containing additives. These phosphorus-containing

1. Vircol 82

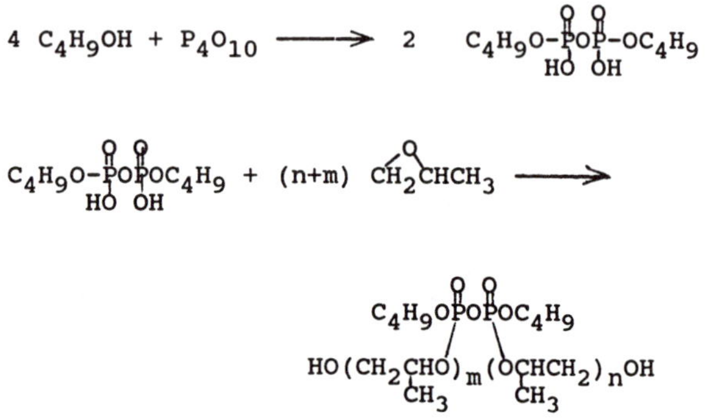

$$4 \ C_4H_9OH + P_4O_{10} \longrightarrow 2 \quad C_4H_9O-\overset{\overset{O}{\|}}{\underset{HO}{P}}\overset{\overset{O}{\|}}{\underset{OH}{P}}-OC_4H_9$$

$$C_4H_9O-\overset{\overset{O}{\|}}{\underset{HO}{P}}\overset{\overset{O}{\|}}{\underset{OH}{P}}OC_4H_9 + (n+m) \ \overset{O}{CH_2}CHCH_3 \longrightarrow$$

$$\begin{array}{c} C_4H_9O\overset{\overset{O}{\|}}{P}O\overset{\overset{O}{\|}}{P}OC_4H_9 \\ HO(CH_2\underset{CH_3}{\overset{|}{C}}HO)_m(O\underset{CH_3}{\overset{|}{C}}HCH_2)_nOH \end{array}$$

2. Fyrol 6

$$(C_2H_5O)_2\overset{\overset{O}{\|}}{P}H + CH_2O + HN(CH_2CH_2OH)_2 \longrightarrow$$

$$(C_2H_5O)_2\overset{\overset{O}{\|}}{P}CH_2N(CH_2CH_2OH)_2$$

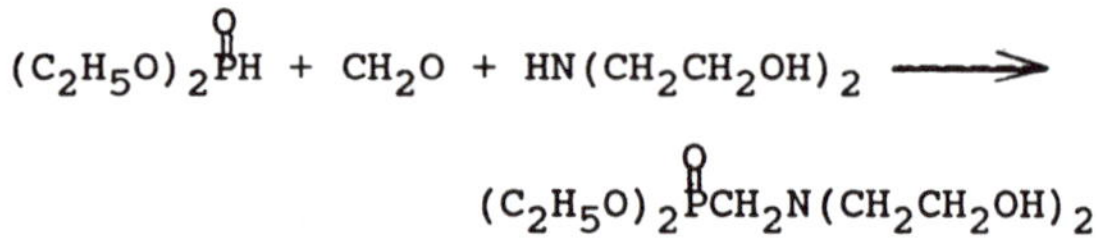

Figure 1. Vircol 82 and Fyrol 6. Two commercial flame retardants for rigid urethane foams. (1, 2)

aldehydes and ketones can be made available through a number of well established chemical routes, such as those shown in Table 1. What was surprising to us at the time this research was begun, was that although these reactive, phosphorus-containing carbonyl compounds were known, little had been reported directed to their use to prepare functional phosphorus-containing plastics additives.

The first phosphorus-containing aldehyde described in the literature that is a neutral non-reactive compound, in the sense mentioned above, is described by N. Dawson and A. Berger (4), who conducted an Arbusov reaction between triethyl phosphite and the diethyl acetal of bromoacetaldehyde, followed by a hydrolysis of the resulting phosphorus-containing acetal, to release the free phosphorus-containing aldehyde. This synthesis is illustrated in Eq. 1 of Table 1. Another approach to preparing a phosphorus-containing aldehyde is that of I. F. Lutsenko and M. Kirilov (5), who added phosphorus pentachloride to vinyl acetate to form an adduct which, when treated with sulfur dioxide, is converted to the corresponding phosphonyl chloride. Treating this phosphonyl chloride with ethanol yields the ester, shown in Eq. 2, Table 1. Hydrolysis releases the phosphorus-containing aldehyde.

The first completely esterified phosphorus-containing ketone is described by R. G. Harvey (7) who used a synthesis route first described by G. Kamai and V. A. Kukhtin (6) for the preparation of phosphorus-containing aldehydes. This process entails the reaction of a trialkyl phosphite, such as triethyl phosphite, with an alpha-beta unsaturated carbonyl compound, such as methyl vinyl ketone, to yield a postulated cyclic intermediate. When treated with an alcohol, the cyclic intermediate is converted into an acetal or ketal. Hydrolysis with water liberates the phosphorus-containing carbonyl compound. This synthetis is illustrated in Eq. 3 of Table 1.

The first reported preparations of neutral phosphorus-containing aldehydes, where the phosphorus atom is linked to the aldehyde containing moiety through an oxygen atom is described by F. Ramirez et al (9) in 1969 and G. Sturtz and J. L. Kraus (8) in 1970. These preparations are illustrated in Eq. 5 and 4 of Table 1.

A review, by A. I. Razumov, B. G. Liorber, V. V. Moskva and M. I. Sokolov (10) describes much of this earlier work.

Examples of phosphorus-containing aldehydes and ketones prepared during our study are shown in Table 2. These were prepared using the methods described either by R. G. Harvey (7) or G. Sturtz and J. L. Kraus (8). The preparations are described in detail in reference (3). As can be seen, some of these compounds were isolated as acetals or ketals.

The first functional derivatives that we prepared of these

Table 1

Phosphorus-Containing Aldehydes and Ketones
Typical Preparations

1. $(C_2H_5O)_3P$ + $BrCH_2CH(OC_2H_5)_2$ $\xrightarrow[\text{2) } H_2O]{\text{1) } \triangle}$ $(C_2H_5O)_2\overset{O}{\overset{\|}{P}}CH_2\overset{O}{\overset{\|}{C}}H$ + C_2H_5Br +

2 C_2H_5OH

[1]SOURCE: Reference 4.

2. PCl_5 + $CH_2=CHO\overset{O}{\overset{\|}{C}}CH_3$ $\xrightarrow[\text{2) } C_2H_5OH]{\text{1) } SO_2}$ $(C_2H_5O)_2\overset{O}{\overset{\|}{P}}CH_2CHClO\overset{O}{\overset{\|}{C}}CH_3$

$(C_2H_5O)_2\overset{O}{\overset{\|}{P}}CH_2\overset{O}{\overset{\|}{C}}H$ $\xleftarrow{}$ H_2O

[2]SOURCE: Reference 5.

3. $(C_2H_5O)_3P$ + $CH_2=CH\overset{O}{\overset{\|}{C}}-CH_3$ $\xrightarrow{}$ $\begin{matrix} CH_3\overset{}{C}-O \\ \| \\ HC-CH_2 \end{matrix}{\Large >}P(OC_2H_5)_3$ $\xrightarrow[\text{2) } H_2O]{\text{1) } ROH}$

$(C_2H_5O)_2\overset{O}{\overset{\|}{P}}CH_2CH_2\overset{O}{\overset{\|}{C}}CH_3$

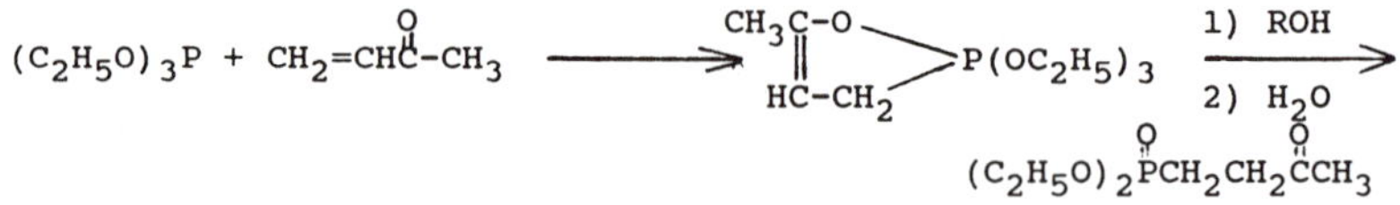

[3]SOURCE: References 6 and 7.

4. $[(CH_3O)_2\overset{O}{\overset{\|}{P}}O^-]$ $[(CH_3)_4N^+]$ + $ClCH_2\overset{O}{\overset{\|}{C}}H$ $\xrightarrow{}$ $(CH_3O)_2\overset{O}{\overset{\|}{P}}OCH_2\overset{O}{\overset{\|}{C}}H$ + $(CH_3)_4NCl$

[4]SOURCE: Reference 8.

5. 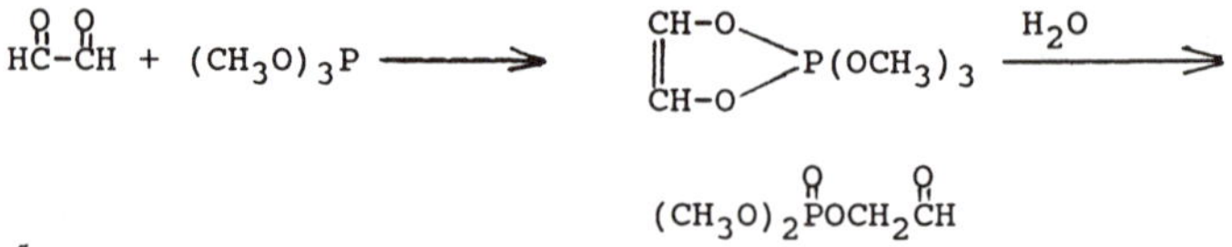 $H\overset{O}{\overset{\|}{C}}-\overset{O}{\overset{\|}{C}}H$ + $(CH_3O)_3P$ $\xrightarrow{}$ $\begin{matrix} CH-O \\ \| \\ CH-O \end{matrix}{\Large >}P(OCH_3)_3$ $\xrightarrow{H_2O}$

$(CH_3O)_2\overset{O}{\overset{\|}{P}}OCH_2\overset{O}{\overset{\|}{C}}H$

[5]SOURCE: Reference 9.

phosphorus-containing aldehydes and ketones were the adducts of a dialkyl phosphonate, such as diethyl phosphonate, to a phosphorus-containing aldehyde or ketone, such as dimethyl (3-oxo-1-butyl)phosphonate to produce 2-[(diethyl) phosphono]-4-[(dimethyl) phosphono]-2-butanol, an undistillable oil. This reaction is shown in Eq. 1 of Fig 2. This reaction was first described by P. Tavs (11), who added diethyl phosphonate to 4-(diethyl)-2-oxobutane. This reaction leads to phosphorus-rich monofunctional additives which, when allowed to react into a polymer system, become non-migratory. However, in most polymer systems di- or polyfunctional additives are preferred, since monofunctional additives can act as chain growth stoppers. Examples of compounds of this type that were prepared in this program are shown in Table 3, Ex. 1 and Ex. 2.

A second type of derivative of phosphorus-containing aldehydes and ketones was prepared from a Mannich-type condensation of a phosphorus-containing carbonyl compound, such as 3-(diethyl phosphono)butanal with diethyl phosphonate and diethanol amine, to form a phosphorus-rich diol, 1-[(diethyl)phosphono]-1-[N,N-di(hyroxyethyl)amino]-3-[(diethyl)phosphono] butane as shown in Eq. 2 of Fig. 2. Examples of the preparation of such diols are shown as Ex 3 and 4 of Table 3.

A third type of derivative was the reaction product of a phosphorus-containing aldehyde with formaldehyde, analogous to the reaction of acetaldehyde with formaldehyde to form polymethanol derivatives. In the case of the phosphorus-containing aldehydes, this leads to a trifunctional additive. This reaction is illustrated in Fig. 3.

As an example, when phosphorus-containing aldehydes were added to paraformaldehyde in the presence of calcium oxide and water, an exothermic reaction occurs which leads to the formation of phosphorus-containing trimethanol propanes. Of the two examples shown in Fig. 3, Compound I is a crystalline solid. Since Compound II could not be crystallized, it is suspected of being contaminated with the ether III.

A fourth type of derivative, the condensation of secondary amines with a phosphorus-containing aldehydes, in some cases leads to tetrafunctional polyols, as shown in Equation 2 of Fig. 4. This reaction proceeds well with phosphonoaldehydes, as long as the oxo group is not beta to the phosphorus atom. In this latter case, as shown in Eq. 1, enamines form (12, 13, 14). The formation of enamines is indicated by a strong carbon-carbon double bond absorption at 1610 cm.$^{-1}$. It was favored even in the presence of a potential trapping agent, diethyl phosphonate. Unfortunately, even when the tetrafunctional diamines formed, the compounds are low in phosphorus content and therefore they lead away from one of our primary goals, compounds having a high phosphorus content.

Finally, a polymeric form of 3-(dimethyl phosphono)

Table 2

Phosphorus-Containing Aldehydes and Ketones

COMPOUND	PROCESS	BOILING POINT, C
$(CH_3O)_2\overset{O}{P}CH_2CH_2\underset{CH_3}{C}(OCH_3)_2$	a	not distilled
$(CH_3O)_2\overset{O}{P}CH_2CH_2\overset{O}{C}CH_3$	a	104-105 (0.4 mm)
$(CH_3O)_2\overset{O}{P}\underset{CH_3}{C}HCH_2CH(OCH_3)_2$	a	110-115 (1.5-1.8 mm)
$(CH_3O)_2\overset{O}{P}\underset{CH_3}{C}HCH_2\overset{O}{C}H$	a	85-89 (0.05-0.10 mm)
$(C_2H_5O)_2\overset{O}{P}CH_2CH_2\overset{O}{C}CH_3$	a	110 (0.05 mm)
$(C_2H_5O)_2\overset{O}{P}\underset{CH_3}{C}HCH_2\overset{O}{C}H$	a	91-96 (0.08 mm)
$(C_2H_5O)_2\overset{O}{P}OCH_2\overset{O}{C}CH_3$	b	101-103 (0.12 mm)

SOURCE: References 6 and 8.

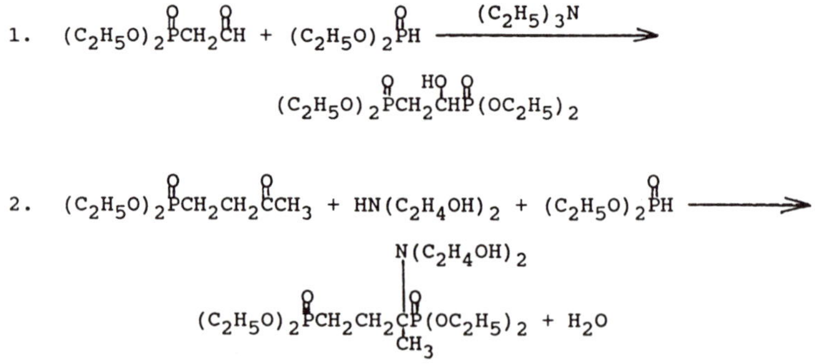

1. $(C_2H_5O)_2\overset{O}{P}CH_2\overset{O}{C}H + (C_2H_5O)_2\overset{O}{P}H \xrightarrow{(C_2H_5)_3N}$

   $(C_2H_5O)_2\overset{O}{P}CH_2\overset{HO}{C}H\overset{O}{P}(OC_2H_5)_2$

2. $(C_2H_5O)_2\overset{O}{P}CH_2CH_2\overset{O}{C}CH_3 + HN(C_2H_4OH)_2 + (C_2H_5O)_2\overset{O}{P}H \longrightarrow$

   $\overset{\textstyle N(C_2H_4OH)_2}{\underset{\textstyle (C_2H_5O)_2\overset{O}{P}CH_2CH_2\underset{CH_3}{\overset{|}{C}}\overset{O}{P}(OC_2H_5)_2}{\big|}} + H_2O$

Figure 2. Reactions of phosphorus-containing aldehydes and ketones with dialkyl phosphonates. (3, 11)

Table 3

Compounds Prepared from Phosphorus - Containing Aldehydes and Ketones

R	R^1	%P	
		Theory	Found

1. $(RO)_2\overset{O}{P}\underset{CH_3}{CH}-CH_2\overset{HO}{\underset{}{C}}H\overset{O}{P}(OC_2H_5)_2$

a. CH_3	–	19.5	19.5
b. C_2H_5	–	17.9	17.8

2. $(RO)_2\overset{O}{P}CH_2CH_2\underset{CH_3}{\overset{OH}{C}}\text{—}\overset{O}{P}(OC_2H_5)_2$

a. CH_3	–	19.5	19.5
b. C_2H_5	–	17.9	17.8

3. $(RO)_2\overset{O}{P}\underset{CH_3}{C}HCH_2\overset{N(C_2H_4OH)_2}{\underset{P(OR^1)_2}{CH}}$

a. CH_3	CH_3	16.5	14.8
b. CH_3	C_2H_5	15.3	14.5
c. C_2H_5	C_2H_5	14.3	14.4

4. $(RO)_2\overset{O}{P}CH_2CH_2\underset{CH_3}{\overset{N(C_2H_4OH)_2}{C}}\text{—}\overset{O}{P}(OR^1)_2$

a. CH_3	CH_3	16.5	14.8
b. CH_3	C_2H_5	15.3	13.0
c. CH_2H_5	C_2H_5	14.3	14.5

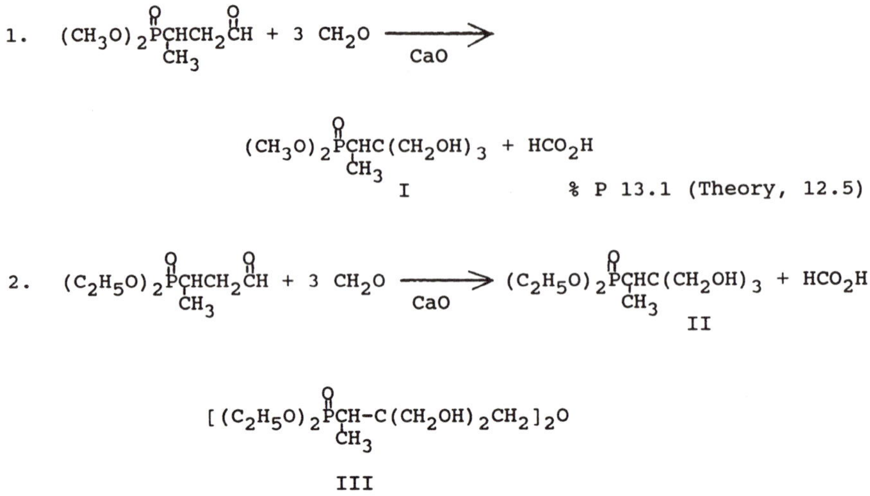

1. $(CH_3O)_2\overset{O}{\underset{CH_3}{P}}\underset{CH_3}{CHCH_2}\overset{O}{C}H$ + 3 CH_2O $\xrightarrow[\text{CaO}]{}$

$(CH_3O)_2\overset{O}{\underset{CH_3}{P}}CHC(CH_2OH)_3$ + HCO_2H

I   % P 13.1 (Theory, 12.5)

2. $(C_2H_5O)_2\overset{O}{\underset{CH_3}{P}}CHCH_2\overset{O}{C}H$ + 3 CH_2O $\xrightarrow[\text{CaO}]{}$ $(C_2H_5O)_2\overset{O}{\underset{CH_3}{P}}CHC(CH_2OH)_3$ + HCO_2H

II

$[(C_2H_5O)_2\overset{O}{\underset{CH_3}{P}}CH-C(CH_2OH)_2CH_2]_2O$

III

Figure 3. Reactions of phosphorus-containing aldehydes with formaldehyde. (3)

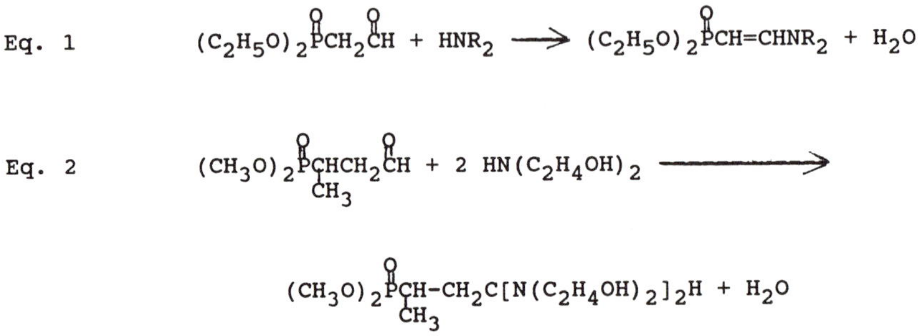

Eq. 1 $(C_2H_5O)_2\overset{O}{P}CH_2\overset{O}{C}H$ + HNR_2 $\longrightarrow$ $(C_2H_5O)_2\overset{O}{P}CH=CHNR_2$ + H_2O

Eq. 2 $(CH_3O)_2\overset{O}{\underset{CH_3}{P}}CHCH_2\overset{O}{C}H$ + 2 $HN(C_2H_4OH)_2$ $\xrightarrow{\hspace{3cm}}$

$(CH_3O)_2\overset{O}{\underset{CH_3}{P}}CH-CH_2C[N(C_2H_4OH)_2]_2H$ + H_2O

Figure 4. Reactions of phosphorus-containing aldehydes with secondary amines. (3, 12, 13, 14)

proplonaldehyde was prepared from the action of trimethyl phosphite on acrolien.

Since most of these products were viscous, non-distillable liquids, purification was difficult. However, based on chemical analysis, the reactions do proceed with sufficient efficiency to provide products of acceptable purity and in good yield. The pathways of these reactions were followed by several techniques, including the use of infrared spectroscopy to follow the disappearance of the carbonyl and P-H bonds and, where possible, the appearance of the new bonds associated with the products.

These additives were tested as flame retardants is rigid urethane foams. The results of one set of these tests are shown in Table 4, where the flame retarding performance of these additives is compared with that of Fyrol 6. In this test, the limiting oxygen index test (L. O. I. or Oxygen Index), the additive is incorporated into a rigid urethane foam formulation at a level equal to 20% of the polyol requirement. The ratio of polyisocyanate to polyol is adjusted to account for the hydroxy-content of the phosphorus-containing additive. A small piece of the rigid foam is ignited in a gas stream which contains various mixtures of oxygen and nitrogen. The ratio of oxygen to nitrogen is varied until the minimum percentage of oxygen that allows sustained combustion was determined. This minimum percentage of oxygen is the limiting oxygen index. The higher the value, the more flame retardant is the material. The data in Table 4 show that all of these additives improved the flame retardance of the rigid urethane foam, but not always in a direct relation to the phosphorus content.

One of the important commercial flame retardants for plastics, the phosphorus-containing oligomer Fyrol 99, is prepared by the condensation of tris -(2-chloroethyl) phosphate, as shown in Fig. 5. In this reaction, the reagent is heated to elevated temperatures and ethylene dichloride is eliminated. The early discoverers of this reaction, V. V. Korshak et al (15) used reaction temperatures in the range of 240-280°C. At these high temperatures, the reaction was difficult to control and it often led to an acidic, highly discolored product. If not carefully watched, gelled materials could form. It was later found (16), that by the use of a nucleophilic catalyst, such as sodium carbonate, the condensation reaction temperature can be reduced to the 140-200°C range, and the reaction becomes more controllable. On first glance, this reaction leads to a non-reactive oligomer. However, on closer examination, it was found that side reactions lead to the formation groups affording both free and latent acidity. This is shown in the lower equations of Fig. 5. These groups resulted from: a) the autocondensation of end groups of the condensate to form a five membered ring, a ring system that can readily undergo hydrolysis to release acidity, b) the dealkylation of ester groups to form vinyl chloride and to librate free acidity, and c) a complex series of reactions

Table 4

Compounds Prepared from Phosphorus-Containing Aldehydes and Ketones

Oxygen Index Data
Rigid Urethane Foams
20 phr Polyol

Compound	% P	Oxygen Index
1. $(C_2H_5O)_2\overset{O}{\overset{\|}{P}}CH_2N(C_2H_4OH)_2$	12.5	23.0
2. $(CH_3O)_2\overset{O}{\overset{\|}{P}}\underset{\overset{\|}{CH_3}}{C}HCH_2\overset{N(C_2H_4OH)_2}{\overset{\|}{C}}H\overset{O}{\overset{\|}{P}}(OC_2H_5)_2$	13.2	23.0
3. $(CH_3O)_2\overset{O}{\overset{\|}{P}}CH_2CH_2\underset{\overset{\|}{CH_3}}{C}\overset{N(C_2H_4OH)_2}{\overset{\|}{P}}(OC_2H_5)_2$	13.0	22.7
4. $(CH_3O)_2\overset{O}{\overset{\|}{P}}\underset{\overset{\|}{CH_3}}{C}HC(CH_2OH)_3$	13.1	21.0
5. $(CH_3O)_2\overset{O}{\overset{\|}{P}}\underset{\overset{\|}{CH_3}}{C}HCH_2\overset{N(C_2H_4OH)_2}{\overset{\|}{C}}H\overset{O}{\overset{\|}{P}}(OCH_3)_2$	14.8	22.8
6. $[(CH_3O)_2\overset{O}{\overset{\|}{P}}CH_2CH_2\overset{O}{\overset{\|}{C}}H]_n$	15.3	21.1
7. $(CH_3O)_2\overset{O}{\overset{\|}{P}}\underset{\overset{\|}{CH_3}}{C}HCH_2\overset{HO}{\overset{\|}{C}}H\overset{O}{\overset{\|}{P}}(OC_2H_5)_2$	18.1	23.9
8. None	0.0	20.0

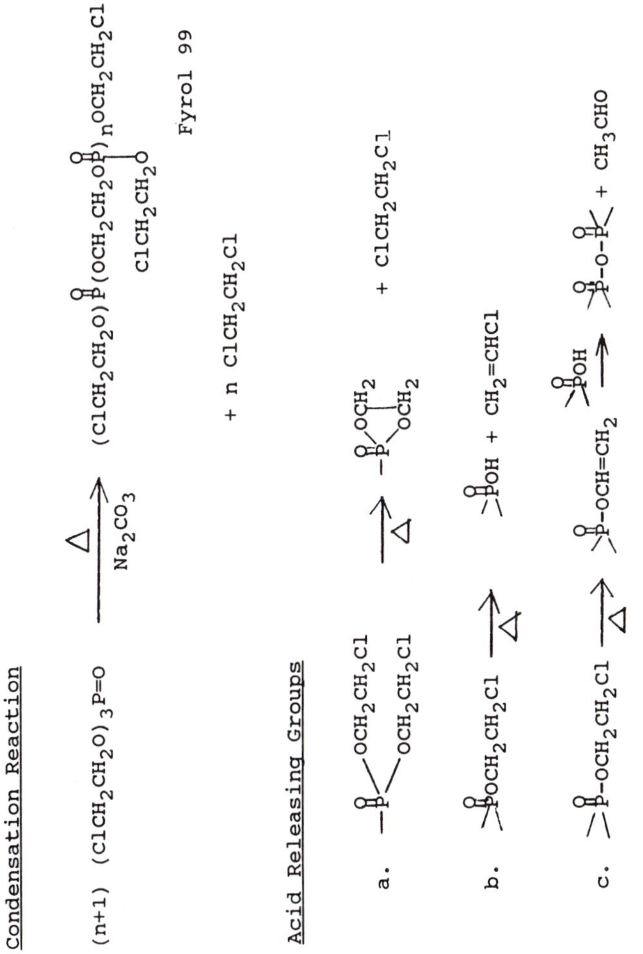

Figure 5. Thermal condensation of tris (2-chloroethyl) phosphate. (15, 16, 17)

leading to the formaton of pyrophosphate linkages in the main chain
of the condensate and to the release of acetaldhyde. The free
acidity is readily removed by direct reaction with ethylene oxide.
However, the anhydride linkages and cyclic rings proved more
troublesome, especially when the oligomer is incorporated in
polymer formulations such as flexible urethane foam formulations.
These formulations can contain amine catalysts, polyols and water.
Since the water and the hydroxy groups of the polyols can react
with these labile entities to release the latent acidity, this
released acidity can adversely affect the polymer-forming reactions
by reacting with the amine and metal salt catalysts added to
promote the polymer formation. Also, this released acidity can
attack the polymer structure, as is sometimes seen as discoloration
and friability of the polymer. Thus, it was necessary to tame this
oligomer by removing these labile, acid-releasing groups and
thereby to make the oligomeric additive compatible with the polymer
system.

A series of techniques was developed to remove both the free
acidity and the latent acidity. These techniques are illustrated
in Fig. 6 and Fig. 6a. Ideally, the addition of ethylene oxide
should neutralize any free acidity, as shown in Eq. 1, and the
addition of water or alcohols to the oligomer should open the
five-membered rings and attack the anhydride linkages to release
all latent acidity (18). An olefin oxide, such as ethylene oxide,
should then be able to remove this released acidity. These
reactions are shown in Eq. 2 of Fig. 6 and Eq. 3 of Fig. 6a.
However, since most of these acid releasing reactions are not
instantaneous, the taming of this oligomer was found to be somewhat
more difficult than described. Many of the problems were solved
when, after treating the oligomer with water or ethanol, a Lewis
acid catalyst, such as titanium chloride or stannous octoate, was
added during the subsequent ethylene oxide addition (20). This is
illustrated in Eq. 4 of Fig. 6a. The trials associated with the
finding of a solution to this problem are described in detail in
reference 17 and in the series of patents shown in references
18-21. It will be noted that these treatments to prevent acid
formation converted the oligomer from a non-reactive oligomer to
one containing reactive hydroxy groups.

One suggested pathway cited in the above studies for the
removal of pyrophosphate linkages from this oligomeric condensate
was that ethylene oxide could react directly to insert itself into
the pyrophosphate bond (17). This is illustrated in Eq. 2c of Fig.
6. This suggestion was based in part upon the remediation
chemistry and in part upon infrared monitoring of the reduction in
the concentration of the pyrophosphate bonds resulting from the
remediation processing. However, the reduction in concentration or
the disappearance of the anhydride groups is not unequivocal
evidence that ethylene oxide inserts into the anhydride bond. In
this oligomeric system there are other pathways for the opening of
the anhydride linkages. Also, there was no clear identification of

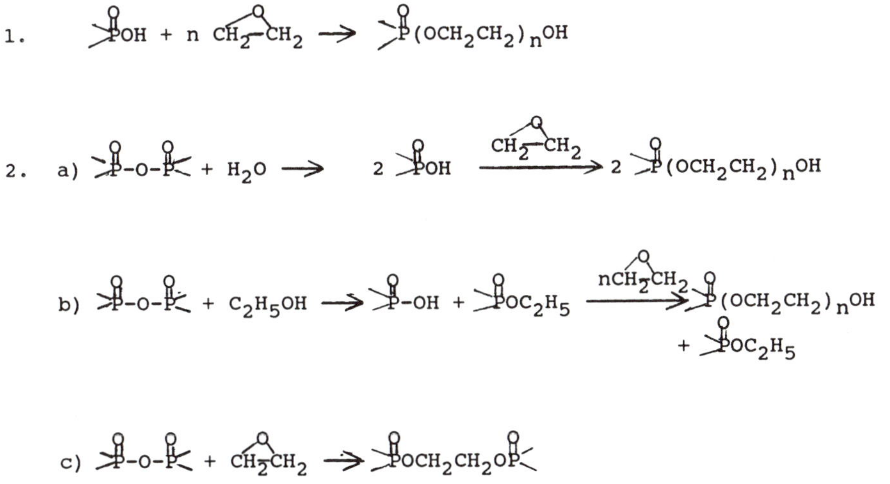

1. $\overset{O}{\underset{}{\overset{\parallel}{P}}}OH + n \ CH_2\text{—}CH_2 \rightarrow \overset{O}{\underset{}{\overset{\parallel}{P}}}(OCH_2CH_2)_nOH$

2. a) $\overset{O}{\underset{}{\overset{\parallel}{P}}}\text{-O-}\overset{O}{\underset{}{\overset{\parallel}{P}}} + H_2O \rightarrow 2 \ \overset{O}{\underset{}{\overset{\parallel}{P}}}OH \xrightarrow{CH_2\text{—}CH_2} 2 \ \overset{O}{\underset{}{\overset{\parallel}{P}}}(OCH_2CH_2)_nOH$

   b) $\overset{O}{\underset{}{\overset{\parallel}{P}}}\text{-O-}\overset{O}{\underset{}{\overset{\parallel}{P}}} + C_2H_5OH \rightarrow \overset{O}{\underset{}{\overset{\parallel}{P}}}\text{-OH} + \overset{O}{\underset{}{\overset{\parallel}{P}}}OC_2H_5 \xrightarrow{nCH_2CH_2} \overset{O}{\underset{}{\overset{\parallel}{P}}}(OCH_2CH_2)_nOH$
   $+ \overset{O}{\underset{}{\overset{\parallel}{P}}}OC_2H_5$

   c) $\overset{O}{\underset{}{\overset{\parallel}{P}}}\text{-O-}\overset{O}{\underset{}{\overset{\parallel}{P}}} + CH_2CH_2 \rightarrow \overset{O}{\underset{}{\overset{\parallel}{P}}}OCH_2CH_2O\overset{O}{\underset{}{\overset{\parallel}{P}}}$

Figure 6. Processes for removing acid and acid-forming groups from condensed <u>tris</u> (2-chloroethyl) phosphate. (17, 18, 19, 20, 21)

3. a) $-\overset{O}{\underset{}{\overset{\parallel}{P}}}\overset{OCH_2}{\underset{OCH_2}{\big\langle}}\big| + H_2O \rightarrow -\overset{O}{\underset{}{\overset{\parallel}{P}}}\overset{OCH_2CH_2OH}{\underset{OH}{\big\langle}} \xrightarrow{CH_2CH_2} -\overset{O}{\underset{}{\overset{\parallel}{P}}}\overset{OCH_2CH_2OH}{\underset{(OCH_2CH_2)_nOH}{\big\langle}}$

   b) $-\overset{O}{\underset{}{\overset{\parallel}{P}}}\overset{OCH_2}{\underset{OCH_2}{\big\langle}}\big| + C_2H_5OH \rightarrow -\overset{O}{\underset{}{\overset{\parallel}{P}}}\overset{OCH_2CH_2OH}{\underset{OCH_2CH_3}{\big\langle}}$

4. $\overset{O}{\underset{}{\overset{\parallel}{P}}}\text{-OH}$

$\begin{matrix} \overset{O}{\underset{}{\overset{\parallel}{P}}}\text{-O-}\overset{O}{\underset{}{\overset{\parallel}{P}}} \\ \overset{O}{\underset{}{\overset{\parallel}{P}}}\overset{OCH_2}{\underset{OCH_2}{\big\langle}}\big| \end{matrix} \quad \xrightarrow[\substack{or \\ C_2H_5OH}]{H_2O} \quad \xrightarrow[\substack{TiCl_4 \\ or \\ O \\ Sn(OCC_7H_{15})_2}]{CH_2CH_2} \quad \text{neutral esters}$

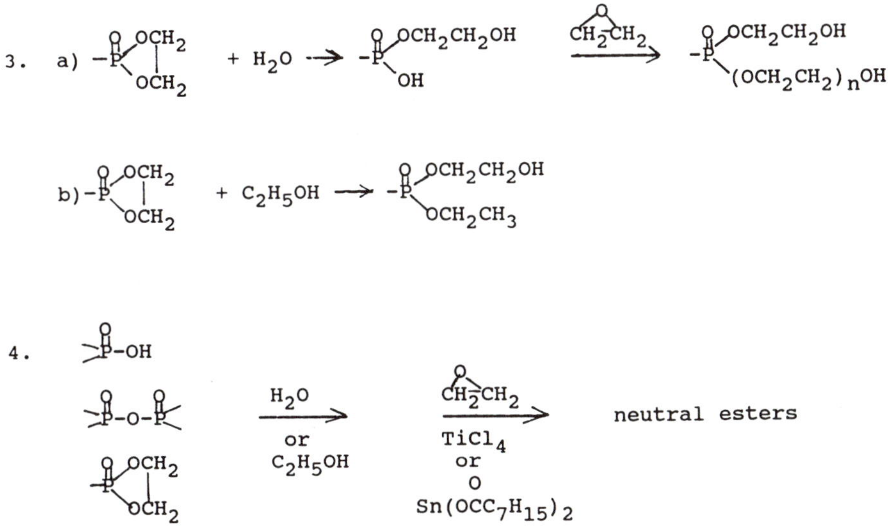

Figure 6a. Processes for removing acid and acid-forming groups from condensed <u>tris</u> (2-chloroethyl) phosphate. (17, 18, 19, 20, 21)

the formation of the products expected from an insertion reaction
of ethylene oxide into the anhydride linkage.

The insertion of ethylene oxide into a phosphorus anhydride
bond has been invoked by several authors. Examples are shown in
Fig. 7. The first example of this reaction is shown in the work of
W. H. Woodstock (22), who effected the reaction of ethylene oxide
with P_4O_{10} to produce a clear, acidic, water soluble liquid
that polymerized on standing to form a solid, resembling art gum.
No yields are given and the source of the acidity is not described.
Also, there is no characterization of the bonding formed in the
product of this reaction. Finally, not all of the phosphorus (V)
oxide is reacted in the Woodstock examples. This in no way impugns
the insertion pathway but, with the formation of an acidic product,
it does suggest reaction pathways other than a direct insertion
reaction may be more prominent. Another reference to the insertion
of an alkylene oxide into a phosphoric anhydride bond is described
by A. J. Papa (23). This reference suggests the insertion of
propylene oxide into the phosphoric anhydride bond of a symmetrical
dialkyl pyrophosphoric acid. Thus, he suggests that the major
components of Vircol 82 may not include significant amounts of the
anhydride linkage. As stated earlier (17), it was suggested that
ethylene oxide inserts into the anhydride linkages formed in a side
reaction during the condensation of tris (2-chloroethyl)
phosphate. However, in none of these two latter cases is insertion
the only pathway available for removal of the anhydride linkage
and, perhaps because of the difficulty in analysiing such reaction
mixtures, in neither case is a product of such a reaction
unequivocally identified.

What we consider as the first unequivocal evidence for such an
insertion reaction has now been demonstrated by D. A. Bright and A.
M. Aaronson (24). Under carefully controlled conditions, they
added ethylene oxide to tetraphenyl pyrophosphate to form and to
identify as the sole reaction product ethylene glycol bis
(diphenyl phosphate). This is illustrated in Fig. 8. The reaction
proceeds at 70 °C when ethylene oxide is added to molten
tetraphenyl pyrophosphate during a 10 hour period. Pyridine was
found to be an effective as a catalyst. The yield is 87.4%; m.p.,
38-40 °C; purity, 97.2 by HPLC.

A similar reaction of tetraphenyl pyrophosphate with propylene
oxide leads to an 80% yield of the corresponding insertion product,
as an oily liquid. The reaction time was six hours at 70 °C.;
purity, 96.7% by HPLC. Magnesium chloride and stannous octoate
were also shown to be effective catalysts for this insertion
reaction.

The next area which we investigated involves a reaction that
is really a variation of the condensation reaction of tris
(2-chloroethyl) phosphate. If the chlorine-containing group is on
one molecule, and the second molecule is a chlorine-free phosphorus
ester, the condensation reaction can be conducted to release and
alkyl chloride and to form a new phosphorus-containing ester. This

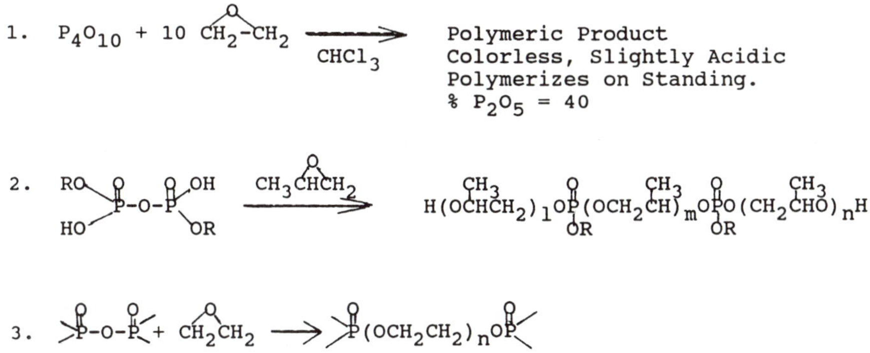

1. P_4O_{10} + 10 CH$_2$-CH$_2$ $\xrightarrow{\text{CHCl}_3}$ Polymeric Product
Colorless, Slightly Acidic
Polymerizes on Standing.
% P_2O_5 = 40

2.

3.

Figure 7. Insertion of alkylene oxides into a pyrophosphate bond. (22, 23, 17)

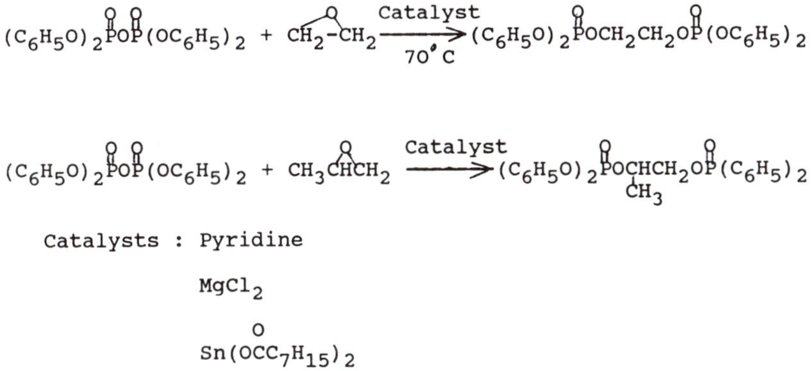

$(C_6H_5O)_2\overset{O}{P}O\overset{O}{P}(OC_6H_5)_2$ + CH$_2$-CH$_2$ $\xrightarrow[\text{70° C}]{\text{Catalyst}}$ $(C_6H_5O)_2\overset{O}{P}OCH_2CH_2O\overset{O}{P}(OC_6H_5)_2$

$(C_6H_5O)_2\overset{O}{P}O\overset{O}{P}(OC_6H_5)_2$ + CH$_3$CHCH$_2$ $\xrightarrow{\text{Catalyst}}$ $(C_6H_5O)_2\overset{O}{P}OCHCH_2O\overset{O}{P}(C_6H_5)_2$
                                                                              CH$_3$

Catalysts : Pyridine

MgCl$_2$

$\overset{O}{Sn(OCC_7H_{15})_2}$

Figure 8. Insertion of alkylene oxides into a pyrophosphate bond. (24)

is demonstrated in Fig. 9, Eq. 1, where dimethyl methylphosphonate can react with ethyl chloroacetate to yield mono- or di-carbethoxymethyl methylphosphonates. In a similar manner, as is shown in Eq. 2, trimethyl phosphate can react with ethyl chloroacetate to liberate methyl chloride and to form mono-, di-, and tri- carbethoxymethyl phosphates. By careful control of the ratio of reagents, one can direct these reaction to yield predominantly a mono-, di- or (in the case of phosphates) a tri-carbalkoxymethyl- derivative.

The first reported application of this reaction is described by A. N. Pudovik et al.(25) who reacted the ethyl esters of bromoacetic acid and chloroacetic acid with various phosphonate esters. His group used molar ratios of the phosphonate and the haloacetate, and obtained a mixture of the mono- and di-carbethoxy alkylphosphonates.

In our laboratory, a nucleophilic catalyst was used to conduct the condensation reaction. A typical reaction entailed heating five moles of dimethyl methylphosphonate with 11.2 moles of ethyl chloroacetate at 175 °C in the presence of a catalytic amount of tetramethylammonium chloride catalyst. The reaction was followed by the amount of methyl chloride collected in a cold trap. After about eight hours, 9.7 moles of methyl chloride is collected. After working up the product, 4.15 moles of bis (carbethoxymethyl) methylphosphonate was isolated by distillation. B.p., 148 °C at 0.3 torr.; the ¹H-nmr spectrum was in accord with the given structure; the ³¹P nmr spectrum was -35 ppm relative to ortho-phosphoric acid; %P, 11.7, theory, 11.5. (26)

A similar preparation was made using methyl chloroacetate. Yield, 312 g (67%); b.p., 136-140 at 0.6 torr; ¹H-nmr spectrum was in accord with the assigned structure.

The preparation of methyl (carbomethoxymethyl) methyl phosphonate was conducted with methyl chloroacetate and an excess of dimethyl methylphosphonate, using sodium carbonate as a catalyst (27).

The reaction of three moles of ethyl chloroacetate with one mole of trimethyl phosphate led to the formation of tris (carbethoxymethyl) phosphate. B.p., 195°C at 0.2 torr (27).

There was no evidence that the esters of chloroacetic acid condensed with themselves.

An alternate process for producing such esters was reported by K. D. Collins (28), whereby diphenyl (carbomethoxymethyl) phosphate was prepared from the reaction of diphenyl phosphorochloridate with methyl glycolate in the presence of a tertiary amine, such as pyridine or lutidine. This is shown in Eq. 3.

The bis (carboalkoxymethyl) methylphosphonates are difunctional esters, which can be converted to linear polyesters when heated with a diol, such as ethylene glycol, at 180°C in the presence of a catalyst, such as stannous octoate (27). This is shown in Fig. 10, Eq. 1. There is no evidence for the alcoholysis of the phosphorus ester bonds. An alternate preparation, the reaction of dimethyl methylphosphonate with bis chloroacetate ester of ethylene glycol in the presence of tetramethyl ammonium chloride led to a similar linear polyester. This is also illustrated in Eq. 2.

The above cited diol, when incorporated into a flexible polyurethane foam formulation at a 5.6% level (10 phr based on the polyol) produced a self-extinguishing urethane foam, based on the Motor Vehicles Safety Standard 302 Flammability Test.

In addition to using these phosphorus-containing ester compounds in synthetic polymer formulations, these esters, when converted to their corresponding amides, are effective in aminoplast resin formulations, particularly those suitable for imbedding the additive on cotton.

The conversion of these (carbethoxymethyl) methylphosphonates to the corresponding amides is shown in Fig. 11, Eq. 1. In this example, methyl (carbomethoxymethyl) methylphosphonate was dissolved in a cold, 4.43 molar solution of anhydrous ammonia in methanol. After being allowed to warm to room temperature, the reaction mixture was heated to 46°C for two hours, and the solvent removed by distillation. The residue, methyl (carbamoylmethyl) methylphosphonate, was isolated as a solid. This solid product was recrystallized from ethanol; m.p., 50-5°C (27). This compound was also prepared by V. E. Shishkin et al (29) by heating methyl (2-ethoxy-2-iminoethyl) methylphosphonate hydrochloride at 50-60°C. This is shown in Eq. 2.

Bis (carbamoylmethyl) methylphosphonate was prepared in a similar fashion from O bis (carbomethoxymethyl) methylphosphonate and methanolic solution of ammonia.

The preparation of dimethyl (carbamoylmethyl) phosphate and bis (carbamoylmethyl) methyl phosphate from ammonia and the corresponding ester precursor are illustrated in Eq. 3a and 3b of Fig. 11. Details of their preparation are described in reference 30.

Of the several reactive flame retardants for cotton, 0,0-dimethyl (N-hydroxymethyl) carbamoylethylphosphonate, sold commercially as Pyrovatex CP, has proven to be one of the more effective (31-33). When imbedded on cotton in conjunction with other aminoplasts, it forms a permanent bond to the cotton, a bond that can withstand at least 50 launderings. At the appropriate loading, it will render the fabric flame retardant, in accord with the Federal Flammability Standard of July 27, 1971. The

1. $CH_3\overset{O}{P}(OCH_3)_2 + ClCH_2\overset{O}{C}OC_2H_5 \xrightarrow[\text{Catalyst}]{} CH_3\overset{O}{P}(OCH_3)(OCH_2\overset{O}{C}OC_2H_5) + CH_3Cl$

   $+ 2\ ClCH_2\overset{O}{C}OC_2H_5 \xrightarrow[\text{Catalyst}]{} CH_3\overset{O}{P}(OCH_2\overset{O}{C}OC_2H_5)_2 + 2CH_3Cl$

2. $(CH_3O)_3P{=}O + ClCH_2\overset{O}{C}OC_2H_5 \xrightarrow[\text{Catalyst}]{} (CH_3O)_2\overset{O}{P}OCH_2\overset{O}{C}OC_2H_5 + CH_3Cl$

   $+ 2\ ClCH_2\overset{O}{C}OC_2H_5 \xrightarrow[\text{Catalyst}]{} CH_3O\overset{O}{P}(OCH_2\overset{O}{C}OC_2H_5)_2 + 2CH_3Cl$

   $+ 3\ ClCH_2\overset{O}{C}OC_2H_5 \xrightarrow[\text{Catalyst}]{} O{=}P(OCH_2\overset{O}{C}OC_2H_5)_3 + 3CH_3Cl$

3. $(C_6H_5O)_2\overset{O}{P}Cl + HOCH_2\overset{O}{C}OCH_3 + C_5H_5N \longrightarrow$

   $(C_6H_5O)_2\overset{O}{P}OCH_2\overset{O}{C}OCH_3 + [C_5H_5NH^+][Cl^-]$

Figure 9. Condensation of ethyl chloroacetate and phosphorus esters. Preparation of phosphorus esters of ethyl glycolate. (25, 26, 27, 28)

1. $n\ CH_3\overset{O}{P}(OCH_2\overset{O}{C}OC_2H_5)_2 + n\ HOCH_2CH_2OH \xrightarrow{\triangle} [-O-\overset{O}{C}CH_2O\overset{O}{\underset{CH_3}{P}}OCH_2\overset{O}{C}-OCH_2CH_2-]_n$

   $+ 2n\ C_2H_5OH$

2. $n\ ClCH_2\overset{O}{C}OCH_2CH_2O\overset{O}{C}CH_2Cl + n\ CH_3\overset{O}{P}(OCH_3)_2 \xrightarrow{\triangle}$

   $-[O\overset{O}{C}CH_2O\overset{O}{\underset{CH_3}{P}}OCH_2\overset{O}{C}OCH_2CH_2]-_n + 2n\ CH_3Cl$

Figure 10. Polycarboxyalkyl phosphonates. Permanent flame retardant for flexible urethane foams. (26)

preparation of this compound is shown in Fig. 12, Eq. 1. A representative co-condensation reaction of this compound with aminoplasts is shown in Eq. 2.

In a screening test, where methyl (carbamoylmethyl) methylphosphonate was incorporated into an aqueous textile padding bath containing trimethoxymethylmelamine, and cotton flannel cloth was padded with this solution, dried and the residual aminoplasts cured at 177°C for two hours, the fabric was found to be permanently flame retardant. This is illustrated in Fig. 12a, Eq. 3. An important characteristic of flame retarded textiles, the "hand" or feel of the cloth, was satisfactory; it did not become stiff. Similar results were obtained when padding cotton flannel with padding solutions containing either dimethyl (carbamoylmethyl) phosphate or methyl bis (carbamoylmethyl) phosphate. Surprisingly, bis (carbamoylmethyl) methylphosphonate failed to provide permanent flame retardance (27,30).

In an attempt to prepare functional compounds having a higher phosphorus content, recent studies have concentrated on the derivatives of dialkyl hydroxymethylphosphonate and dialkyl (2-hydroxyethyl) phosphate. Some of this research is illlustrated in Fig. 13.

In the case of the reaction of diethyl hydroxymethylphosphonate with phosphoryl chloride, as shown in Eq. 1 of Fig. 13, this reaction results in the formation of tris (diethyl phosphonomethyl) phosphate (34). Although this is not a functional compound, it is not volatile and it has a phosphorus content of 22.6%. The methyl ester analog should have a phosphorus content of 26.7%.

The alcoholysis of diethyl phosphonate with diethyl hydroxy-methylphosphonate, shown in Eq. 2 of Fig. 13, leads to the formation of bis (diethyl phosphonomethyl) phosphonate, a functional compound with a phosphorus content of 25.1% (35). The tetramethyl ester should have a phosphorus content of 28.5%. These compounds, with reactive P-H bonds, should be able to undergo many of the reactions of diethyl phosphonate to yield phosphorus-containing functional additives of high phosphorus content. An example is the addition of bis (diethyl phosphonomethyl) phosphonate to formaldehyde to form bis (diethyl phosphonomethyl) hydroxymethylphosphonate.

Diethyl 2-hydroxyethyl phosphate was prepared from the reaction of diethyl hydrogen phosphate with ethylene oxide. This is shown in Eq 3a of Fig. 13. An average of 1.3 moles of ethylene oxide added per mole of the diethyl hydrogen phosphate. Similar results were obtained when using pure dibutyl hydrogen phosphate (36). Gas chromatography of the silinated compounds obtained from the reaction of dibutyl hydrogen phosphate and ethylene oxide shows compounds with one, two and three ethylene oxide adducts are formed in this reaction. The reaction of alkylene oxides with

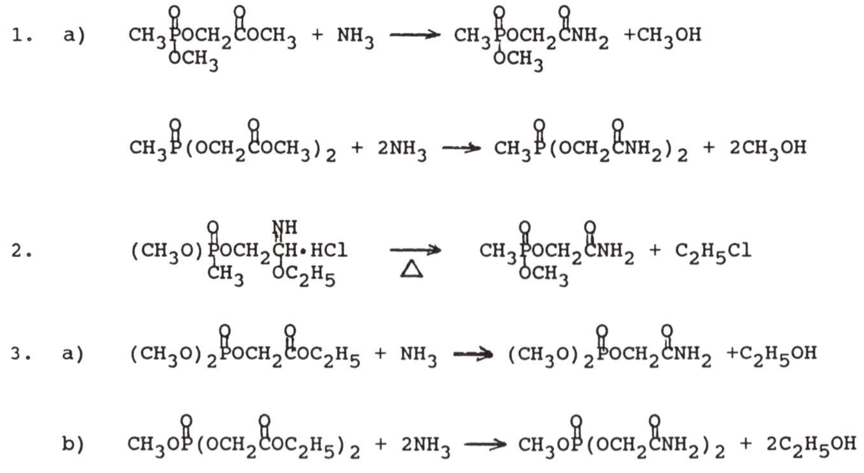

1. a) $CH_3\overset{O}{\underset{OCH_3}{P}}OCH_2\overset{O}{C}OCH_3 + NH_3 \longrightarrow CH_3\overset{O}{\underset{OCH_3}{P}}OCH_2\overset{O}{C}NH_2 + CH_3OH$

$CH_3\overset{O}{P}(OCH_2\overset{O}{C}OCH_3)_2 + 2NH_3 \longrightarrow CH_3\overset{O}{P}(OCH_2\overset{O}{C}NH_2)_2 + 2CH_3OH$

2. $(CH_3O)\overset{O}{\underset{CH_3}{P}}OCH_2\overset{NH}{\underset{OC_2H_5}{C}}\cdot HCl \xrightarrow{\Delta} CH_3\overset{O}{\underset{OCH_3}{P}}OCH_2\overset{O}{C}NH_2 + C_2H_5Cl$

3. a) $(CH_3O)_2\overset{O}{P}OCH_2\overset{O}{C}OC_2H_5 + NH_3 \longrightarrow (CH_3O)_2\overset{O}{P}OCH_2\overset{O}{C}NH_2 + C_2H_5OH$

    b) $CH_3O\overset{O}{P}(OCH_2\overset{O}{C}OC_2H_5)_2 + 2NH_3 \longrightarrow CH_3O\overset{O}{P}(OCH_2\overset{O}{C}NH_2)_2 + 2C_2H_5OH$

Figure 11. Preparation of phosphonoxy carboxamides.
Potential flame retardants for cotton fabrics(27,29,30).

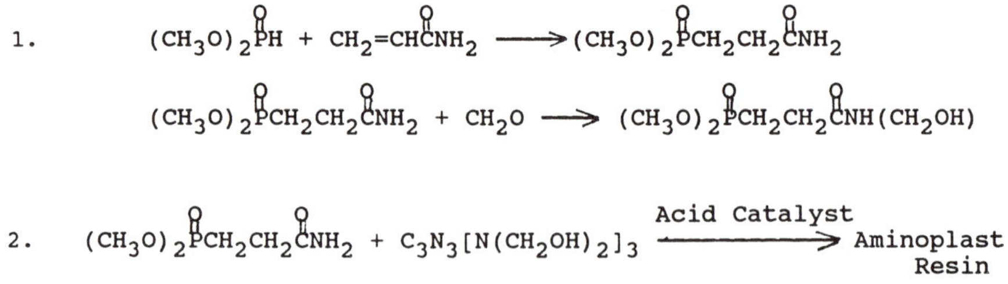

1. $(CH_3O)_2\overset{O}{P}H + CH_2=CH\overset{O}{C}NH_2 \longrightarrow (CH_3O)_2\overset{O}{P}CH_2CH_2\overset{O}{C}NH_2$

$(CH_3O)_2\overset{O}{P}CH_2CH_2\overset{O}{C}NH_2 + CH_2O \longrightarrow (CH_3O)_2\overset{O}{P}CH_2CH_2\overset{O}{C}NH(CH_2OH)$

2. $(CH_3O)_2\overset{O}{P}CH_2CH_2\overset{O}{C}NH_2 + C_3N_3[N(CH_2OH)_2]_3 \xrightarrow{\text{Acid Catalyst}}$ Aminoplast Resin

Figure 12. Pyrovatex CP. Commercial flame retardant for cotton
fabrics. (31, 32, 33)

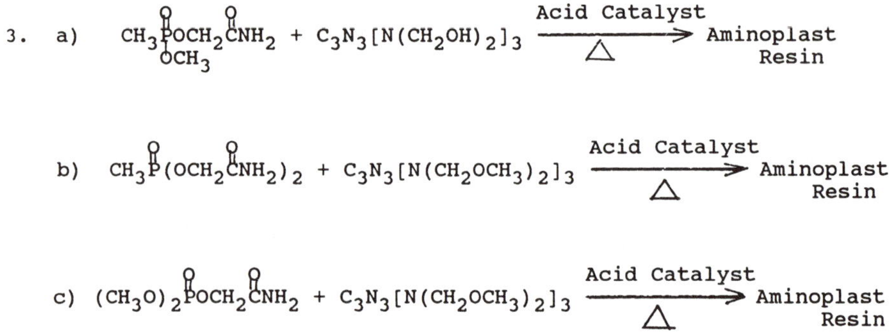

3. a) $CH_3\overset{O}{\underset{OCH_3}{P}}OCH_2\overset{O}{C}NH_2 + C_3N_3[N(CH_2OH)_2]_3 \xrightarrow[\Delta]{\text{Acid Catalyst}}$ Aminoplast Resin

    b) $CH_3\overset{O}{P}(OCH_2\overset{O}{C}NH_2)_2 + C_3N_3[N(CH_2OCH_3)_2]_3 \xrightarrow[\Delta]{\text{Acid Catalyst}}$ Aminoplast Resin

    c) $(CH_3O)_2\overset{O}{P}OCH_2\overset{O}{C}NH_2 + C_3N_3[N(CH_2OCH_3)_2]_3 \xrightarrow[\Delta]{\text{Acid Catalyst}}$ Aminoplast Resin

Figure 12a. Flame retardants for cotton fabrics. (27, 30)

1. $3 \ (C_2H_5O)_2\overset{O}{P}CH_2OH + POCl_3 \xrightarrow{\text{Base}} [(CH_2H_5O)_2\overset{O}{P}CH_2O]_3\overset{O}{P} + 3 \ \text{Base} \cdot HCl$

2. $2 \ (C_2H_5O)_2\overset{O}{P}CH_2OH + (C_2H_5O)_2\overset{O}{P}H \longrightarrow [(C_2H_5O)_2\overset{O}{P}CH_2O]_2\overset{O}{P}H + 2 \ C_2H_5OH$

Figure 13. New phosphorus-containing plastics additives. (35)

3. a) $(C_2H_5O)_2\overset{O}{P}OH + n \ \overset{O}{CH_2}CH_2 \longrightarrow (C_2H_5O)_2\overset{O}{P}(OCH_2CH_2)_nOH$

   b) $6(C_2H_5O)_2\overset{O}{P}(OCH_2CH_2)_nOH + P_4O_{10} \longrightarrow$

   $2[(C_2H_5O)_2\overset{O}{P}(OCH_2CH_2)_nO]_2\overset{O}{P}OH \ +$

   I

   $2[(C_2H_5O)_2\overset{O}{P}(OCH_2CH_2)_nO]\overset{O}{P}(OH)_2$

   II

   c) I & II + $(r+s+t) \ \overset{O}{CH_2}-CH_2 \longrightarrow$

   $[(C_2H_5O)_2\overset{O}{P}(OCH_2CH_2)_nO]_2\overset{O}{P}(OCH_2CH_2)_rOH$

   $+$

   $[(C_2H_5O)_2\overset{O}{P}(OCH_2CH_2)_nO]\overset{O}{P}\overset{\displaystyle -(OCH_2CH_2)_sOH}{\diagdown}_{(OCH_2CH_2)_tOH}$

Figure 13a. New phosphorus-containing plastics additives. (36)

phosphorus-containing acids is an often used procedure in industrial laboratories to form neutral esters of phosphorus acids. Unfortunately, little is published about the reaction mechanism, the methods to control the direction of the reaction, the methods to control the product formed, or the nature of the polyoxyalkylene products formed. Perhaps the availability of techniques such as silination and gas chromatography will encourage investigators to reexamine this field of research.

The impure diethyl hydroxyethyl phosphate was reacted with phosphorus (V) oxide, as shown in Fig. 13, Eq. 3b, and the resulting mixture of phosphoric acids was neutralized with ethylene oxide to yield a clear, non-volatile oil with a phosphorus content of 16.5%. The reaction of these phosphorus acids with ethylene oxide is illusrated in Eq. 3c. This product mixture has a functionality similar to that of Vircol 82.

One last item that has little to do with polymer additives.

During the summer of 1981, J. L. Mills, of Texas Technical University, joined our laboratory for a summer research program. One of his assignments was to look at methods for preparing phosphorus (III) oxide and to conduct research on the chemistry of this phosphorus oxide. In the course of this work, Mills considered the use of mild oxidants and mild oxidatation conditions for the oxidation of elemental phosphorus. Some of the oxidants he proposed or tried were triphenylphosphine oxide, dimethyl sulfoxide and trimethylamine oxide. These reactions are illustrated in Fig. 14. To my knowledge, if he tried these reactions he did not succeed in getting these latter three compounds to act as an oxidants for elemental phosphorus. However, as he left our laboratory, he suggested that iodine, since it readily oxidizes elemental phosphorus, may act as a catalyst for such oxidation reactions. Many years later the attempts to oxidize elemental (white) phosphorus with triphenylphosphine oxide and with dimethyl sulfoxide were repeated. In the case of dimethyl sulfoxide, when dimethyl sulfoxide was added to a suspension of white phosphorus in xylene and the system was warmed to reflux while stirring, no evidence for a reaction was observed; the elemental phosphorus remained a separate phase. When this experiment was repeated in the presence of a crystal of iodine, a moderate exotherm was noted at 40-44°C, as the elemental phosphorus melted. When the reaction was completed, dimethyl sulfide had formed and the oxidized phosphorus was present as a separate gummy phase. Hydrolysis of the oxidized phosphorus led to a mixture of ortho- and pyrophosphoric acid. This is shown in Eq. 2. Thus, the oxidation reaction had oxidized the phosphorus to the pentavalent state. Unfortunately, the reaction did not stop at the trivalent state of phosphorus. This is perhaps consistent with the findings of R. W. Light and R. T. Paine (38), who used sulfur dioxide to oxidize hexamethylphosphorous triamide to form hexamethylphosphoric triamide and reduced sulfur, as shown in Eq. 5. The sulfur dioxide had been reduced at least to the divalent state (and, considering

1. $P_4 + (C_6H_5)_3PO \xrightarrow[\text{to } 137^\circ C]{I_2}$ No Reaction

2. $P_4 + CH_3\overset{O}{\overset{\|}{S}}CH_3 \xrightarrow[\text{to } 137^\circ C]{}$ No Reaction

$\xrightarrow[40-45^\circ C]{I_2} [P_4O_{10}] + CH_3SCH_3 \xrightarrow{H_2O} H_3PO_4 +$

$H_4P_2O_7 + CH_3SCH_3$

3. $P_4 + (CH_3)_3NO \xrightarrow{I_2}$?

4. $(C_6H_5)_3PO + (C_6H_5O)_3P \xrightarrow[365^\circ C]{} (C_6H_5)_3P + (C_6H_5O)_3PO$

5. $P[N(CH_3)_2]_3 + SO_2 \longrightarrow OP[N(CH_3)_2]_3 + SP[N(CH_3)_2]_3 +$ Polymeric
Products

Figure 14. Use of elemental phosphorus as a reducing agent.
(37, 38, 39, 40)

the by-product, perhaps originally to the zero valent state).
Thus, it is likely the iodine reacts to form a trivalent (or at
least a non-zero valent) phosphorus intermediate, which in turn is
easily oxidized to the pentavalent state (40).

It is intriguing to consider the mode of reduction that
elemental phosphorus might take if, in the absence or presence of
iodine, it is effective in reducing nitro and nitroso compounds
(Eq. 3). What type of reduced nitrogen compounds would form?
Also, this might be an elegant method for reducing the N-oxides of
heterocyclic compounds. These N-oxides are often formed in
heterocylic compounds, such as pyridine, to alter the orientation
of substitution.

When reaction of elemental phosphorus with triphenyl phosphine
oxide was tried, no reduction of the triphenyl phosphine oxide was
observed at temperatures up to that of refluxing xylene, 137
°C, with or without the presence of the catalytic amount of
iodine. Perhaps, had a higher boiling solvent been used, one that
allowed one to reach temperatures closer to those achieved by J. N.
Gardner and J. Kochling (39) who reduced triphenyl phosphine oxide
to triphenyl phosphine at 365°C, using triphenyl phosphite as
the reducing agent, a similar reduction might have been effected.

Literature Cited

1. C. L. Harowitz, (to Mobile Oil Corp.) U.S.P. 3,525,705
    (Aug. 25, 1970) (to Virginia-Carolina Chemical Corp.)
    Brit. 954,792 (April 8, 1964)
2. T. M. Beck and E. N. Walsh, (to Stauffer Chemical Co.)
    U.S.P. 3,235,517 (Feb. 16, 1966)
3. R. L. Muntz and E. N. Walsh, (to Stauffer Chemical Co)
    U.S.P. 4,067,932 (Jan. 10, 1978)
4. N. Dawson and A. Burger, J. Am. Chem. Soc., 74 ,5312 (1952)
5. I. F. Lutsenko and M. Kirilov, Doklady Akad. Nauk SSSR,
    132 , 842 (1960)
6. G. Kamai and V. A. Kukhtin, Doklady Akad. Nauk SSSR, 112 ,
    868 (1957)
7. R. G. Harvey, Tetrahedron, 1966 , 2561
8. G. Sturtz and J. L. Kraus, Compt. rend., C271 , 744 (1970)
9. F. Ramirez, S. L. Glaser, A. J. Bigler and J. F. Pilot,
    J. Am. Chem. Soc., 91 , 496 (1969)
10. A. I. Razumov, B. G. Liorber, V. V. Moskva and M. P.
    Sokolov, Russian Chemical Reviews, 42 , (7) 538 (1973)
11. P. Tavs, Chem. Ber., 100 , 1571 (1967)
12. V. V. Moskva, A. I. Razumov, Z. Ya. Sazonova and T. V.
    Zykova, Zhur. Obshch. Khim., 41 , 1874 (1971)
13. A. I. Razumov, B. G. Liorber, M. P. Sokolov and T. V.
    Zykova, Zhur. Obshch. Khim., 41 , 2106 (1971)
14. N. Collignon, G. Fabre, J. M. Varlet and Ph. Savignac,
    Phosphorus and Sulfur, 10 , 1002 (1981)
15. V. V. Korshak, I. A. Gribova and V. K. Shitikov, Izvest.
    Akad. Nauk SSSR, 1958 , 210
16. E. D. Weil, (to Stauffer Chemical Co.) U.S.P. 3,896,187
    (July 22, 1975); U.S.P. 4,005,034 (Jan. 25, 1977); U.S.P.
    4,013,814 (Mar. 22, 1977)
17. E. D. Weil, R. B. Fearing and F. Jaffe, J. Fire Retardant
    Chem., 9 ,39 (1982)
18. E. D. Weil, (to Stauffer Chemical Co.) U.S.P. 3,891,727
    (June 24, 1975)
19. K. S. Shim and E. N. Walsh, (to Stauffer Chemical Co.)
    U.S.P. 3,959,414 (May 25, 1976); U.S.P. 3,959,415
    (May 25, 1976); U.S.P. 3,962,374 (June 8, 1976); U.S.P.
    3,965,217 (June 22, 1976)
20. E. N. Walsh, F. Jaffe, M. L. Honig, K. S. Shim and M. E.
    Brokke (to Stauffer Chemical Co.) U.S.P. 4,012,463
    (Mar. 15, 1977)
21. M. L. Honig and E. N. Walsh, (to Stauffer Chemical Co.)
    U.S.P. 3,954,917 (May 4, 1976)
22. W. H. Woodstock (to Victor Chemical Works) U.S.P. 2,568,784
    (Sep. 25, 1951)
23. A. J. Papa, Ind. Eng. Chem. Prod. Res. Devel. 9 (4), 478
    (1971)
24. D. A. Bright and A. M. Aaronson, (to AKZO Chemicals, Inc.)
    U. S. Pat. 5,041,596 (Aug. 20, 1991)

25. A. N. Pudovik, A. A. Muratova, T. I. Konnova, T. Feoktistova and L. N. Levtova, Zhur. Obshch. Khim., 30 , 2624 (1960)
26. E. N. Walsh and M. L. Honig (to Stauffer Chemical Co.) U.S.P. 4,044. 074 (Aug. 23, 1979); U.S.P. 4,142,904 (Mar. 6, 1979)
27. E. N. Walsh and R. B. Fearing, (to Stauffer Chemical Co.) U.S.P. 4,162,279 (July 24, 1979); U.S.P. 4,177,300 (Dec. 4, 1979)
28. K. D. Collins (to Research Corporation) U.S.P. 3,199,360 (Nov. 11, 1975)
29. V. E. Shishkin, Yu. M. Yukhno, B. I. No and N. M. Mamutova, Zhur. Obshch. Khim. 46 , 2233 (1975)
30. E. N. Walsh and T. M Hardy, (to Stauffer Chemical Co.) U.S.P. 4,292,036 (Sep. 29, 1981)
31. S. A. Zahir, (to Ciba Limited) U.S.P. 3,374,292 (Mar. 19, 1967)
32. H. Nachbur and A. Maeder, (to Ciba Limited) U.S.P. 3,634,422 (Jan. 11, 1972)
33. H. Nachbur, J. Kern and A. Maeder, (to Ciba Limited) U.S.P. 3,669,725 (June 13, 1972)
34. T. A. Hardy and E. N. Walsh, (to Stauffer Chemical Co.) U.S.P. 4,697,030 (Sep. 29, 1987)
35. T. A. Hardy and E. N. Walsh, (to AKZO America, Inc.) U.S.P. 4,808,744 (Feb. 28, 1987)
36. T. A. Hardy and E. N. Walsh, (to AKZO America, Inc.) U.S.P. 4,820,854 (Apr. 11, 1989)
37. J. R. Mills, Program Report, Aug. 7, 1984.
38. R. W. Light and R. T. Paine, Phosphorus and Sulfur, 1980 , 255
39. J. N. Gardner and J. Kochling, (to Hoffmann La-Roche, Inc.) U.S.P 3,847,999 (Nov. 12, 1974)
40. J. N. Denis and A. Krief, J. Chem. Soc. Chem. Comm. 1980 , 544

RECEIVED November 27, 1991

INDEXES

Author Index

Affiliation Index

Subject Index

Production: Peggy D. Smith
Indexing: Debra Steiner
Acquisition: Cheryl Shanks

Printed and bound by Maple Press, York, PA

Other ACS Books

Chemical Structure Software for Personal Computers
Edited by Daniel E. Meyer, Wendy A. Warr, and Richard A. Love
ACS Professional Reference Book; 107 pp;
clothbound, ISBN 0–8412–1538–3; paperback, ISBN 0–8412–1539–1

Personal Computers for Scientists: A Byte at a Time
By Glenn I. Ouchi
276 pp; clothbound, ISBN 0–8412–1000–4; paperback, ISBN 0–8412–1001–2

Biotechnology and Materials Science: Chemistry for the Future
Edited by Mary L. Good
160 pp; clothbound, ISBN 0–8412–1472–7; paperback, ISBN 0–8412–1473–5

Polymeric Materials: Chemistry for the Future
By Joseph Alper and Gordon L. Nelson
110 pp; clothbound, ISBN 0–8412–1622–3; paperback, ISBN 0–8412–1613–4

The Language of Biotechnology: A Dictionary of Terms
By John M. Walker and Michael Cox
ACS Professional Reference Book; 256 pp;
clothbound, ISBN 0–8412–1489–1; paperback, ISBN 0–8412–1490–5

Cancer: The Outlaw Cell, Second Edition
Edited by Richard E. LaFond
274 pp; clothbound, ISBN 0–8412–1419–0; paperback, ISBN 0–8412–1420–4

Practical Statistics for the Physical Sciences
By Larry L. Havlicek
ACS Professional Reference Book; 198 pp; clothbound; ISBN 0–8412–1453–0

The Basics of Technical Communicating
By B. Edward Cain
ACS Professional Reference Book; 198 pp;
clothbound, ISBN 0–8412–1451–4; paperback, ISBN 0–8412–1452–2

The ACS Style Guide: A Manual for Authors and Editors
Edited by Janet S. Dodd
264 pp; clothbound, ISBN 0–8412–0917–0; paperback, ISBN 0–8412–0943–X

Chemistry and Crime: From Sherlock Holmes to Today's Courtroom
Edited by Samuel M. Gerber
135 pp; clothbound, ISBN 0–8412–0784–4; paperback, ISBN 0–8412–0785–2

For further information and a free catalog of ACS books, contact:
American Chemical Society
Distribution Office, Department 225
1155 16th Street, NW, Washington, DC 20036
Telephone 800–227–5558